T0338607

An Introduction to Gravity

Einstein's theory of gravity can be difficult to introduce at the undergraduate level, or for self-study. One way to ease its introduction is to construct intermediate theories between the previous successful theory of gravity, Newton's, and our modern theory, Einstein's general relativity. This textbook bridges the gap by merging Newtonian gravity and special relativity (by analogy with electricity and magnetism), a process that both builds intuition about general relativity, and indicates why it has the form that it does. This approach is used to motivate the structure of the full theory, as a nonlinear field equation governing a second-rank tensor with geometric interpretation, and to understand its predictions by comparing it with the – often qualitatively correct – predictions of intermediate theories between Newton's and Einstein's. Suitable for a one-semester course at junior or senior level, this student-friendly approach builds on familiar undergraduate physics to illuminate the structure of general relativity.

Joel Franklin is a professor in the Physics Department of Reed College, Oregon. His research focuses on mathematical and computational methods with applications to classical mechanics, quantum mechanics, electrodynamics, general relativity, and its modifications. He is also the author of textbooks on *Advanced Mechanics and General Relativity*, *Computational Methods for Physics*, *Classical Field Theory*, and *Mathematical Methods for Oscillations and Waves*, all published by Cambridge University Press.

An Introduction to Gravity

JOEL FRANKLIN

Reed College, Oregon

CAMBRIDGE
UNIVERSITY PRESS

Shaftesbury Road, Cambridge CB2 8EA, United Kingdom

One Liberty Plaza, 20th Floor, New York, NY 10006, USA

477 Williamstown Road, Port Melbourne, VIC 3207, Australia

314–321, 3rd Floor, Plot 3, Splendor Forum, Jasola District Centre, New Delhi – 110025, India

103 Penang Road, #05-06/07, Visioncrest Commercial, Singapore 238467

Cambridge University Press is part of Cambridge University Press & Assessment, a department of the University of Cambridge.

We share the University's mission to contribute to society through the pursuit of education, learning and research at the highest international levels of excellence.

www.cambridge.org
Information on this title: www.cambridge.org/highereducation/isbn/9781009389709
DOI: 10.1017/9781009389693

First published 2024

A catalogue record for this publication is available from the British Library.

Library of Congress Cataloging-in-Publication Data
Names: Franklin, Joel, 1975– author.
Title: An introduction to gravity / Joel Franklin, Reed College, Oregon.
Description: Cambridge, United Kingdom ; New York, NY, USA : Cambridge University Press, 2024. | Includes bibliographical references and index.
Identifiers: LCCN 2023046090 (print) | LCCN 2023046091 (ebook) | ISBN 9781009389709 (hardback) | ISBN 9781009389693 (ebook)
Subjects: LCSH: Gravity. | Relativity (Physics) | Gravitation.
Classification: LCC QB334 .F73 2024 (print) | LCC QB334 (ebook) | DDC 531/.14–dc23/eng/20231031
LC record available at https://lccn.loc.gov/2023046090
LC ebook record available at https://lccn.loc.gov/2023046091

ISBN 978-1-009-38970-9 Hardback

Additional resources for this publication at www.cambridge.org/franklin-gravity

Contents

Preface

Einstein's theory of gravity, general relativity, is over a century old, as established and tested as any fundamental physical theory. It has been around long enough for teachers to develop pedagogically sound approaches to its introduction, at both the undergraduate and graduate level. There are many excellent books on the subject, among my favorites (and ones that informed much of this book) are by Schutz [43], Hartle [23], d'Inverno [10], Wald [51], and the encyclopedic classic by Misner, Thorne, and Wheeler [35]. I recommend these to your attention as primary resources for learning the subject, and delightful references for review and reflection. What, then, is the point of the current offering?

It is striking that gravity, as an undergraduate course, is not taught with the same frequency as, say, electricity and magnetism (E&M). The two are similar in that they deal with a single, specific interaction, making them different from subjects like classical and quantum mechanics, or solid-state physics, that span multiple physical interactions, and which involve a menagerie of techniques for solving their fundamental equations. A course on E&M must also introduce students to techniques for solving its relevant equations, but the relatively narrower set of physical ideas, tightly focused around a single force, can be an advantage in the classroom. Similarly, gravity is a subject with only one interaction (not a force, as it turns out), and as with E&M, there are really only two fundamental objects of study: the field equation that connects the sources to the field, and an equation of motion that describes how particles move in response to that field. Many of the techniques used to teach E&M can be usefully applied to gravity. Even the side benefits of a subject like E&M, like the field-theoretic intuition that is built up naturally as one learns the subject, is shared by an introduction to gravity. So why is gravity set aside in the undergraduate curriculum?

Attempting to answer this question was the starting point for this book. One reason that gravity is difficult to teach is that the field equation, Einstein's equation, involves mathematical objects that are not as familiar as the vector calculus required to learn E&M (and which benefits from the physical intuition that E&M provides). Another is that the interpretation of the field in gravity, the metric that tells us how distance is measured, makes it difficult to understand the observable predictions of the theory. We are used to thinking about a fixed, Minkowski spacetime with forces that act on particles, causing them to move. In general relativity, there is no force – particles move along length-minimizing curves (geodesics) in a spacetime that is determined by the metric, a solution of the field equation given some source configuration. That metric can be expressed in any coordinate system, and figuring out the physical content of its geodesics in an unfamiliar set of coordinates, and in a spacetime that is necessarily very different from Minkowski spacetime, is much harder than using the Lorentz force in Newton's second law to predict motion given an electric and magnetic field.

To address these challenges, I decided to use the question of why Newtonian gravity is "not enough," to both highlight gravitational physics beyond Newtonian gravity (which can be used to introduce some of the more exotic features of general relativity in a familiar setting), and serve as a motivator for the full theory (a motivator both in the sense of why general relativity has to be the way it is, and in the sense of generating student interest, so that they see the value in learning enough geometry to appreciate the subject). Newtonian gravity is virtually identical to electrostatics, and students are very familiar with Newtonian gravity's interpretation and predictions: Mass generates a gravitational force that acts on other masses. It is natural to wonder if the similarity between Newtonian gravity and electrostatics extends to dynamics – does gravity share the relativistic structure of the full set of Maxwell's equations? The question provides a target for discussion, and the setup work involved in answering it is relevant in understanding the linearized limit of general relativity, itself important in building physical intuition about the full theory. There are also differences between gravity and E&M, notably the mass sourcing in gravity, and the opportunity for more general forms of energy to act as a source. Focusing on the similarities and differences between the two theories gives students a framework for thinking about general relativity and its predictions, prior to encountering the less familiar geometric language in which the story of gravity is told. My goal was to create a relatively coherent narrative, starting from the familiar and working incrementally by asking physically motivated questions about the merger of gravity and relativity until the full theory emerges almost as a matter of course. This approach is meant to ease the difficulties with both the formal statement of the theory, and the understanding of its observable predictions by providing a set of familiar footholds along the way to which students can return and regroup as they move along.

Outline

I'll give a brief outline of the book, which will also serve as an informal example syllabus for the course that I teach at Reed College.[1]

---

[1]   Reed College offers a general relativity course every other year. When I teach that course, my audience is junior and senior students who have had a semester of electricity and magnetism at the level of Griffiths [21], and seen special relativity in a sophomore "modern physics" course.

The fundamental question driving the course is "Why can't you take Newtonian gravity, add special relativity, and get a viable relativistic theory of gravity?" So the first thing to do is review Newtonian gravity, and special relativity (both of which my students have seen prior to the class), and begin the process of combining the two, pointing out new predictions along the way. A side effect of this careful attempt to marry gravity and special relativity is that almost everything that shows up in the full theory: black holes, bending of light, even gravitational radiation, is in place, qualitatively, before any really new ideas are introduced. Chapter 1 represents that discussion and is used to build intuition about the subject, while also exploring the deficiencies of the resulting, incomplete theory.

Hopefully, the very physically concrete and accessible Chapter 1 provides motivation for the tool-building Chapter 2. The latter is essentially a review of vector calculus (similar to the review that introductions to E&M provide) set in a new notation. Here I develop the minimal set of machinery, in a flat spacetime setting, that will be needed to present Einstein's equation. Students learn about tensors, their transformation, and are introduced to the metric as a length-measuring device. The focus is on tensors as building blocks of physical theories that are "generally covariant," the same in all coordinate systems. Since those equations typically involve derivatives, and partial derivatives do not always have tensor behavior, I introduce the covariant derivative, together with connections, and the rest of the pieces necessary to discuss geodesics and parallel transport, all in the familiar, Minkowski, setting.

By the end of Chapter 2 (and together with the physics content of Chapter 1), we are led, quite naturally, to a target theory of gravity that is nonlinear, with a second-rank tensor as its field, one that has an almost immediate geometric interpretation. That target theory is then developed in Chapter 3, with only the Riemann tensor needed to set the scene. I present the standard "derivation" of Einstein's equation by comparison of the trajectories of nearby falling masses with geodesic deviation. My hope is that at this point, there is a sense of inevitability that drives students along. Before solving Einstein's field equation, I make contact with the incomplete "gravito-electo-magnetic" theory from Chapter 1 by working through the linearized limit of both the field equation and geodesic equation of motion. This is meant to give students a way to understand the physics predicted by the full theory. The linearized limit of gravity is also a nice one for comparison with E&M, and I discuss its solutions in that context.

Chapter 4 explores the Schwarzschild solution. I make use of symbolic computational packages to ease the calculation of the Ricci tensor, so that we can focus instead on the content of the field equation, and its solution. With this first nontrivial vacuum solution in place, I go through the usual set of geodesic observations, coordinate transformations, and experimental tests, noting the qualitative similarity to the predictions made in Chapter 1 while also highlighting the quantitative differences. There are both linearized geodesic calculations, and numerical ones, and students are invited to explore geodesics in spacetimes like Kerr using these techniques.

Staying in vacuum, plane waves and radiation are the topics in Chapter 5. I work only in the linearized limit of general relativity, and so I rely heavily on E&M as an example theory. Almost all of the development of gravitational radiation is done in parallel with descriptions of electromagnetic radiation. This is true in building vacuum solutions from (far away) sources using Green's functions and limits that define radiation. But the parallel approach is also useful in thinking about the detection of radiation by looking at the particle motion resulting from interaction with the field. Comparison and contrast between gravity and E&M are the main pedagogical tactics.

Moving from vacuum to "material" in the Chapter 6, I discuss continuum mechanics in both Newtonian and relativistic settings. Descriptions of the stress tensor sources for gravity, and their conservation, benefit from a review of electromagnetic sources. As topics, stellar interiors and a brief introduction to cosmology form the bulk of the applications, but I take the "in material" opportunity to present wormholes and the Alcubierre warp drive, both of which require (exotic) sources. These last topics represent a nice inversion of Einstein's equation, in which a metric with desired properties is built, and the stress tensor required to generate that metric is then characterized (as opposed to starting with a source and finding the field, the usual direction).

Chapter 7 is somewhat indulgent given my background and interests, and could easily be omitted – my ultimate goal is to round out the general relativistic solutions that were foreshadowed in Chapter 1 by solving Einstein's equation for a spherically symmetric central mass that also contains charge. That source is extended, like the ones in Chapter 6, but rather than study it in that setting, I tried to add a little value by introducing field actions for scalar and vector fields. Then adding in the Einstein–Hilbert action allows for a discussion of universal coupling and a concrete place to think about the role of the metric in scalar/vector field equations together with the role of scalar/vector field sources in Einstein's equation. Once we see how any theory necessarily couples to gravity, I develop the Reissner–Nordstrøm solution, as a final spacetime for study.

There are three appendices that I use only for reference when I teach the course. Appendix A is a bare bones review of special relativity, which, again, my students have seen (usually multiple times). For completeness, I include Appendix B on the Runge–Kutta method, which can be used to solve for geodesic trajectories in any spacetime of interest. While there are numerical differential equation solvers built in to many programming language libraries by now, it's always a good idea to understand what they are doing to avoid identifying a numerical artifact as a physical feature. It is also the case that the fourth-order Runge–Kutta method is widely used and requires only about ten lines of code to implement, making it worth the minimal effort. Lastly, while the curvature that is relevant in general relativity is somewhat different from the curvature of curves and surfaces, there is some intuition that can be built by thinking about the latter, and so I have a short introduction to those curvature ideas as Appendix C.

Notes

That overview completed, I'll note some personal idiosyncrasies to help avoid distraction when reading the text and carrying out the exercises:

1. Notation: I use fairly standard tensor notation, and have tried to highlight those places where I stray, temporarily, for the sake of clarity. In general, Greek indices occur when sums are over the full spacetime or when I do not need to distinguish between time and space components. Roman indices are used to refer to the spatial components of a tensor that has both time and space components. When dealing with vectors that have components and basis in place, I use bold typeface, for example, $\mathbf{E} = E^x\hat{\mathbf{x}} + E^y\hat{\mathbf{y}} + E^z\hat{\mathbf{z}}$. I use blackboard bold for matrices (or, if focusing on a matrix-like table representing a tensor, I'll use an indexed object). When writing out a matrix as part of an assignment or equality, I use the notation $\doteq$ rather than $=$, a reminder that these are representations of an object rather than the object itself, a habit picked up from Sakurai's quantum mechanics text (J. J. Sakurai, *Modern Quantum Mechanics Revised Edition*, Addison-Wesley, 1994).

   General relativity is a subject that contains a lot of symbolic information, with tensor indices dangling here and there, and other reminders cluttering expressions. One place where some economy can be found is in indicating functional dependence. When you have a tensor $T^\mu(t, x, y, z)$ that depends on the coordinates, I tend to write $T^\mu(x)$ with x standing in for all the coordinates. If I want to highlight a spacetime split in functional dependence, I use $T^\mu(\mathbf{r}, t)$ where $\mathbf{r}$ represents the spatial pieces (sometimes I use $\mathbf{x}$ to do the same thing, depending on what other symbols are nearby). There are times when indicating the dependence is unwieldy, as in expressions like $S[A^\mu(x)]$, the "action depending on the field A^μ, itself a function of the coordinates," and I usually omit the innermost set of dependence reminders. When the functional dependence is not the point, I omit it altogether, so you'll see things like $\dot{x}^\alpha$ standing in for $\dot{x}^\alpha(t)$.

2. References: I have tried to include original sources where appropriate, together with some interesting papers specifically aimed at undergraduate physics majors. Beyond that, and as a general rule, most of the topics in this book can be found in the books on gravity listed above (and many others), all of which have much more extensive references for advanced study.

3. Exercises: To quote from Mary Boas's excellent *Mathematical Methods in the Physical Sciences* (3rd ed. John Wiley & Sons, 2006): (replace "mathematics" with "physics" in the current setting) "To use mathematics effectively in applications, you need not just knowledge but *skill*. Skill can be obtained only through practice... The *only* way to develop the skill necessary to use this material in your later courses is to practice by solving many problems. Always study with pencil and paper at hand. Don't just

read through a solved problem – try to do it yourself! Then solve similar ones from the problem set for that section ... " Sage advice, and while I have not included problems at the end of sections, I do reference problems within the text itself. You should work those problems when you run across them in your reading, and do the rest when you get to the end of the chapter.

Acknowledgments

I learned general relativity from Professor Stanley Deser, who was a wonderful mentor to me. Much of the approach used in this book is informed by my interactions with him. He impressed upon me the utility of thinking about difficult physics by analogy with simpler physics as a way to make progress. I have attempted to do that (and using one of his favorite vehicles for comparison, E&M) in this book and hope that he would have enjoyed it. My postdoctoral adviser Professor Scott Hughes taught me the importance of making clear and concrete physical predictions in general relativity. That has never been easy (for me) to do, but it sure is fun when you can pull it off, and well worth the effort of trying. Professor David Griffiths remains the clearest and most creative expositor of physics that I have ever encountered, producing an infuriating standard that is, nevertheless, something to shoot for. He provided commentary and suggestions for this text, for which I am extremely grateful. Of course, it would be even better if he would just write a book on gravity ... Finally, I'd like to thank the Physics Department at Reed College – my colleagues and our students in the department are an inspiration.

1 Newtonian Gravity

Newton's law of universal gravitation was the accepted theory of gravity until general relativity supplanted it. Because of its predictive power in a nonrelativistic limit, Newtonian gravity retains an important place in the physics curriculum. It provides quantitative accuracy for many cases of interest. One of its first feats was providing a theoretical description of Kepler's (observational) laws. Yet there is a problem with this simplified form of gravity: Its description relies on the concept of force. This is a problem because the modern theory of gravity, while it reduces to a theory that looks like Newtonian gravity in approximation, is fundamentally different. General relativity provides a force-free gravitational interaction. It may not appear to matter much, but a reliance on Newtonian gravity pollutes our description of physics almost from the very start. We talk about the "four forces of nature," a simple and lovely idea that suggests that all interparticle relationships can be boiled down to four types. This way of thinking, in part, also invites reduction. The electromagnetic and weak forces, for example, can be combined into one. Can further simplification be found? Is it possible that all of these forces are manifestations of a single interaction, just viewed through different lenses, at different energy scales?

No. At least, not without an intermediate idea that would allow the electroweak and strong forces to be combined with gravity, which can be viewed as a geometric effect. That shift from force to geometry is not an incremental one that can be easily bridged. One focus of this chapter is to establish the structure of the gravitational field. By thinking about bringing Newtonian gravity in line with special relativity we can see the structural elements that are required. In particular, we will show that a theory of gravity (1) must be nonlinear, (2) must involve a second-rank tensor (as both source and field), and then in Chapter 2, we will finish the job by showing that (3) only the symmetric part of that tensor is physically relevant, and at that point, one can view the theory as a geometric one. So we will see that the target theory, which will end up being general relativity, is highly constrained from the start by special relativity. I want to highlight this shift away from gravity as a force as much as possible, especially since the "gravitational force" is so ingrained in our collective educational toolbox. But, to develop the implications of special relativity for a gravitational interaction, it will be easiest to work from Newtonian gravity, modifying it and adding to it as we consider more and more implications of special relativity. That morphing will occur very much within a force framework, and much of it will be done by comparison with electricity and magnetism, another force-based theory of interaction. Consider yourself warned, then, that while we will use familiar descriptions of gravity, our end point in this chapter and the next are that those familiar descriptions cannot hold.

The chapter begins with a description of Newtonian gravity as a field theory, meaning that we will be focused on the divergence and curl of the gravitational field, relating it to its sources and characterizing its vector geometry. Much of this will be a review, both of vector calculus and of the usual observations about a gravitational field, like the orbital motion of massive bodies in the presence of the field, the existence of a gravitational potential, analogous to an electrostatic one, and the energy required to build configurations of mass, energy that is then stored in the field itself, and so on.

After this review, we'll think about the implications of adding "a little" special relativity. In particular, what happens if we use $E = mc^2$ to relate the massive sources that we know from Newtonian gravity to energetic sources? This shift, from mass to energy, will also allow us to consider the response of particles that have energy but no mass to gravitational fields. This opens up a number of interesting possibilities. We can consider the bending of light by a gravitational field and also the gravitational field that is generated by a beam of light. We can also compute the gravitational contribution from static electric fields and find the gravitational field that comes from a massive point charge. Just as electric fields have an associated energy density, gravitational fields also carry energy and therefore must act as sources for themselves. That idea already establishes that a theory of gravity must be nonlinear, as the self-coupling energy density term goes like the field-squared, just as it does for an electric or magnetic field.

We'll make further progress towards our goal by considering the gravitational field in different inertial frames related by a Lorentz boost. Just as one can show that a magnetic field must exist by demanding that the predictions of two frames match a manifestation of "the laws of physics are the same in all inertial frames," the tenet of special relativity, you can also establish that there must be a "gravitomagnetic" field that is sourced by moving mass. Once the gravitomagnetic field is in place, we get "the rest" of the "Maxwell" equations easily. And so, without appealing to general relativity at all, we will see that, for example, gravitational radiation is already present as a theoretical prediction with even a naive marriage of Newtonian gravity and special relativity. Many of the extensions that we discuss in this chapter are qualitatively correct in full GR, often differing only by factors of two and four. But please don't forget that the ultimate structure of general relativity is very different from these force-field observations.

1.1 Two Observations and Their Consequences

All of Newtonian gravity can be built from two fundamental observations. The first is the experimentally verifiable form of the force between two masses m_1 and m_2. For m_1 at vector location $\mathbf{r}_1$ and m_2 at location $\mathbf{r}_2$ as shown in Figure 1.1, the force on the first mass due to the second is

$$\mathbf{F}_{12} = -\frac{Gm_1m_2}{R_{12}^2}\hat{\mathbf{R}}_{12}, \tag{1.1}$$

where $G \approx 6.67 \times 10^{-11}\,\mathrm{N\,m^2/kg^2}$ is the gravitational constant. The force has magnitude proportional to the product of the masses and acts along the line connecting the centers of mass of each. The minus sign out front tells us that the force is attractive, tending to pull the first mass towards the second. We have defined the vector $\mathbf{R}_{12} = \mathbf{r}_1 - \mathbf{r}_2$ to be the vector pointing from $\mathbf{r}_2$ to $\mathbf{r}_1$, with associated unit vector $\hat{\mathbf{R}}_{12}$. The force acting on mass two due to mass one is just the negative of the force acting on mass one due to mass two: $\mathbf{F}_{21} = -\mathbf{F}_{12}$.

Suppose you have several masses $\{m_i\}_{i=1}^N$ at locations $\{\mathbf{r}_i\}_{i=1}^N$. A second observation about the gravitational force is that the net force acting on a mass m at location $\mathbf{r}$ is the sum of the individual forces from each of the N masses. Referring to Figure 1.2, the total force on m is

$$\mathbf{F} = -\sum_{i=1}^N \frac{Gmm_i}{R_i^2}\hat{\mathbf{R}}_i, \qquad \mathbf{R}_i \equiv \mathbf{r} - \mathbf{r}_i, \tag{1.2}$$

where the individual separation vectors are $\mathbf{R}_i \equiv \mathbf{r} - \mathbf{r}_i$, pointing from the ith mass to m, and $\hat{\mathbf{R}}_i$ is the associated unit vector. This idea, that the net force is the vector sum of each mass's force contribution, is known as "superposi-

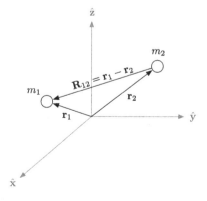

Fig. 1.1   Mass m_1 is located at $\mathbf{r}_1$, mass m_2 at $\mathbf{r}_2$. The vector $\mathbf{R}_{12} = \mathbf{r}_1 - \mathbf{r}_2$ is the vector that points from $\mathbf{r}_2$ to $\mathbf{r}_1$.

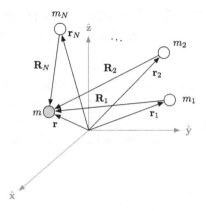

Fig. 1.2 A set of masses $\{m_i\}_{i=1}^{N}$ at locations $\{\mathbf{r}_i\}_{i=1}^{N}$ exert gravitational forces on a mass m at $\mathbf{r}$. The observation is that the net force on m is the sum of the individual forces.

tion." It is a familiar concept, but one that we highlight must be established experimentally, superposition is not a given.

In the force expression from (1.2), the mass of the target particle, m, appears in each term of the sum. We can define the gravitational field to be the force per unit mass acting at location $\mathbf{r}$, similar to the electric field's force per unit charge definition. Let $\mathbf{g} = \mathbf{F}/m$, so that for the setup in Figure 1.2, the gravitational field at $\mathbf{r}$ is

$$\mathbf{g}(\mathbf{r}) = -\sum_{i=1}^{N} \frac{Gm_i}{R_i^2} \hat{\mathbf{R}}_i \qquad (1.3)$$

with units of acceleration. As in E&M, we will focus on the field. To recover the force, just multiply the field by m: $\mathbf{F} = m\mathbf{g}$. The mass m at $\mathbf{r}$ will play the role of a "test mass" just as we often think of a positive "test charge" when working with the electric and magnetic fields.

We can generalize the field in (1.3) to apply to continuous distributions of mass. If we have a mass density function $\rho(\mathbf{r}')$ that gives the mass per unit volume at the point $\mathbf{r}'$, then in a small volume $d\tau'$ surrounding the point $\mathbf{r}'$, there is mass $dm' = \rho(\mathbf{r}')\,d\tau'$. Referring to Figure 1.3, the contribution of that little patch of mass to the gravitational field at $\mathbf{r}$ is

$$d\mathbf{g} = -\frac{Gdm'}{R^2}\hat{\mathbf{R}} = -\frac{G\rho(\mathbf{r}')d\tau'}{R^2}\hat{\mathbf{R}}, \qquad \mathbf{R} \equiv \mathbf{r} - \mathbf{r}', \qquad (1.4)$$

and then using superposition, we can sum up all those pieces to arrive at a continuous form of (1.3),

$$\mathbf{g}(\mathbf{r}) = \int d\mathbf{g} = -\int \frac{G\rho(\mathbf{r}')}{R^2}\hat{\mathbf{R}}\,d\tau', \qquad (1.5)$$

where the volume integral is over all space, with the understanding that $\rho(\mathbf{r}') = 0$ where there is no source mass. In Cartesian coordinates, the volume element is $d\tau' = dx'dy'dz'$; we are integrating over the primed source coordinates.

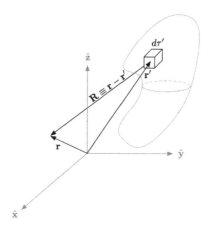

Fig. 1.3 A continuous distribution of mass described by the mass density $\rho(\mathbf{r}')$, which gives the mass per unit volume at $\mathbf{r}'$. The mass in the small $d\tau'$ volume shown is $dm' = \rho(\mathbf{r}')d\tau'$. That mass will contribute to the gravitational field at the point $\mathbf{r}$.

The integral expression for the field in (1.5) can in principle be used to find the gravitational field at any point $\mathbf{r}$ for any given distribution $\rho(\mathbf{r}')$ that is localized (and for which the integral exists). It encapsulates our two observations, the vector form of the field due to a point source and superposition. But it is often useful to have differential equations relating $\mathbf{g}$ to its source ρ. We can obtain these by taking the divergence and curl of $\mathbf{g}(\mathbf{r})$ directly from (1.5). Taking the curl, which slips through the integral (the curl has derivatives with respect to $\mathbf{r}$ in it, whereas the integral is over the dummy variables $\mathbf{r}'$),

$$\nabla \times \mathbf{g}(\mathbf{r}) = -G \int \rho(\mathbf{r}') \nabla \times \left(\frac{\hat{\mathbf{R}}}{R^2}\right) d\tau'. \tag{1.6}$$

We can evaluate the curl of $\hat{\mathbf{R}}/R^2 = \mathbf{R}/R^3$,[1]

$$\nabla \times \frac{\mathbf{R}}{R^3} = \frac{\nabla \times \mathbf{r}}{R^3} + \nabla\left(\frac{1}{R^3}\right) \times \mathbf{r} - \frac{\nabla \times \mathbf{r}'}{R^3} - \nabla\left(\frac{1}{R^3}\right) \times \mathbf{r}', \tag{1.7}$$

and it is clear that $\nabla \times \mathbf{r} = 0$ from the definition of the curl. The curl of $\mathbf{r}'$ is zero because $\mathbf{r}' = x'\hat{\mathbf{x}} + y'\hat{\mathbf{y}} + z'\hat{\mathbf{z}}$ doesn't depend on x, y, or z, which is what ∇ acts on. The gradient of R^{-3} is

$$\nabla\left[\left((x-x')^2 + (y-y')^2 + (z-z')^2\right)^{-3/2}\right] = -3\frac{\mathbf{R}}{R^5}. \tag{1.8}$$

Using the relation from (1.8) in (1.7) gives

$$\nabla \times \frac{\mathbf{R}}{R^3} = -3\frac{\mathbf{R} \times \mathbf{r}}{R^5} + 3\frac{\mathbf{R} \times \mathbf{r}'}{R^5} = -3\frac{\mathbf{R} \times \mathbf{R}}{R^5} = 0. \tag{1.9}$$

---

[1] The relevant product rule for curls acting on a function a and vector function $\mathbf{B}$ is

$$\nabla \times (a\mathbf{B}) = a(\nabla \times \mathbf{B}) + (\nabla a) \times \mathbf{B}.$$

This result makes good geometric sense, since the curl measures the extent to which a vector field "curls" around a point, and the vector field $\mathbf{R}/R^3$ points radially away from the point $\mathbf{r}'$, it doesn't curl around anything. The gravitational field is made up of a sum of these divergent pieces, and (1.9) used in (1.6) makes it is clear that $\nabla \times \mathbf{g}(\mathbf{r}) = 0$.

The divergence of the gravitational field is

$$\nabla \cdot \mathbf{g}(\mathbf{r}) = -G \int \rho(\mathbf{r}') \nabla \cdot \left(\frac{\hat{\mathbf{R}}}{R^2} \right) d\tau'. \tag{1.10}$$

We have to be careful with the divergence inside the integrand here, $\nabla \cdot (\mathbf{R}/R^3)$. If we just calculate it directly,[2]

$$\nabla \cdot \left(\frac{\mathbf{R}}{R^3} \right) = -3 \frac{\mathbf{R}}{R^5} \cdot \mathbf{R} + \frac{3}{R^3} = 0. \tag{1.11}$$

This result is surprising since the vector field diverges strongly from the point $\mathbf{r}'$. The calculation in (1.11) is valid at all points except for $\mathbf{r}'$ itself, so the divergence is indeed zero away from the source point. But it is possible that the divergence *at* $\mathbf{r}'$ is nonzero (since at this point we are dividing zero by zero), and we hope that this is the case to save our intuitive expectation. It is not immediately obvious how to probe the point $\mathbf{r}'$ using the derivative operator ∇, but we can remove that derivative by integrating over a ball of radius ϵ centered at the point $\mathbf{r}'$ and using the divergence theorem.

To carry out the integral, move the origin of the coordinate system to $\mathbf{r}'$, so that $\mathbf{R} = \mathbf{r}$, a vector pointing from the origin to any point in three dimensions (after we evaluate the integral, we will move the origin back to $\mathbf{r}'$). For $B(\epsilon, 0)$, the ball of radius ϵ centered at the origin, the divergence theorem gives

$$\int_{B(\epsilon,0)} \nabla \cdot \frac{\hat{\mathbf{r}}}{r^2} d\tau = \oint_{\partial B(\epsilon,0)} \frac{\hat{\mathbf{r}}}{r^2} \cdot d\mathbf{a}, \tag{1.12}$$

with $r^2 = \epsilon^2$ at the surface of the ball. That surface has area element $d\mathbf{a} = \epsilon^2 \sin\theta d\theta d\phi \hat{\mathbf{r}}$ in spherical coordinates. Performing the surface integral,

$$\oint_{\partial B(\epsilon,0)} \frac{\hat{\mathbf{r}}}{r^2} \cdot d\mathbf{a} = \int_0^{2\pi} \int_0^{\pi} d\theta d\phi = 4\pi. \tag{1.13}$$

Now we know that $\nabla \cdot (\hat{\mathbf{r}}/r^2)$ is zero everywhere except at the origin, with indeterminate value there, but a value that can be integrated to 4π. That is practically the definition of the Dirac delta function,[3] and we conclude that

---

[2] Here we use the product rule for function a and vector $\mathbf{B}$,

$$\nabla \cdot (a\mathbf{B}) = \nabla a \cdot \mathbf{B} + a \nabla \cdot \mathbf{B}.$$

[3] The one-dimensional Dirac delta function is defined by its integral behavior: $\delta(x) = 0$ unless $x = 0$ where it is infinite but such that

$$\int_{-\infty}^{\infty} \delta(x) dx = 1.$$

The three-dimensional form is a product: $\delta(\mathbf{r}) = \delta(x)\delta(y)\delta(z)$. Try Problem 1.1 and Problem 1.2 for a review of the Dirac delta function.

$\nabla \cdot (\hat{\mathbf{r}}/r^2) = 4\pi\delta^3(\mathbf{r})$ or, moving the origin back,

$$\nabla \cdot \left(\frac{\hat{\mathbf{R}}}{R^2}\right) = 4\pi\delta^3(\mathbf{r} - \mathbf{r}'). \tag{1.14}$$

Returning to (1.10) with (1.14), we learn that the divergence of the gravitational field is

$$\nabla \cdot \mathbf{g}(\mathbf{r}) = -G \int \rho(\mathbf{r}')4\pi\delta^3(\mathbf{r} - \mathbf{r}') \, d\tau' = -4\pi G\rho(\mathbf{r}). \tag{1.15}$$

Our original two observations about the gravitational force have become the pair of partial differential equations for the gravitational field:

$$\boxed{\nabla \cdot \mathbf{g}(\mathbf{r}) = -4\pi G\rho(\mathbf{r}), \qquad \nabla \times \mathbf{g}(\mathbf{r}) = 0,} \tag{1.16}$$

providing a force on a particle of mass m at $\mathbf{r}$ of $\mathbf{F} = m\mathbf{g}(\mathbf{r})$. You should compare these equations with the similar ones that govern the electrostatic field

$$\nabla \cdot \mathbf{E}(\mathbf{r}) = \frac{\rho_e(\mathbf{r})}{\epsilon_0}, \qquad \nabla \times \mathbf{E}(\mathbf{r}) = 0, \tag{1.17}$$

for $\rho_e(\mathbf{r})$ the charge density. The force on a charge q at $\mathbf{r}$ is $\mathbf{F} = q\mathbf{E}(\mathbf{r})$. The comparison gives us a way to formally map electrostatic results to gravitational ones by taking $\epsilon_0^{-1} \to -4\pi G$ and $\rho_e \to \rho$.

1.2 The Field Equations

Both the divergence and curl carry useful information about the field $\mathbf{g}$. If we integrate the equation governing the divergence of $\mathbf{g}$ over an arbitrary volume Ω and use the divergence theorem, we return to an integral form,

$$\int_\Omega \nabla \cdot \mathbf{g}(\mathbf{r}) \, d\tau = -4\pi G \int_\Omega \rho(\mathbf{r}) \, d\tau \longrightarrow \oint_{\partial\Omega} \mathbf{g}(\mathbf{r}) \cdot d\mathbf{a} = -4\pi G M_{\text{enc}}, \tag{1.18}$$

where M_{enc} is the mass enclosed by the volume. The gravitational field flux through the surface enclosing Ω is proportional to the mass enclosed, much like Gauss's law. We can use the integral form to find the gravitational field due to highly symmetric distributions of mass.

Example 1.1 (Spherical Shell of Mass). What is the gravitational field inside and outside a sphere of radius R with mass M uniformly distributed over its surface? Assume that the sphere is centered at the origin of our coordinate system. Since the source is spherically symmetric, it is reasonable to take the gravitational field to be spherically symmetric[4] which, for a vector field, means that the magnitude should depend only on the distance to the origin, with direction that is radial: $\mathbf{g}(\mathbf{r}) = g(r)\hat{\mathbf{r}}$.

Take a "Gaussian surface" that is a sphere of radius $r < R$. There is no mass enclosed by that sphere, $M_{\text{enc}} = 0$. The surface integral in (1.18) is easy to evaluate from the assumed spherically symmetric form,

$$\oint \mathbf{g}(\mathbf{r}) \cdot d\mathbf{a} = g(r)4\pi r^2, \tag{1.19}$$

and putting these two pieces together in (1.18), we learn that $g(r)4\pi r^2 = 0$, so that $\mathbf{g}(\mathbf{r}) = 0$ for points inside the spherical shell.

If our Gaussian surface has radius $r > R$, then all that changes is the mass enclosed, $M_{\text{enc}} = M$. Now we have $g(r)4\pi r^2 = -4\pi GM$ giving $g(r) = -GM/r^2\hat{\mathbf{r}}$. We have recovered the usual "shell theorem" result: Inside the sphere, there is no field, and outside the field is that of a point source of mass M sitting at the origin.

Example 1.2 (Infinite Line of Mass). As another case where we can use this gravitational Gauss's law, take an infinite uniform line of mass lying along the $\hat{\mathbf{z}}$ axis. The mass per unit length of the line is λ, what is the gravitational field a distance s from the line? This time, we will take $\mathbf{g}(\mathbf{r})$ to have cylindrical symmetry, and its magnitude depends only on s, the distance to the axis, with direction radially away from the axis, $\mathbf{g}(\mathbf{r}) = g(s)\hat{\mathbf{s}}$ (where s is the distance to the $\hat{\mathbf{z}}$ axis, and $\hat{\mathbf{s}}$ is its direction of increase). For the Gaussian surface, take a cylinder of radius s and height ℓ centered on the line. The mass enclosed is $M_{\text{enc}} = \lambda\ell$, and the surface integral, easily evaluated in cylindrical coordinates, is

$$\oint \mathbf{g}(\mathbf{r}) \cdot d\mathbf{a} = g(s)\ell 2\pi s. \tag{1.20}$$

When used in (1.18), $g(s)\ell 2\pi s = -4\pi G\lambda\ell \rightarrow \mathbf{g}(\mathbf{r}) = -2G\lambda/s\hat{\mathbf{s}}$.

This is a good place to see how the mapping $\epsilon_0^{-1} \rightarrow -4\pi G$ can be useful in generating gravitational results from electrostatic ones. For an infinite line of electric charge with charge density λ_e, the electric field is $\mathbf{E}(\mathbf{r}) = \lambda_e/(2\pi\epsilon_0 s)\hat{\mathbf{s}}$. We would predict, from the map, that

$$\mathbf{E}(\mathbf{r}) = \frac{\lambda_e}{2\pi\epsilon_0 s}\hat{\mathbf{s}} \longrightarrow \mathbf{g}(\mathbf{r}) = -\frac{2G\lambda}{s}\hat{\mathbf{s}}, \tag{1.21}$$

confirming our result above.

We can keep going, of course, finding the gravitational fields associated with infinite cylinders, sheets and blocks of mass. You should try solving Prob-

---

[4] Why is that? There is nothing in the equation $\nabla \cdot \mathbf{g} = -4\pi G\rho$ that indicates that the symmetry of the source should be reflected in the field. But a partial differential equation by itself is not enough to find the field. There are also boundary conditions to consider, and these inherit the source symmetries and propagate those symmetries to the field.

lem 1.3, Problem 1.4, and Problem 1.5 using both the gravitational Gauss' law
and the electrostatic to gravitational map.

So much for the divergence of the gravitational field and its integral form.
How about the curl? Pick a surface S with boundary ∂S and integrate $\nabla \times \mathbf{g} = 0$ over it using the curl theorem to turn the surface integral into a boundary
integral

$$0 = \int_S \left(\nabla \times \mathbf{g}(\mathbf{r}) \right) \cdot d\mathbf{a} = \oint_{\partial S} \mathbf{g}(\mathbf{r}) \cdot d\boldsymbol{\ell}. \tag{1.22}$$

If we multiply this equation by m, then it says that the work done around
a closed loop is zero, the gravitational force is conservative. Then it can be
developed from a potential energy function U via the gradient. While this is
a familiar statement of conservation, we can work directly with the curl of
$\mathbf{g}$ itself. The most general vector field that has curl vanishing everywhere is
the gradient of a scalar, so $\nabla \times \mathbf{g}(\mathbf{r}) = 0 \rightarrow \mathbf{g}(\mathbf{r}) = -\nabla \varphi(\mathbf{r})$. The minus
sign is traditional (think of $\mathbf{E}(\mathbf{r}) = -\nabla V(\mathbf{r})$ from E&M), and since we are
dealing with the field, φ is an energy per unit mass (just as V is energy per unit
charge) called the "gravitational potential."[5] The potential energy is related to
the potential by a factor of m, $U = m\varphi$, just as the field $\mathbf{g}$ is related to the
force.

The gravitational potential allows us to combine the content of the pair in
(1.16) into a single equation. By construction, $\nabla \times \mathbf{g}(\mathbf{r}) = 0$, so we only have
to worry about the divergence, $\nabla \cdot (-\nabla \varphi(\mathbf{r})) = -4\pi G \rho(\mathbf{r})$, or

$$\boxed{\nabla^2 \varphi(\mathbf{r}) = 4\pi G \rho(\mathbf{r}),} \tag{1.23}$$

which is Poisson's equation. The linearity of Poisson's equation tells us that
superposition holds here: adding sources adds their contribution to the potential. That allows us to develop an integral solution to (1.23) by building up
combinations of its point source solution. A point source sitting at the origin
has mass density $\rho(\mathbf{r}) = m\delta^3(\mathbf{r})$. How should we think about solving (1.23)
for this source? Since the source is spherically symmetric, we expect φ to be as
well (again in order to impose boundary conditions), and a function is spherically symmetric if it depends only on the distance to the origin: $\varphi(\mathbf{r}) = \varphi(r)$.
Under this assumption, the Laplacian simplifies to $\nabla^2 \varphi(\mathbf{r}) = (r\varphi(r))''/r$ with
primes denoting r derivatives. If we work away from the source, at points with
$r > 0$, then Poisson's equation reads

$$\frac{d^2}{dr^2} \left(r\varphi(r) \right) = 0 \longrightarrow \varphi(r) = \frac{\alpha}{r} + \beta \tag{1.24}$$

for constants of integration α and β. To set those constants, we rely on boundary conditions. One generally implicit one is that the potential should vanish

---

[5] The units of φ are already tantalizing: energy per unit mass is $(m/s)^2$, a speed squared. In
Newtonian gravity, there is no natural speed that can be made out of the available constants, G and
m, so there is nothing to set a scale for the gravitational potential. If we were working in a special
relativistic setting, though, we *would* have a natural speed, the speed of light c.

at spatial infinity, $\varphi(\mathbf{r}) \to 0$ as $r \to \infty$, requiring that $\beta = 0$.[6] The other boundary condition comes from $r = 0$, the source of the delta function in the field. Integrating (1.23) with the delta source over a ball of radius ϵ centered on the mass and again using the divergence theorem on $\nabla^2\varphi = \nabla \cdot (\nabla\varphi)$,

$$\oint_{\partial B(\epsilon,0)} \nabla\left(\frac{\alpha}{r}\right) \cdot d\mathbf{a} = 4\pi Gm \longrightarrow -4\pi\alpha = 4\pi Gm, \qquad (1.25)$$

and $\alpha = -Gm$. Thus the potential for a point source at the origin is $\varphi = -Gm/r$, and the associated field is $\mathbf{g} = -\nabla\varphi = -Gm/r^2\hat{\mathbf{r}}$, a good check.

We can move the source to an arbitrary location $\mathbf{r}'$, and then the potential just has $r \to |\mathbf{r} - \mathbf{r}'| \equiv \mathbf{R}$. The solution to a partial differential equation, like Poisson's, for a point source described by a delta function is called the "Green's function solution" and is denoted (problematically for our present purposes) $G(\mathbf{r}, \mathbf{r}')$ for a source at $\mathbf{r}'$. We have just found the Green's function for the gravitational Poisson problem,[7]

$$G(\mathbf{r}, \mathbf{r}') = -\frac{G}{R}, \qquad \mathbf{R} \equiv \mathbf{r} - \mathbf{r}'. \qquad (1.26)$$

The Green's function is useful in theories that support superposition, linear PDEs, since we can build solutions for arbitrary source distributions. Referring to Figure 1.3, the potential at $\mathbf{r}$ due to the sum of point contributions from $\rho(\mathbf{r}')$ is

$$\varphi(\mathbf{r}) = \int G(\mathbf{r}, \mathbf{r}')\rho(\mathbf{r}')\,d\tau', \qquad (1.27)$$

where again the integral is over all space, and ρ is taken to be a function that is zero outside of the actual mass distribution. This integral solution is built in much the same way as (1.5), and rather than review the construction, we will content ourselves to check that $\varphi(\mathbf{r})$ satisfies (1.23). Hitting both sides of (1.27) with the Laplacian and noting that $\nabla^2 G(\mathbf{r}, \mathbf{r}') = 4\pi G\delta^3(\mathbf{r} - \mathbf{r}')$ by definition,

$$\nabla^2\varphi(\mathbf{r}) = \int \left[\nabla^2 G(\mathbf{r}, \mathbf{r}')\right]\rho(\mathbf{r}')\,d\tau'$$
$$= \int \left[4\pi G\delta^3(\mathbf{r} - \mathbf{r}')\right]\rho(\mathbf{r}')\,d\tau' = 4\pi G\rho(\mathbf{r}). \qquad (1.28)$$

The computational advantage in working with $\varphi(\mathbf{r})$ instead of $\mathbf{g}(\mathbf{r})$ is simplicity: a single function rather than the field, which is described by three functions. But it can also be used to get at different types of quantities than the field. As an example, let's compute the energy required to build a discrete distribution of mass, like the one shown in Figure 1.4. There are N masses $\{m_i\}_{i=1}^N$ positioned at $\{\mathbf{r}_i\}_{i=1}^N$. To find the work required to build the distribution, we'll

[6]   This is conventional, and comes more from the notion of a gauge choice than any physical requirement. The potential energy $m\varphi$ has an undetectable offset that can be chosen however we like (only energy differences matter). If the field itself were nonzero at spatial infinity, we would have an actual observational problem, since there are no forces with infinite range.

[7]   The Green's function traditionally omits the point mass m, so that $G(\mathbf{r}, \mathbf{r}')$ satisfies $\nabla^2 G(\mathbf{r}, \mathbf{r}') = 4\pi G\delta^3(\mathbf{r} - \mathbf{r}')$. I have highlighted the argument of the Green's function to avoid confusion with the gravitational constant G, which appears on the right in (1.26).

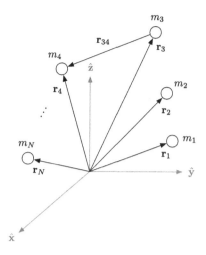

Fig. 1.4   A set of masses $\{m_i\}_{i=1}^N$ at locations $\{\mathbf{r}_i\}_{i=1}^N$. The vector pointing from the ith mass to the jth is $\mathbf{r}_{ij} \equiv \mathbf{r}_j - \mathbf{r}_i$, with $\mathbf{r}_{34}$ shown.

start with a set of N masses located out at spatial infinity and compute the minimum work required to bring each one from infinity to its final resting location. For the first mass, we don't do any work moving it from infinity to $\mathbf{r}_1$, since there is no gravitational force acting on it. Calling W_i the work required to bring in the ith mass in the presence of the gravitational field of the $j = 1 \to i-1$ masses, we have $W_1 = 0$. To bring in the second mass, we have to do work against the gravitational field of the first mass (hence the minus sign out front),

$$W_2 = -m_2 \int_\infty^{\mathbf{r}_2} \nabla\varphi(\mathbf{r}) \cdot d\boldsymbol{\ell} = -m_2\big(\varphi_1(\infty) - \varphi_1(\mathbf{r}_2)\big) = m_2\varphi_1(\mathbf{r}_2), \quad (1.29)$$

where $\varphi_1(\mathbf{r}) = -Gm_1/|\mathbf{r}_2 - \mathbf{r}_1|$ is the potential due to m_1, and it vanishes at spatial infinity, so we have gotten rid of that term in (1.29). Bringing in the third mass, we do work against the gravitational field of both the first and second masses, and it is clear that we will get $W_3 = m_3\varphi_1(\mathbf{r}_3) + m_3\varphi_2(\mathbf{r}_3)$, considering each one separately (via superposition). The process continues, and the first four or so demonstrate the pattern

$$\begin{aligned} W_2 &= m_2\varphi_1(\mathbf{r}_2), \\ W_3 &= m_3\varphi_1(\mathbf{r}_3) + m_3\varphi_2(\mathbf{r}_3), \\ W_4 &= m_4\varphi_1(\mathbf{r}_4) + m_4\varphi_2(\mathbf{r}_4) + m_4\varphi_3(\mathbf{r}_4). \end{aligned} \quad (1.30)$$

The total energy required is the sum of these individual pieces, $W = W_1 + W_2 + W_3 + \cdots + W_N$. Imagine arranging these sums from the columns shown in (1.30),

$$\begin{aligned} W &= \sum_{k=2}^N m_k\varphi_1(\mathbf{r}_k) + \sum_{k=3}^N m_k\varphi_2(\mathbf{r}_k) + \sum_{k=4}^N m_k\varphi_3(\mathbf{r}_k) + \cdots \\ &= \sum_{j=1}^N \sum_{k=j+1}^N m_k\varphi_j(\mathbf{r}_k). \end{aligned} \quad (1.31)$$

We can get an equivalent expression by organizing the sums by the rows of (1.30),

$$W = m_2 \sum_{k=1}^{2-1} \varphi_k(\mathbf{r}_2) + m_3 \sum_{k=1}^{3-1} \varphi_k(\mathbf{r}_3) + m_4 \sum_{k=1}^{4-1} \varphi_k(\mathbf{r}_4) + \cdots$$
$$= \sum_{j=1}^{N} \sum_{k=1}^{j-1} m_j \varphi_k(\mathbf{r}_j). \tag{1.32}$$

From the form of the potential,

$$m_k \varphi_j(\mathbf{r}_k) = -\frac{G m_k m_j}{|\mathbf{r}_k - \mathbf{r}_j|} = -\frac{G m_j m_k}{|\mathbf{r}_j - \mathbf{r}_k|} = m_j \varphi_k(\mathbf{r}_j), \tag{1.33}$$

so we could also write the final double sum in (1.32) as

$$W = \sum_{j=1}^{N} \sum_{k=1}^{j-1} m_j \varphi_k(\mathbf{r}_j) = \sum_{j=1}^{N} \sum_{k=1}^{j-1} m_k \varphi_j(\mathbf{r}_k) \tag{1.34}$$

matching the index placement of the terms in (1.31). Finally, note that we can split up the sum,

$$W = \sum_{j=1}^{N} \sum_{k=1}^{j-1} m_k \varphi_j(\mathbf{r}_k) = \sum_{j=1}^{N} \sum_{k=1\neq j}^{N} m_k \varphi_j(\mathbf{r}_k) - \underbrace{\sum_{j=1}^{N} \sum_{k=j+1}^{N} m_k \varphi_j(\mathbf{r}_k)}_{=W \text{ from } (1.31)}, \tag{1.35}$$

and we can write the total work done in a nicely symmetric fashion, summing over all particles (but one) in each sum at the expense of a $1/2$. All that has really happened here is a double-counting,

$$W = \frac{1}{2} \sum_{j=1}^{N} \sum_{k=1, k\neq j}^{N} m_k \varphi_j(\mathbf{r}_k) = \frac{1}{2} \sum_{j=1}^{N} \sum_{k=1, k\neq j}^{N} m_j \varphi_k(\mathbf{r}_j). \tag{1.36}$$

The sum over k in the second equality represents the total potential due to all the masses other than the jth one, and we could call that $\varphi(\mathbf{r}_j)$ (remembering to omit the infinity that comes from including the contribution from m_j itself),

$$W = \frac{1}{2} \sum_{j=1}^{N} m_j \varphi(\mathbf{r}_j). \tag{1.37}$$

Using the final form in (1.37), we can extend the concept of "work done to build a configuration of mass" to the continuum. Referring to the mass density in Figure 1.3, the mass m_j will become $dm' = \rho(\mathbf{r}') d\tau'$, and the sum over all masses $\{m_j\}_{j=1}^{N}$ is a sum over all space, so for the continuum limit,

$$W = \frac{1}{2} \int \varphi(\mathbf{r}') \rho(\mathbf{r}') d\tau', \tag{1.38}$$

where the function φ describes the potential due to the distribution ρ, and the integral is over all space, appropriately cut off by ρ. We can rewrite the integrand entirely in terms of the potential using (1.23) "solved" for $\rho =$

$(\nabla^2 \varphi)/(4\pi G)$, and put that ρ in the above integrand (with ∇' the gradient with respect to the primed coordinates),

$$W = \int \left[\frac{1}{8\pi G} \varphi(\mathbf{r'}) \nabla' \cdot \left(\nabla' \varphi(\mathbf{r'}) \right) \right] d\tau'. \tag{1.39}$$

Finally, integration by parts[8] can be applied. The domain Ω is "all space," and $\partial\Omega$ is a surface out at spatial infinity where φ vanishes (one of our boundary conditions), so we can just flip the derivative acting on $\nabla'\varphi(\mathbf{r'})$ onto $\varphi(\mathbf{r'})$, picking up a minus sign in the process:

$$W = \int_\Omega \left[-\frac{1}{8\pi G} \nabla' \varphi(\mathbf{r'}) \cdot \nabla' \varphi(\mathbf{r'}) \right] d\tau'. \tag{1.40}$$

The term in brackets can be interpreted as a function that tells us the "energy density" at a point $\mathbf{r'}$. This energy per unit volume gets multiplied by the infinitesimal volume $d\tau'$, and we add to get the total W. As a function, the energy density stored in the gravitational field is

$$u(\mathbf{r}) = -\frac{1}{8\pi G} \nabla \varphi(\mathbf{r}) \cdot \nabla \varphi(\mathbf{r}) = -\frac{g(\mathbf{r})^2}{8\pi G}. \tag{1.41}$$

Going back to the electrostatic energy density $u(\mathbf{r}) = 1/2\epsilon_0 E(\mathbf{r})^2$ and making the usual replacement, we recover the expression in (1.41), just an easy check at this point. The main difference between the electrostatic energy density and the gravitational one is the sign. For a net positive or negative charge density, bringing in additional charge involves an electric force that is repulsive, so we do work *pushing* the charge into place, leading to a positive energy density. In the gravitational setting, where the force is always attractive, we are pulling on the charge to keep it moving at constant speed instead of racing towards the partially built mass distribution. The total energy stored and its density are negative.

Here we have used the potential to develop the expression for energy stored in a gravitational field, an interesting notion, and one that is harder to derive using the field directly. That is an example of the type of information that can be developed using the potential, but its real utility is its role in energy conservation and the determination of radial and other types of particle motion.

---

[8] The divergence theorem gives, for functions f and g, and a domain Ω with boundary $\partial\Omega$:

$$\int_\Omega \nabla \cdot (f\nabla g)\, d\tau = \oint_{\partial\Omega} f\nabla g \cdot d\mathbf{a}.$$

Meanwhile, the product rule has $\nabla \cdot (f\nabla g) = \nabla f \cdot \nabla g + f\nabla^2 g$, and we can use this on the left above,

$$\int_\Omega \nabla f \cdot \nabla g\, d\tau + \int_\Omega f\nabla^2 g\, d\tau = \oint_{\partial\Omega} f\nabla g \cdot d\mathbf{a}.$$

If f or ∇g vanish on the boundary of Ω, then

$$\int_\Omega \nabla f \cdot \nabla g\, d\tau = -\int_\Omega f\nabla^2 g\, d\tau.$$

1.3 Radial Infall

We turn from finding fields given sources to the question: How do particles move in the presence of a gravitational field? Newton's second law, written in terms of the potential φ for some configuration of source mass reads

$$m\ddot{\mathbf{r}}(t) = m\mathbf{g}(\mathbf{r}(t)) = -m\big(\nabla\varphi(\mathbf{r})\big)\big|_{\mathbf{r}=\mathbf{r}(t)}. \tag{1.42}$$

If we dot $\dot{\mathbf{r}}(t)$ into both sides of this equation, then we can write the left-hand side as a total time derivative of what we recognize as the kinetic energy. On the right, we also have a time derivative from the chain rule:

$$\frac{d}{dt}\left(\frac{1}{2}m\dot{\mathbf{r}}(t)\cdot\dot{\mathbf{r}}(t)\right) = \frac{d}{dt}\big(-m\varphi(\mathbf{r}(t))\big), \tag{1.43}$$

a differential equation that we can solve by integrating both sides, picking up a constant of integration along the way:

$$\frac{1}{2}m\dot{\mathbf{r}}(t)\cdot\dot{\mathbf{r}}(t) = -m\varphi(\mathbf{r}(t)) + E \longrightarrow E = \frac{1}{2}m\dot{\mathbf{r}}(t)\cdot\dot{\mathbf{r}}(t) + m\varphi(\mathbf{r}(t)). \tag{1.44}$$

The aptly named constant E represents the total energy of the particle and consists of the familiar kinetic and potential energy $U = m\varphi$.

As a simple starting point for investigating motion, consider a spherically symmetric central body of mass M and radius R with its center at the origin. That sets up a gravitational potential $\varphi(\mathbf{r}) = -GM/r$. A particle of mass m starts from rest at some initial distance r_0 from the central sphere. How long does it take for the particle to run into the surface of the sphere? Since the gravitational force acts along the line connecting the center of the sphere to the particle, this is really a one-dimensional problem. In spherical coordinates, the only direction that matters is the radial one. Since the particle starts from rest at $r(0) = r_0$, we know the constant $E = -GMm/r_0$, and then at any other time,

$$E = -\frac{GMm}{r_0} = \frac{1}{2}m\dot{r}(t)^2 - \frac{GMm}{r(t)}. \tag{1.45}$$

The particle mass m can be canceled throughout, typical of gravitational problems, and the radial velocity can be isolated:

$$\dot{r}(t) = \pm\sqrt{2GM\left(\frac{1}{r(t)} - \frac{1}{r_0}\right)}. \tag{1.46}$$

Which sign should we take here? Since the particle is falling *in*, its distance to the origin shrinks in time, and $\dot{r}(t) < 0$. Picking the minus sign, we can separate the equation,

$$-\frac{dr}{\sqrt{2GM(1/r - 1/r_0)}} = dt, \tag{1.47}$$

and integrate both sides. To keep the solution general, at least at first, we'll integrate from $r_0 \to r$ on the left and from $0 \to t$ on the right, the easy side,

$$\int_r^{r_0} \frac{d\bar{r}}{\sqrt{2GM(1/\bar{r} - 1/r_0)}} = t. \tag{1.48}$$

Integrating on the left,

$$\sqrt{\frac{r_0}{2GM}} \left[\sqrt{r(r_0 - r)} + r_0 \tan^{-1}\left(\sqrt{-1 + \frac{r_0}{r}} \right) \right] = t. \tag{1.49}$$

In this form, we can easily find the time T associated with $r = R$, the time at which the particle crashes into the surface. What is the speed of the particle when this happens? We can get the value directly from (1.46),

$$\dot{r}(T) = -\sqrt{2GM\left(\frac{1}{R} - \frac{1}{r_0} \right)}. \tag{1.50}$$

This expression has no limit, there is nothing stopping the particle from traveling at a speed greater than the speed of light. We haven't used the correct relativistic dynamics, so this comes as no surprise (try out Problem 1.9 if you want to see the relativistic resolution). Nevertheless, there are interesting things to think about here. Suppose the particle starts out at spatial infinity, $r_0 \to \infty$. Then its speed upon impact is $v_T = \sqrt{2GM/R}$. Newton's second law is time reversible, so we also know, given M and R, that if we send a particle from the central sphere's surface with this same speed, then the particle will "just" make it to spatial infinity, where it will be at rest. That special initial launch speed is called the "escape speed," since it is the minimum speed needed to leave the gravitational environment of the central mass. Any more than this, and you'd end up out at spatial infinity moving with some constant nonzero speed. Here we have obtained this speed circuitously, but in Problem 1.10, you will do it quickly and directly from (1.45).

It is amusing that we can already make our first prediction, combining gravity with special relativity. Suppose we take the escape speed idea and turn it around. For a sphere of mass M, what is the maximum radius such that the escape speed is the speed of light? Setting $v_T = c$ and solving for R, we get

$$R = \frac{2GM}{c^2}. \tag{1.51}$$

A massive body with radius less than this would have a surface from which light itself could not escape. This was known by Laplace [32] and Michell [33] in the 1700s, and Laplace called such objects "dark stars." In general relativity, these have been relabeled "black holes," but the idea (although not the mechanism)[9] is the same. In the relativistic context, this critical size is known as the "Schwarzschild radius," and we write $R_M \equiv 2GM/c^2$.

---

[9]   There is a major loophole in the argument here....

Example 1.3 (Infall Reparameterization). We can use the integral in (1.48) to write a general expression for radial infall parameterized by an angle ψ. Define the constant $A \equiv \sqrt{2GM/r_0}$, and let $r = r_0 \cos^2 \theta$ where $\theta = 0$ is associated with r_0 and θ itself pins down r. Then (1.48) becomes (using $\bar{\theta}$ as the integration variable)

$$-\int_\theta^0 r_0 \big(1 + \cos(2\bar{\theta})\big) d\bar{\theta} = At \longrightarrow r_0 \Big(\theta + \frac{1}{2}\sin(2\theta)\Big) = At. \quad (1.52)$$

Letting $\theta = \psi/2$,

$$t = \frac{r_0}{2}\sqrt{\frac{r_0}{2GM}}(\psi + \sin\psi). \quad (1.53)$$

The total time it takes to reach $r = 0$, where $\theta = \pi/2 \to \psi = \pi$, is

$$T = \frac{r_0}{2}\sqrt{\frac{r_0}{2GM}}\pi. \quad (1.54)$$

We have the pair of variables $r(\psi)$ and $t(\psi)$ parameterized by $\psi = 0 \to \pi$, and these can be used to describe the trajectory in a spacetime diagram (see Section A.3 for the definition). In terms of ψ, $ct(\psi)$ and $r(\psi)$ are, using $R_M = 2GM/c^2$,

$$\begin{aligned} ct(\psi) &= \frac{r_0}{2}\sqrt{\frac{r_0}{R_M}}(\psi + \sin\psi), \\ r(\psi) &= r_0 \cos^2(\psi/2) = \frac{r_0}{2}(1 + \cos\psi). \end{aligned} \quad (1.55)$$

Notice that this solution can also be run backwards in time to tell a complete "up and down" story: At $t = -T$, a particle leaves $r = 0$ heading "up," at time $t = 0$, it reaches its maximum height r_0, where it is momentarily at rest, and then the particle falls back into $r = 0$ at time T. All that changes is we have to run time from $-T \to T$, which automatically changes the sign of $\dot{r}(t)$ from negative (for infall) to positive. The full trajectory from $t = -T \to T$ is shown in Figure 1.5.

1.3.1 Tidal Forces

We can use the radial infall idea to think about the effect of gravitational forces on extended bodies. Suppose you have a rectangular volume of noninteracting material ("dust") in the gravitational field of the Earth as shown in Figure 1.6. Each particle in the volume experiences an acceleration inward towards the center of the Earth, setting up a squeezing of the sides of the rectangle. At the same time, the bottom of the volume has acceleration that is greater than the top, so there is also a stretching force. This simultaneous stretching and squeezing is typical of gravity, and the combination is known as the "tidal force" – you'll work out the relative accelerations in Problem 1.12.

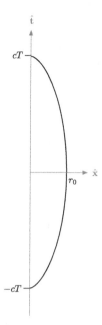

Fig. 1.5 A particle starts from $r = 0$ at time $t = -T$ with T in (1.54), moving with positive initial velocity. It travels out to its maximum "height" r_0, where it is at rest at $t = 0$, and then falls back into the origin at T.

1.4 Scattering and Bound Trajectories

In addition to the conservation of energy statement in (1.44), we can establish that the angular momentum of a test mass is conserved by spherically symmetric gravitational fields, $\mathbf{g}(\mathbf{r}) = g(r)\hat{\mathbf{r}}$. Crossing $\mathbf{r}(t)$ into both sides of (1.42) and noting that $\mathbf{r} \times (g(r)\hat{\mathbf{r}}) = 0$, we find that

$$\mathbf{r}(t) \times \left(m\ddot{\mathbf{r}}(t)\right) = 0 = \frac{d}{dt}\left(\mathbf{r}(t) \times \left(m\dot{\mathbf{r}}(t)\right)\right), \tag{1.56}$$

which shows that the angular momentum $\mathbf{L} = \mathbf{r}(t) \times (m\mathbf{v}(t))$ is a constant of the motion. Writing the velocity vector in spherical coordinates,

$$\mathbf{v}(t) \equiv \dot{\mathbf{r}}(t) = \dot{r}(t)\hat{\mathbf{r}} + r(t)\dot{\theta}(t)\hat{\boldsymbol{\theta}} + r(t)\sin\left(\theta(t)\right)\dot{\phi}(t)\hat{\boldsymbol{\phi}}, \tag{1.57}$$

where the unit vectors are themselves time-dependent, owing to their position dependence (but are still orthogonal at every point), we can evaluate the angular momentum vector in spherical coordinates,

$$\mathbf{L} = \mathbf{r} \times (m\mathbf{v}) = mr(t)^2\dot{\theta}(t)\hat{\boldsymbol{\phi}} - mr(t)^2\sin\left(\theta(t)\right)\dot{\phi}(t)\hat{\boldsymbol{\theta}}. \tag{1.58}$$

Since $\mathbf{L}$ is constant, we can align the $\hat{\mathbf{z}}$ axis with it initially, isolating its direction, $\mathbf{L} = L_z\hat{\mathbf{z}}$. Equating the spherical form from (1.58) with $L_z\hat{\mathbf{z}}$ written in

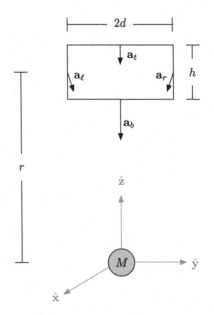

A spherical central body of mass M at the origin sets up a gravitational field. A macroscopic massive object with height h and width $2d$ has its center a distance r away from the center of the sphere. Different pieces of the massive object experience different accelerations, with $\mathbf{a}_t$ at the top, a smaller (in magnitude) $\mathbf{a}_b$ at the bottom, and equal magnitude but different direction accelerations for the left, $\mathbf{a}_\ell$, and right, $\mathbf{a}_r$, sides. All acceleration vectors point towards the center of the sphere.

spherical coordinates,

$$mr(t)^2\dot{\theta}(t)\hat{\boldsymbol{\phi}} - mr(t)^2 \sin\big(\theta(t)\big)\dot{\phi}(t)\hat{\boldsymbol{\theta}} = L_z \cos\theta(t)\hat{\mathbf{r}} - L_z \sin\theta(t)\hat{\boldsymbol{\theta}}, \quad (1.59)$$

we can obtain expressions for our constants of motion. Dotting $\hat{\mathbf{r}}$ into both sides of the above gives $L_z \cos\theta(t) = 0$, and if we want $\mathbf{L}$ to be nonzero, then we must have $\theta(t) = \pi/2$, a constant. That puts the motion of the particle in the xy plane (as expected). Because $\theta(t)$ is constant, $\dot{\theta}(t) = 0$, and the equality from (1.59) becomes

$$-mr(t)^2\dot{\phi}(t)\hat{\boldsymbol{\theta}} = -L_z\hat{\boldsymbol{\theta}} \longrightarrow L_z = mr(t)^2\dot{\phi}(t), \qquad (1.60)$$

which we recognize as the z component of angular momentum.

Then we have two constants of integration, E and L_z. Using $\dot{\phi}(t) = L_z/(mr(t)^2)$ in the expression for E from (1.44), we can write the energy conservation as a one-dimensional kinetic term involving $\dot{r}(t)$ and an effective potential that depends only on $r(t)$:

$$\begin{aligned}
E &= \frac{1}{2}m\big(\dot{r}(t)^2 + r(t)^2\dot{\phi}(t)^2\big) + m\varphi\big(r(t)\big) \\
&= \frac{1}{2}m\dot{r}(t)^2 + \underbrace{\left(\frac{L_z^2}{2mr(t)^2} + m\varphi\big(r(t)\big)\right)}_{\equiv U_{\text{eff}}}.
\end{aligned} \qquad (1.61)$$

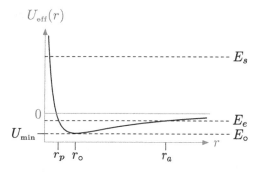

A sketch of an effective potential energy function. Near $r = 0$, the function is dominated by the $1/r^2$ term. As r gets large, the $-1/r$ term takes over, and in between, we could have a minimum at r_o. Three different energies are shown, E_o, E_e, and E_s associated with "bound" (circular, elliptical) and "scattering" (hyperbolic) trajectories, respectively.

Specializing to the potential outside of a sphere of mass centered at the origin, $\varphi = -GM/r$, let's sketch $U_{\text{eff}}(r)$ as a one-dimensional potential energy and see what it tells us about the possible motion of particles.

Near zero, the $1/r^2$ term dominates, so the potential energy is positive and approaching infinity. On the other end, as $r \to \infty$, the energy is negative and decaying as $1/r$. There are a few options, but in general the effective potential looks something like the one shown in Figure 1.7. In that figure, there is a minimum effective energy at r_o, and you will develop expressions for r_o and U_{min} in Problem 1.13. There are some typical energies overlaid on the effective potential, and we can think of these as one-dimensional energies, so that motion is prohibited when $E < U_{\text{eff}}$, where the kinetic $1/2m\dot{r}(t)^2$ would need to be negative. Then we can tell the story of the motion for each energy shown. The lowest energy trajectory is at E_o, and there is only one point where motion can occur, at r_o. Remember that we are really talking about two-dimensional motion in the xy plane, so a single value of $r(t) = r_o$ corresponds to circular motion with $\dot{\phi}(t) = L_z/(mr_o^2)$. Moving up, at E_e, the particle must move back and forth between r_p and r_a, an elliptical orbit here, with turning points at these locations. Finally, E_s represents a "scattering" trajectory: The particle comes in from spatial infinity, "hits" the potential energy curve where it must have zero speed, and then goes back out to infinity (hyperbolic motion in this spherical source context). It is interesting to note that *none* of these trajectories represents particle capture by the central body, meaning that there are no trajectories where a particle comes in from infinity and then runs into the central body. For a central spherical mass of finite radius, a particle could run into the surface, but for a point source, none of the trajectories shown here would ever reach zero. The only way for a mass m to run into a point source is pure, $L_z = 0$, radial infall as in the last section.

1.4.1 Circular Motion

Let's think about circular motion first, for $E = U_{\text{min}}$. That minimum occurs at

$$\frac{dU_{\text{eff}}(r)}{dr}\bigg|_{r=r_{\text{o}}} = 0 \longrightarrow r_{\text{o}} = \frac{L_z^2}{GMm^2}, \tag{1.62}$$

and the value of the potential energy there is

$$U_{\text{min}} = U(r_{\text{o}}) = -\frac{G^2M^2m^3}{2L_z^2}. \tag{1.63}$$

By tuning the constant angular momentum value L_z we can achieve circular orbits at any radius. A circular orbit is "stable" if it occurs, as shown in Figure 1.7, at a minimum (as opposed to a maximum, which does not appear in Figure 1.7), as all of these do. However, we will encounter effective potentials that do have maxima later on, and it will be important to determine whether a circular orbit is stable or unstable. Stability here is really a question of longevity, since if a circular orbit is unstable, then you wouldn't "see" particles orbiting at that radius; they would have been perturbed off of the trajectory by some other physical interaction.

1.4.2 Elliptical Motion

Moving on to elliptical motion associated with the energy E_e in Figure 1.7, we have to do some work to get a solution here, and we can start by changing from $r(t)$ in (1.61) to $\rho(t) \equiv 1/r(t)$, motivated by the form of the gravitational potential: it is better to deal with the linear $-GMm\rho(t)$ than with $-GMm/r(t)$. If $r(t) = 1/\rho(t)$, then $\dot{r}(t) = -\dot{\rho}(t)/\rho(t)^2$, and (1.61) becomes

$$E_e = \frac{1}{2}m\frac{\dot{\rho}(t)^2}{\rho(t)^4} + \left(\frac{L_z^2}{2m}\rho(t)^2 - GMm\rho(t)\right). \tag{1.64}$$

To determine the shape of the trajectory, we can switch to a geometric parameterization, expressing ρ as a function of ϕ instead of t. We will lose the temporal evolution information but obtain a simple ODE that can be solved to get the elliptical trajectory quickly. From the constant angular momentum we know that $\dot{\phi}(t) = L_z/(mr(t)^2) = (L_z/m)\rho(t)^2$ and

$$\frac{d\rho(\phi(t))}{dt} = \frac{d\rho(\phi)}{d\phi}\frac{d\phi}{dt} = \frac{d\rho(\phi)}{d\phi}\frac{L_z}{m}\rho(\phi)^2. \tag{1.65}$$

Using this form for $\dot{\rho}(t)$ back in the energy conservation equation,

$$E_e = \frac{1}{2}\frac{L_z^2}{m}\left(\frac{d\rho(\phi)}{d\phi}\right)^2 + \frac{1}{2}\frac{L_z^2}{m}\rho(\phi)^2 - GMm\rho(\phi). \tag{1.66}$$

In the radial infall section, we solved (1.45) for $\dot{r}(t)$. In that setting, when we took the square root, we knew that only the negative piece contributed (that was *infall*). Here the trajectory has turning points, so if we solved (1.66) for $\rho(\phi)$, we'd have to switch signs to cover the entire trajectory, which leads to more complicated bookkeeping. Fortunately, as you will show in Problem 1.17, we can eliminate the sign ambiguity by taking the ϕ derivative. If we take the ϕ derivative of (1.66) and solve for the second derivative of $\rho(\phi)$, then we obtain the delightfully familiar

$$\frac{d^2\rho(\phi)}{d\phi^2} = -\rho(\phi) + \frac{GMm^2}{L_z^2} \tag{1.67}$$

with equally familiar solution

$$\rho(\phi) = A\cos(\phi) + B\sin(\phi) + \frac{GMm^2}{L_z^2}. \tag{1.68}$$

The constants A and B can be set from the initial conditions. Looking back once again at Figure 1.7, we expect there to be identifiable turning points at r_p and r_a. Those points are determined by the intersection of E_e and the effective potential, points where $\dot{r}(t) = 0$, and we can solve for them explicitly from (1.61):

$$r_{\pm} = -\frac{GMm}{2E_e} \pm \sqrt{\frac{L_z^2}{2mE_e} + \left(\frac{GMm}{2E_e}\right)^2} \tag{1.69}$$

(keeping in mind that $E_e < 0$). Using terminology appropriate for the Sun as the central body, the point of closest approach is called the "perihelion," with $r_p = |r_-|$, whereas the point furthest from the central body is the "aphelion," which we'll denote $r_a = r_+$. We can orient the orbit so that at $\phi = 0$, the particle is at r_p. At that point, we know that the derivative of ρ is zero (since $\dot{r} = 0$ there, so is $\dot{\rho}$, and then from (1.65), the ϕ derivative of ρ is also zero), so that $B = 0$. For A, we have

$$\rho(0) = A + \frac{GMm^2}{L_z^2} = \frac{1}{r_p} \longrightarrow A = \frac{L_z^2 - GMm^2 r_p}{L_z^2 r_p}. \tag{1.70}$$

Finally, we can replace reference to L_z and m with their expressions in terms of the orbital parameters E_e, r_p, and r_a obtained from (1.69),

$$L_z = \pm E_e\sqrt{\frac{2}{MG}r_a r_p(r_a + r_p)}, \qquad m = -\frac{E_e}{GM}(r_a + r_p), \tag{1.71}$$

where the $\pm$ on L_z indicates clockwise or counterclockwise motion. Using these, $r(\phi) = 1/\rho(\phi)$ is

$$\begin{aligned} r(\phi) &= \frac{1}{((r_a - r_p)/(2r_a r_p))\cos\phi + (r_a + r_p)/(2r_a r_p)} \\ &= \frac{2r_a r_p/(r_a + r_p)}{1 + ((r_a - r_p)/(r_a + r_p))\cos\phi}. \end{aligned} \tag{1.72}$$

This is the radial form of an ellipse with eccentricity e and semilatus rectum p defined by

$$e = \frac{r_a - r_p}{r_a + r_p}, \qquad p = \frac{2r_a r_p}{r_a + r_p}, \tag{1.73}$$

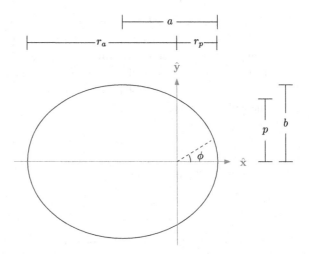

Fig. 1.8   An elliptical trajectory for a particle orbiting a spherical central body at the origin. The ellipse has
closest approach point at r_p, and furthest point from the central sphere at r_a. The particle crosses the
$\hat{\mathbf{y}}$ axis at the semilatus rectum p. We can relate these horizontal and vertical distances to the
semimajor axis a and semiminor axis b.

where e tells us how circular the ellipse is (with $e = 0$ corresponding to a
circle), and p gives a measure of the vertical extent of the ellipse. We can
sketch the ellipse: from the points $r(0) = r_p$, $r(\pi) = r_a$, and $r(\pi/2) =
2r_a r_p/(r_a + r_p) = p$ we get the picture shown in Figure 1.8. There are many
ways to characterize elliptical motion; another set of horizontal and vertical
length descriptions are the semimajor axis, a and semiminor axis b, also shown
in Figure 1.8. These can be related to r_a, r_p, and p (see Problem 1.19).

1.4.3 Kepler's Laws

The elliptical orbit, in the context of the Sun as the central body, expresses
Kepler's first law: Planets travel in elliptical orbits with the Sun at a focus.
Kepler's second law says that "equal areas are swept out in equal times." We
can establish this from the constant angular momentum vector's magnitude.
Referring to Figure 1.9, consider the infinitesimal triangle made out of $\mathbf{r}(t)$
and $\dot{\mathbf{r}}(t)dt$ with area

$$dA = \frac{1}{2}\left|\mathbf{r}(t) \times \dot{\mathbf{r}}(t)dt\right| = \frac{1}{2}\frac{L_z}{m}dt \longrightarrow \frac{dA}{dt} = \frac{L_z}{2m}, \qquad (1.74)$$

and the change in area per unit time is a constant.

Kepler's third law says that the orbital period is proportional to $a^{3/2}$. To set
up the calculation that establishes this, we'll go all the way back to (1.61),
write it in terms of a and b (first, we'll write L_z and m in terms of r_p and r_a,

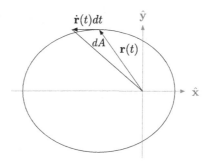

An orbiting body sweeps out the infinitesimal area dA in a time dt. That area is half of the parallelogram area spanned by $\mathbf{r}(t)$ and $\dot{\mathbf{r}}(t)dt$, which can be computed using the cross product of the two, itself proportional to the angular momentum magnitude.

then replace those using the result of Problem 1.19), and then solve for $\dot{r}(t)$:

$$\dot{r}(t) = \pm \frac{1}{r(t)} \sqrt{\frac{GM}{a} \left(-b^2 + 2ar(t) - r(t)^2 \right)}. \tag{1.75}$$

To find the period, we only need to consider a portion of the trajectory, and we can pick a piece with definite sign for $\dot{r}(t)$. The orbiting particle starts at r_p with $\dot{r} = 0$. Its radial distance from the central body increases until we get to r_a, and this represents half of the period of motion. Taking the plus sign in (1.75) ($\dot{r}$ is increasing), we can separate just as we did in the radial infall case to get

$$\frac{r\,dr}{\sqrt{(GM/a)(-b^2 + 2ar - r^2)}} = dt. \tag{1.76}$$

Integrating on the left from $r_p \to r_a$ and on the right from $0 \to T/2$,

$$\frac{\pi a^{3/2}}{\sqrt{MG}} = \frac{T}{2} \longrightarrow T = \frac{2\pi a^{3/2}}{\sqrt{MG}}, \tag{1.77}$$

which is the third law. Keep in mind that Kepler's laws are direct observations of the motion of the planets; he had no idea *why* the planets moved as they did. Newton, with his inverse square force law published in 1687, got all three observational facts right off the bat. It would be a long time before people found an observational need for any modifications of Newton's theory.

1.4.4 Hyperbolic Motion

Finally, let's work out the trajectories of the "scattering" motion, the one associated with E_s in Figure 1.7. These can be developed geometrically from (1.67) as a special case, since nothing there required bound motion. Starting with the solution to (1.67) in the case where there is no central body, that is, $M = 0$, we expect a linear trajectory. If we take $A = 0$ and set $B = 1/\bar{R}$ for some length scale $\bar{R}$ in (1.68), then as a function of ϕ, the x and y locations of the particle are

$$x(\phi) = \frac{\cos\phi}{\rho(\phi)} = \bar{R}\cot\phi, \qquad y(\phi) = \frac{\sin\phi}{\rho(\phi)} = \bar{R}, \qquad (1.78)$$

which does indeed describe a straight line going from $x = -\infty \to \infty$ with constant $y = \bar{R}$ as ϕ goes from π to 0. Now let's put back the central mass; our solution is

$$\rho(\phi) = \frac{\sin\phi}{\bar{R}} + \frac{GMm^2}{L_z^2}. \qquad (1.79)$$

What happens? The previously straight trajectory bends about the central mass. At $\phi = \pi$ and 0, we have $y = 0$, whereas x goes from $-L_z^2/(GMm^2)$ to $L_z^2/(GMm^2)$, so let $x_0 \equiv L_z^2/(GMm^2)$. The point of closest approach is at $\phi = \pi/2$, where of course $x = 0$, and

$$y(\phi = \pi/2) = \frac{1}{1/\bar{R} + 1/x_0}. \qquad (1.80)$$

As advertised in Figure 1.7, the particle starts at infinity, comes in to a finite distance from the central body at which point its radial velocity is zero, and then heads back out to infinity. The particle is at spatial infinity for $\rho(\phi_\infty) = 0$, which occurs when $\sin\phi_\infty = -\bar{R}/x_0$, a negative value for the angle ϕ. Our initial "launch" angle for the mass that moves along this trajectory is, then, $\theta = -\phi_\infty = \sin^{-1}(\bar{R}/x_0)$. We can also define the "scattering" angle ψ to be the angle of asymptotic deflection between the initial and final velocity vectors. The trajectory and angles θ and ψ are shown in Figure 1.10.

We highlight in these trajectories the lack of "capture" by the central body. For a point source mass M, there is no way for the family of trajectories described by (1.79) to pass through the origin; they are all bent and will miss the source (again, ignoring the possible finite source radius for a particle to run

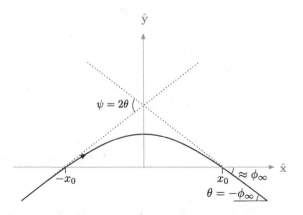

Fig. 1.10 A particle moves from left to right along the trajectory shown, crossing the x axis at $\pm x_0$, where $x_0 \equiv L_z^2/(GMm^2)$. The particle is infinitely far away from the central body (located at the origin) when $\rho(\phi_\infty) = 0$, giving $\phi_\infty = -\sin^{-1}(\bar{R}/x_0)$, and then the "launch" angle $\theta = -\phi_\infty = \sin^{-1}(\bar{R}/x_0)$.

into). This was also true for the orbital motion of the previous section. Only zero angular momentum infall trajectories can reach the source, in general. That will not be the case for one of the defining objects of general relativity; black holes capture orbiting particles and can also draw in these otherwise unbound trajectories.

1.5 A Little Special Relativity

What happens if we add a little special relativity to Newtonian gravity? From Einstein's relativistic energy relation

$$E = \frac{mc^2}{\sqrt{1 - v^2/c^2}} \tag{1.81}$$

we recover the famous $E = mc^2$ for a particle at rest. Although energy and mass are *not* the same, their functional equivalence here gives us an interesting entry point. Mass is explicitly the source in the field equation of Newtonian gravity (1.16), but what if we expand our view to include both familiar mass and also the effective mass E/c^2? Where would we see contributions? Almost everywhere! First, we will explore the physics of including the energy density of an electric field for massive charged particles. The electric field clearly acts as a source of gravity, but can it also *respond* to gravitational fields? If we are to take energy as a source seriously, then we must also allow it to be a "test particle," that is, we must also consider the dynamical effects of gravitational fields on massless energy blobs, such as the light beam from a laser. Can light be bent by gravitational fields? Can it orbit a central body the way massive particles can? The answer to all of these questions will be an emphatic "yes," although, as a caveat, the mechanism in general relativity is quite different from the one we will allow in this section. Nevertheless, the qualitative predictions we generate here are all borne out in the full theory and were predicted for Newtonian gravity by Soldner [46] in the early eighteen hundreds.

To start things off, then, we will distinguish between "real" mass densities $\rho(\mathbf{r})$ and energy "mass" densities $\rho_E(\mathbf{r}) = u(\mathbf{r})/c^2$, where $u(\mathbf{r})$ is an energy density. Then our formal update to (1.16) will be

$$\nabla \cdot \mathbf{g}(\mathbf{r}) = -4\pi G\left(\rho(\mathbf{r}) + \frac{u(\mathbf{r})}{c^2}\right), \qquad \nabla \times \mathbf{g}(\mathbf{r}) = 0, \tag{1.82}$$

and, of course, we'll leave the force contribution as $\mathbf{F}(\mathbf{r}) = m\mathbf{g}(\mathbf{r})$ (where we understand that m could also be an effective energetic mass). It is easier to work with the potential (although developing the integral "Gauss's law" for (1.82) is possible, you will do that in Problem 1.21), and the Poisson problem from (1.23) becomes

$$\nabla^2 \varphi(\mathbf{r}) = 4\pi G \left(\rho(\mathbf{r}) + \frac{u(\mathbf{r})}{c^2} \right). \tag{1.83}$$

One interesting element of the relativistic addition is the role of the speed of light c in setting the scale of the potential, which itself has units of speed squared.

With the Pandora's box open, we are also invited to think about the energy density associated with the gravitational field itself; it should act as its own source if we are allowing energy coupling. That will end up raising questions that will take us through the next few chapters. As a prelude, take a look at Problem 1.22 to see the type of issues that arise.

1.5.1 Massive Charged Particles

A point particle of mass M and charge Q sits at the origin. What is the gravitational potential that it generates? There are two independent contributions. First, the usual $\varphi_M = -GM/r$, but there is a new one coming from the second source term in (1.83). The electric field of the charge is $\mathbf{E} = Q/(4\pi\epsilon_0 r^2)\hat{\mathbf{r}}$, and the energy density of that field is

$$u(r) = \frac{1}{2}\epsilon_0 E^2 = \frac{1}{32\epsilon_0}\left(\frac{Q}{\pi r^2}\right)^2. \tag{1.84}$$

Since we already know about the mass M contribution, let's focus on the potential due to just this new energy density, we can always add back in the familiar piece by superposition.

The spherically symmetric field equation we need to solve is

$$\nabla^2 \varphi(r) = \frac{GQ^2}{8\pi\epsilon_0 c^2 r^4}. \tag{1.85}$$

The units on the left are inverse seconds squared, and on the right, we can identify the combination $GQ^2/(\epsilon_0 c^2)$ as having units of m^4/s^2. So just as there is a natural length associated with the mass M, the Schwarzschild radius $R_M \equiv 2GM/c^2$, there is also a natural length that goes along with the charge, $R_Q^2 \equiv GQ^2/(4\pi\epsilon_0 c^4)$.

Using $\nabla^2 \varphi(r) = (r\varphi(r))''/r$ (where prime denotes the derivative with respect to r), we can multiply both sides of (1.85) by r and integrate twice to get

$$r\varphi(r) = Br + A + \frac{c^2 R_Q^2}{4r} \longrightarrow \varphi(r) = B + \frac{A}{r} + \frac{c^2 R_Q^2}{4r^2} \tag{1.86}$$

with constants of integration A and B. Taking $\varphi(r \to \infty) = 0$, we can set $B = 0$, and we recognize A/r as the vacuum solution associated with the mass of the central body which we are accounting for separately, so the contribution to the potential coming from the charge is just the new $1/r^2$ term.[10] Putting this together with the mass term $-GM/r = -c^2 R_M/(2r)$, the gravitational potential is

$$\varphi(r) = c^2\left(-\frac{R_M}{2r} + \frac{R_Q^2}{4r^2}\right). \tag{1.87}$$

Notice the factor of c^2 sitting out front setting the units and overall scale for $\varphi(r)$.

The gravitational field coming from this potential is

$$\mathbf{g} = -\nabla\varphi(r) = \frac{c^2}{2}\left(-\frac{R_M}{r^2} + \frac{R_Q^2}{r^3}\right)\hat{\mathbf{r}}, \tag{1.88}$$

and the force on a mass m is $\mathbf{F} = m\mathbf{g}$. There are a variety of observations we can now make. Perhaps the most striking is that there is a region in which the net gravitational force on a mass could be repulsive, owing to the opposite signs of the two terms on the right in (1.88). The crossover from attractive to repulsive occurs at the radius $r_c = R_Q^2/R_M$. Let's take a look at these length scales for a familiar particle, the electron with mass $M \approx 9.11 \times 10^{-31}$ kg and charge $Q \approx -1.6 \times 10^{-19}$ C. Then

$$R_M \approx 1.4 \times 10^{-57}\,\text{m}, \qquad R_Q = 1.4 \times 10^{-36}\,\text{m}, \qquad \frac{R_Q^2}{R_M} \approx 1.410^{-15}\,\text{m}. \tag{1.89}$$

All of these length scales involve physics beyond classical mechanics, so there is only so much we can say concretely here (the situation becomes more interesting for massive objects with finite size like the proton; see Problem 1.23). If we made a sphere of mass M and radius R that carried charge Q, and had both R_M and R_Q inside R, could we measure the charge of the sphere by watching *neutral* test particle trajectories? Absolutely, and it is interesting to compare and contrast those trajectories with the $Q = 0$ ones we found in Section 1.3 and Section 1.4. Try calculating the time it takes for a neutral particle of mass m to fall into the central body in Problem 1.24. You can also calculate the period of circular motion in this type of setting and compare that with the one from Kepler's law as in Problem 1.25. If you were to look for elliptical orbits here, then you would find that the additional $1/r^2$ piece of the potential leads to precession of elliptical orbits (as do all terms that are added to a $1/r$ potential). So you would certainly know that the central body was charged using only its gravitational interaction with other massive (but uncharged) particles. If you were to add charge to your test particles, so that they had mass m and charge

---

10 There could be a $1/r$ contribution from the charge, as long as the constant multiplying it vanished when $Q = 0$. Absent a more realistic source model that provides realizable boundary conditions for the vacuum solution, there is not much we can say about such $1/r$ contributions, so we have omitted terms of this form for simplicity here. See Problem 1.21 for a concrete example, in which a model particle is specified (and leads to a charge-dependent $1/r$ term in the potential). If it bothers you, then imagine $M(Q)$ as a modified mass term that depends on the charge of the central body.

q, then there would be two contributions from the charge Q of a central body: one due to the gravitational piece and then the usual electrostatic force acting on the charge q.

To be clear, what we have done here is neither Newtonian gravity nor general relativity. The field theory defined by (1.83) is a toy intermediary, although it does share qualitative predictions with the full theory of gravity.

1.5.2 Electromagnetic Fields in Vacuum

For a massive charged particle, the mass source is localized. The gravitational field that mass generates extends beyond the matter distribution, of course, but the source is confined to the particle's location. The charge piece of the gravitational sourcing comes through the charge's electric field energy density, which extends to infinity. Let's remove the material source from the picture and just think about electromagnetic field energy as a gravitational source. We can include magnetic sources; for a magnetic field $\mathbf{B}$, the energy density is $u = B^2/(2\mu_0)$. As a model, we could take our charged sphere from above and spin it, and then a dipole magnetic field develops outside the sphere. We are focusing on the contribution from the electromagnetic fields in (material) vacuum, with $\nabla \times \mathbf{B} = 0$, so out there the magnetic field comes from a magnetic scalar "potential" W with $\mathbf{B} = -\nabla W$ (the minus sign is just to parallel the minus sign in the electric $\mathbf{E} = -\nabla V$).

The equation governing the gravitational potential with these static electric and magnetic energy density sources is

$$\nabla^2 \varphi(\mathbf{r}) = \frac{2\pi G}{c^2} \left[\epsilon_0 \nabla V(\mathbf{r}) \cdot \nabla V(\mathbf{r}) + \frac{1}{\mu_0} \nabla W(\mathbf{r}) \cdot \nabla W(\mathbf{r}) \right], \qquad (1.90)$$

and as it turns out, we can solve this quite generally, again provided that we are in vacuum, so that $\nabla^2 V = 0$ and $\nabla^2 W = 0$. For a potential φ with gradient

$$\nabla \varphi(\mathbf{r}) = \frac{2\pi G}{c^2} \left[\epsilon_0 V(\mathbf{r}) \nabla V(\mathbf{r}) + \frac{1}{\mu_0} W(\mathbf{r}) \nabla W(\mathbf{r}) \right], \qquad (1.91)$$

the divergence of the gradient is

$$\nabla \cdot \nabla \varphi(\mathbf{r}) = \frac{2\pi G}{c^2} \left[\epsilon_0 \nabla V(\mathbf{r}) \cdot \nabla V(\mathbf{r}) + \epsilon_0 V \nabla^2 V(\mathbf{r}) \right.$$
$$\left. + \frac{1}{\mu_0} \nabla W(\mathbf{r}) \cdot \nabla W(\mathbf{r}) + \frac{1}{\mu_0} W \nabla^2 W(\mathbf{r}) \right], \qquad (1.92)$$

which matches (1.90) since the Laplacians are zero (and omitting contributions to $\nabla \varphi$ that have zero divergence). We can "integrate" the gradient in (1.91) to get

$$\varphi(\mathbf{r}) = \frac{\pi G}{c^2} \left[\epsilon_0 V(\mathbf{r})^2 + \frac{1}{\mu_0} W(\mathbf{r})^2 \right]. \qquad (1.93)$$

Fantastic: all you have to do is give me the electric and magnetic potentials, and we get the gravitational potential associated with these for free. Since we already know the gradient of φ from (1.91), the gravitational field is similarly easy,

$$\mathbf{g}(\mathbf{r}) = -\nabla\varphi(\mathbf{r}) = \frac{2\pi G}{c^2}\left[\epsilon_0 V(\mathbf{r})\mathbf{E}(\mathbf{r}) + \frac{1}{\mu_0}W(\mathbf{r})\mathbf{B}(\mathbf{r})\right]. \tag{1.94}$$

We have worked away from the source, here, where the Laplacians of $\mathbf{E}$ and $\mathbf{B}$ vanish. It is possible to add to the field in (1.94) any term whose divergence vanishes at points away from the source of the electric and magnetic fields. Notably, we could add a term of the form $A/r^2\,\hat{\mathbf{r}}$, with delta-function divergence, which is zero for points other than the origin.

Example 1.4 (A Spinning Sphere of Charge). A ball of radius R carries total charge Q uniformly distributed inside of it. The ball is centered at the origin and spins about the z axis with angular speed ω. The electric field and potential outside of the sphere are

$$\mathbf{E} = \frac{Q}{4\pi\epsilon_0 r^2}\hat{\mathbf{r}}, \qquad V = \frac{Q}{4\pi\epsilon_0 r}. \tag{1.95}$$

The magnetic field outside of the configuration is the usual dipolar one with

$$\mathbf{B} = \frac{\mu_0 m}{4\pi r^3}(2\cos\theta\,\hat{\mathbf{r}} + \sin\theta\,\hat{\boldsymbol{\theta}}), \tag{1.96}$$

and the dipole moment for a spinning charged ball is $m = Q\omega R^2/5$ in magnitude. We need to find W such that $\mathbf{B} = -\nabla W$. You should try to generate the appropriate form in Problem 1.27; it ends up being

$$W = -\frac{\mu_0 m}{4\pi r^2}\cos\theta. \tag{1.97}$$

Using the potentials and fields in (1.94), we can find the gravitational field outside of the configuration due to the energy stored in its electric and magnetic fields,

$$\mathbf{g} = \frac{\mu_0 G Q^2}{4\pi r^3}\left[\left(\frac{1}{2} + \frac{\omega^2 R^4\cos^2\theta}{25c^2 r^2}\right)\hat{\mathbf{r}} + \frac{\omega^2 R^4\cos\theta\sin\theta}{50c^2 r^2}\hat{\boldsymbol{\theta}}\right]. \tag{1.98}$$

Example 1.5 (Lasers). As another, even simpler example that does not require the above machinery, think of a highly localized electromagnetic energy density source, a laser. This presents an infinite line of energy density λ that is confined to the laser beam itself. What is the gravitational field associated with such a massless energy source? We already know the answer from Example 1.2. For a laser pointing along the $\hat{\mathbf{z}}$ axis, the gravitational field a distance s from the axis is

$$\mathbf{g} = -\frac{2G\lambda}{s}\hat{\mathbf{s}}, \tag{1.99}$$

and there will be an attractive force exerted on a particle of mass m. Of course, the size of that force is very small (see Problem 1.28).

1.5.3 Gravitational Interaction with Light

A charged massive object generates an "extra" gravitational field because of the energy content of its electric field. Can that energy content also *respond* to gravitational fields? An electric or magnetic field doesn't have mass, but they do carry energy and therefore have effective mass.[11] Let's revisit some of our massive orbital trajectories to see if they apply at all to massless particles. We have already seen how "light" can interact with gravity in discussing escape speeds of c, so there is a radial component to light interaction with spherical masses. Are there circular or elliptical orbits in this toy theory? How about "scattering" solutions, associated with a "bending" of light?

Circular Orbits for Light

For our purposes here, light is a massless particle that travels at speed c. Let's go back to the radius associated with circular orbits (1.62) and see what we get for such a "particle." Let's give light a fictitious mass m for a moment (again, think of it as an effective mass coming from the light's energy density if this temporary bookkeeping device bothers you). Then if the light is to travel in a circle at speed c, it must have $L_z = mr_\circ c$, and then (1.62) becomes

$$r_\circ = \frac{m^2 r_\circ^2 c^2}{GMm^2} \longrightarrow r_\circ = \frac{GM}{c^2}. \tag{1.100}$$

The circular orbit occurs at half the Schwarzschild radius, so that if you had a sphere of mass M with radius $R < GM/c^2$ (so that light is orbiting *outside* the sphere), you'd never know about it anyway, and no one could send a signal to tell you about it since the orbit is occurring inside the "dark star" and no light could escape.

Bending of Light

We again take the view that light has an effective mass m so that we can use expression (1.79) directly. The angular momentum L_z is a constant of the motion, and we can evaluate it when $\phi = \pi/2$, at the closest approach point. Let $\rho(\pi/2) \equiv 1/U$; then $L_z = mUc$, and we'll use the natural length scale for M,

---

[11] We could take the quantum mechanical view and focus on massless photons of frequency ν with energy $E = h\nu$ and effective mass $E/c^2 = h\nu/c^2$, but I prefer not to mix in quantum mechanics at this stage.

the Schwarzschild radius $R_M = 2GM/c^2$, to write (1.79) as

$$\frac{1}{U} = \frac{1}{\bar{R}} + \frac{R_M m^2}{2m^2 U^2 c^2} = \frac{1}{\bar{R}}\left(1 + \frac{\bar{R} R_M}{2U^2}\right) \longrightarrow U = \frac{1}{2}\left(\bar{R} \pm \bar{R}\sqrt{1 - \frac{2R_M}{\bar{R}}}\right).$$
(1.101)

Pick the positive root, which is the one that makes sense in the $R_M \to 0$ limit, where we set up our solution to have closest approach $\bar{R}$. Suppose that $R_M/\bar{R} \ll 1$; then $U \approx \bar{R}$. Then what does it mean to have $R_M/\bar{R} \approx R_M/U \ll 1$? Well, U is the closest point to the central body achieved by the trajectory. For an actual sphere of radius R_S, the smallest U that could be is R_S. For most spherical objects, the Schwarzschild radius is well inside the radius of the sphere, so that $R_M/R_S \ll 1$ automatically. As an example, take the Sun, for which $R_M \approx 3$ km, and the solar radius is roughly $R_S \approx 7 \times 10^5$ km. Then $R_M/R_S \approx 10^{-6}$, so the approximation is justified in that case (and you should try calculating this ratio for the Earth).

The result of this approximation is that we can take $U \approx \bar{R}$, and then $L_z = m\bar{R}c$. The x axis crossing point is

$$x_0 = \frac{L_z^2}{GMm^2} \approx \frac{\bar{R}^2 c^2}{GM} = \frac{2\bar{R}^2}{R_M},$$
(1.102)

and then the angle θ shown in Figure 1.10 is

$$\sin\theta = \frac{\bar{R}}{x_0} \approx \frac{R_M}{2\bar{R}} \longrightarrow \theta \approx \frac{R_M}{2\bar{R}}$$
(1.103)

or half the ratio of the Schwarzschild radius to the closest approach radius, a ratio that we have already established is small, justifying the small angle approximation $\sin\theta \approx \theta$ here.

If we think of the trajectory shown in Figure 1.10 as an actual, continuous beam of light,[12] then our interpretation of θ is as our viewing angle, that is, we see the light coming in at the angle θ. A peculiarity of our intuition with light is that we automatically assume that it is traveling in a straight line. If I ask you to point to the source of light in a room, you point straight towards the bulb, not off at an angle. So if we were asked where the light that we view at angle θ came from, our answer would be informed by a straight line interpretation. Referring to the modified form of Figure 1.10 in Figure 1.11, we would say that the light came from point B. In reality, because of the gravitational bending, the light came from point A. The difference between our perception and the actual emission point is characterized by the "deflection angle" ψ shown in the figure (the same angle we had in Figure 1.10). That angle, for Newtonian gravity, with the assumptions we have made is

$$\psi = 2\theta \approx \frac{R_M}{\bar{R}}.$$
(1.104)

This result, as with all of the ones we have developed for sources and interactions involving energy, is qualitatively correct. Light does get bent by grav-

---

[12] Or the path of a set of light particles emitted at some faraway star.

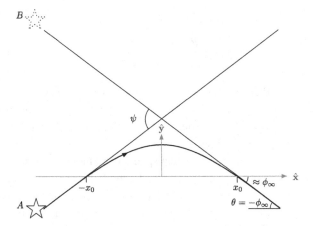

Fig. 1.11 Light is emitted by the star at point A, travels along the trajectory shown, and is received by us at the end of the trajectory, making an angle θ with respect to horizontal. We perceive the light coming from point B, an error described by the deflection angle $\psi = 2\theta$.

itational fields. But in general relativity the deflection angle $\psi = 2R_M/\bar{R}$, so our result here is off by a factor of two with respect to the observational result. Remember that the temporary theory we have here is intermediate between Newtonian gravity (which does not predict bending of light) and general relativity: we do not expect it to yield quantitatively accurate results.

1.6 A Little More Special Relativity

You can review the Lorentz transformation and its implications for both inertial frames and motion in Appendix A, which will refresh some useful results (see [21] and [19] for good introductions to special relativity) – in particular, we'll need length contraction, the transformation rule for the perpendicular (to a boost direction) component of a force, and velocity addition. To set these up, consider two inertial frames L, taken at rest, and $\bar{L}$ moving to the right through L with constant speed v along a shared x axis (this setup is depicted in Figure A.1). For a length $\Delta\bar{x}$ at rest in $\bar{L}$, the measured length in L is $\Delta x = \gamma^{-1}\Delta\bar{x}$ with

$$\gamma \equiv \frac{1}{\sqrt{1 - v^2/c^2}} > 1. \tag{1.105}$$

A mass at rest in $\bar{L}$ experiences a force $\bar{F}_\perp$ acting in a direction perpendicular to the x axis. That same force, measured in L, is $F_\perp = \gamma^{-1}\bar{F}_\perp$, as you can establish in Problem 1.30. Finally, for velocity addition, if an object moves

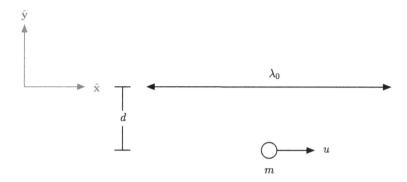

Fig. 1.12 An infinite line of mass is at rest with mass density λ_0. A distance d from the line there is a mass m moving to the right with speed u.

with speed $\bar{v}$ relative to $\bar{L}$, then its speed relative to L is

$$w = \frac{v + \bar{v}}{1 + \bar{v}v/c^2}. \tag{1.106}$$

With these three results in mind, we'll run through a thought experiment that should be familiar from E&M [40]. We'll start with two example configurations to get the pieces in place and then study a combination of the two. An infinite line of mass with mass density λ_0 is lying at rest along the x axis. A distance d away is a mass m moving to the right with speed u, as shown in Figure 1.12. We know the gravitational field associated with the line from Example 1.2, and then the force on the mass m is, in Cartesian coordinates, $2G\lambda_0 m/d\hat{\mathbf{y}}$.

Next, suppose the infinite line of mass is moving with speed s to the right. The mass density changes because of Lorentz contraction. When the line of mass was at rest, the mass contained in an interval $\Delta\bar{x}$ was $\Delta m = \lambda_0\Delta\bar{x}$. The interval is now contracted, and that same Δm mass is now in an interval $\Delta x = \gamma_s^{-1}\Delta\bar{x}$, where γ_s is the boost factor associated with the speed s, $\gamma_s \equiv (1 - s^2/c^2)^{-1/2}$. The new density λ is

$$\lambda = \frac{\Delta m}{\Delta x} = \frac{\lambda_0\Delta\bar{x}}{\gamma_s^{-1}\Delta\bar{x}} = \gamma_s\lambda_0 \tag{1.107}$$

with $\lambda > \lambda_0$, which makes sense; the same mass in a smaller interval means a larger density. Now the force on the mass m is

$$\frac{2G\lambda m}{d}\hat{\mathbf{y}} = \gamma_s\frac{2G\lambda_0 m}{d}\hat{\mathbf{y}}. \tag{1.108}$$

All right, time for the full configuration: an infinite line of mass with rest mass density λ_0 is moving to the right through L with speed v. A distance d away is a mass m moving to the right in L with speed u, as in the top sketch of Figure 1.13. We'll calculate the force in L and in the rest frame of the particle, which we'll call $\bar{L}$. In both frames, the force is perpendicular to the relative motion, so we can compare them using the perpendicular force transformation relation.

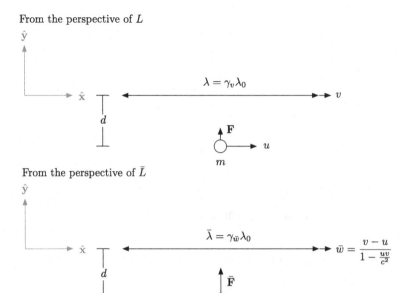

Fig. 1.13 A moving line of mass and moving test mass from two different points of view. In L, the line of mass is moving to the right with speed v, and the test mass is moving with speed u. There the line of mass has mass density $\lambda = \gamma_v \lambda_0$ (the boost factor γ_v has v as the relevant speed). In the rest frame of the mass m, called $\bar{L}$, the mass is at rest, and the line of mass moves with speed $\bar{w}$ obtained by relativistic velocity addition. There the line of mass has mass density $\bar{\lambda} = \gamma_{\bar{w}} \lambda_0$ with $\gamma_{\bar{w}}$ the boost factor attached to $\bar{w}$.

From the point of view of L, the infinite line of mass has mass density $\lambda = \gamma_v \lambda_0$ with $\gamma_v = (1 - v^2/c^2)^{-1/2}$. Then the force, which is perpendicular to the relative motion of the two frames, is

$$\mathbf{F} = \frac{2G\lambda m}{d}\hat{\mathbf{y}} = \gamma_v \frac{2G\lambda_0 m}{d}\hat{\mathbf{y}}. \tag{1.109}$$

In $\bar{L}$, the infinite line of mass is moving with speed $\bar{w}$, obtained through the relativistic velocity addition formula,

$$\bar{w} = \frac{v - u}{1 - uv/c^2}, \tag{1.110}$$

and it has the mass density $\bar{\lambda} = \gamma_{\bar{w}} \lambda_0$ with $\gamma_{\bar{w}} = (1 - \bar{w}^2/c^2)^{-1/2}$. Then the force, as measured in the rest frame of the mass, is

$$\bar{\mathbf{F}} = \frac{2G\bar{\lambda} m}{d}\hat{\mathbf{y}} = \gamma_{\bar{w}} \frac{2G\lambda_0 m}{d}\hat{\mathbf{y}}. \tag{1.111}$$

Now the question is whether or not the relation between perpendicular force components in L and $\bar{L}$ holds. If it did, then the magnitude F of the force in L would be related to the magnitude of the force in $\bar{L}$ by $F = \bar{F}/\gamma_u$, where $\gamma_u = (1 - u^2/c^2)^{-1/2}$ since u is the relative speed of the two frames. Let's compute $\bar{F}/\gamma_u$,

$$\frac{\bar{F}}{\gamma_u} = \frac{\gamma_{\bar{w}}}{\gamma_u}\frac{2G\lambda_0 m}{d} = \frac{\gamma_{\bar{w}}}{\gamma_u\gamma_v}F. \tag{1.112}$$

If $\gamma_{\bar{w}}/(\gamma_u\gamma_v)$ is 1, the correct relation holds, and we are done. But, in fact, you can show in Problem 1.31 that

$$\frac{\gamma_{\bar{w}}}{\gamma_u\gamma_v} = \left(1 - \frac{uv}{c^2}\right), \tag{1.113}$$

so the forces in the two frames do not stand in the correct ratio.

Presumably, we could have run this whole argument with an electric configuration, an infinite line of charge, with a moving charge nearby, etc. What is different in that setting that allows the special relativistic requirement to hold? Of course, it is the presence of the magnetic field and associated force that saves the calculation. That magnetic force is missing from Newtonian gravity, yet the present problem indicates that such a "gravitomagnetic" force is *necessary* in a special relativistic setting. Let's augment our field **g** with a new gravitomagnetic field **b** sourced by moving mass, just as the magnetic field is sourced by moving charge. Working by analogy with E&M, we'll take $\nabla \cdot \mathbf{b} = 0$, and therefore there are no monopole sources for **b**. The curl will be where the moving mass source lives, and we'll take $\nabla \times \mathbf{b} \propto \rho\mathbf{v}$, where ρ is the mass density, and **v** the velocity, telling how that mass is moving at any given location. Using a gravitational form for the Lorentz force will ensure that we have copied the structure faithfully, so let $\mathbf{F} = m\mathbf{v} \times \mathbf{b}$ to mimic $\mathbf{F} = q\mathbf{v} \times \mathbf{B}$ in E&M.

To set the constants of proportionality in the source equation, we could use our current example. But it is worth highlighting a completion of the electrostatic map, $1/\epsilon_0 \to -4\pi G$, to include the magnetic piece. That will also help us check the field **b** for various moving sources of interest against their stationary magnetic counterparts. To update the map, we need to know what to do with the magnetic constant μ_0. The speed of light is shared between the electromagnetic and gravitational theories, and $\mu_0\epsilon_0 = 1/c^2$. The map for $1/\epsilon_0$ can now be applied:

$$\frac{1}{\epsilon_0} = \mu_0 c^2 \longrightarrow -4\pi G \text{ gives } \mu_0 \longrightarrow -\frac{4\pi G}{c^2}. \tag{1.114}$$

Using this in the magnetic $\nabla \times \mathbf{B} = \mu_0\mathbf{J}$, we have

$$\boxed{\nabla \times \mathbf{b}(\mathbf{r}) = -\frac{4\pi G}{c^2}\rho(\mathbf{r})\mathbf{v}(\mathbf{r}).} \tag{1.115}$$

That's a differential form for a gravitational "Ampere's law." To get the integral form, take a surface S with boundary ∂S (oriented according to the right-hand rule: if you curl your fingers around the boundary, then your thumb points in the direction of the surface normal to S) and integrate both sides of (1.115) over the surface using the curl theorem to turn the curl of **b** into an integral of

b over the boundary:

$$\oint_{\partial S} \mathbf{b} \cdot d\boldsymbol{\ell} = -\frac{4\pi G}{c^2} \int_S \rho \mathbf{v} \cdot d\mathbf{a}. \tag{1.116}$$

Example 1.6 (Laser Redux). We calculated the gravitational field of a laser back in Example 1.5 using the infinite line of energy density as a source for the gravitational field. It is now clear that we can also find the gravitomagnetic contribution from the motion of the energy source. To keep the calculation general, take the line of energy to be lying on the z axis and moving up with constant speed v. The source density is described formally by the current density $\mathbf{J} = \rho \mathbf{v} = \lambda \delta(x)\delta(y)v\hat{\mathbf{z}}$. Because the source has the cylindrical symmetry of an electromagnetic current source, we expect $\mathbf{b}(\mathbf{r}) = b(s)\hat{\boldsymbol{\phi}}$. Taking the "Amperian loop" to be the edge of a disk of radius s with unit normal pointing in the $\hat{\mathbf{z}}$ direction (then the direction of its boundary line is $\hat{\boldsymbol{\phi}}$) and applying (1.116),

$$b(s)2\pi s = -\frac{4\pi G}{c^2}\lambda v \longrightarrow \mathbf{b}(\mathbf{r}) = -\frac{2G\lambda v}{c^2 s}\hat{\boldsymbol{\phi}}. \tag{1.117}$$

Now it's time to use (1.114): for an infinite line of charge, λ_e moving up the z axis with speed v (giving current $I = \lambda v$), the magnetic field is

$$\mathbf{B} = \frac{\mu_0 \lambda v}{2\pi s}\hat{\boldsymbol{\phi}} \longrightarrow -\frac{2G\lambda v}{c^2 s}\hat{\boldsymbol{\phi}}, \tag{1.118}$$

a good check. In the case of a laser, the energy density is moving with speed $v = c$, and $\mathbf{b} = -2G\lambda/(cs)\hat{\boldsymbol{\phi}}$.

The presence of the gravitomagnetic field saves the relativistic force transformation between L and $\bar{L}$ from (1.112). The force in $\bar{L}$ does not change, the mass m isn't moving in its own rest frame, so there is no gravitomagnetic force (although there is a nonzero gravitomagnetic field). In L, there is a new contribution to the force on the mass. The gravitomagnetic field at the particle location is, from (1.117), $\mathbf{b} = -(2G/c^2)\lambda v/d\hat{\mathbf{z}}$, and then the gravitomagnetic force is $-(2G/c^2)mu\lambda v/d\hat{\mathbf{y}}$. The net force on the particle is

$$\mathbf{F} = \gamma_v \frac{2G\lambda_0 m}{d}\left(1 - \frac{uv}{c^2}\right)\hat{\mathbf{y}}. \tag{1.119}$$

This is, again, a force perpendicular to the relative motion of L and $\bar{L}$, so we just need to update (1.112) with the new F magnitude:

$$\frac{\bar{F}}{\gamma_u} = \frac{\gamma_{\bar{w}}}{\gamma_u}\frac{2G\lambda_0 m}{d} = \frac{\gamma_{\bar{w}}}{\gamma_u \gamma_v}F\left(1 - \frac{uv}{c^2}\right)^{-1} = F, \tag{1.120}$$

where the final equality comes from (1.113). The gravitomagnetic force has produced compliance with the demands of special relativity.

So far, for static distributions of mass (or energy) and stationary distributions of mass (or energy) current, the field equations encapsulating all of our work extending Newtonian gravity are

$$\boxed{\begin{aligned}
\nabla \cdot \mathbf{g}(\mathbf{r}) &= -4\pi G \rho(\mathbf{r}), & \nabla \times \mathbf{g}(\mathbf{r}) &= 0, \\
\nabla \cdot \mathbf{b}(\mathbf{r}) &= 0, & \nabla \times \mathbf{b}(\mathbf{r}) &= -\frac{4\pi G}{c^2} \mathbf{J}(\mathbf{r}),
\end{aligned}}$$
(1.121)

where $\mathbf{J}(\mathbf{r}) = \rho(\mathbf{r})\mathbf{v}(\mathbf{r})$ is the mass current density. Here the density $\rho(\mathbf{r})$ could be "real mass," or $\rho(\mathbf{r}) = u(\mathbf{r})/c^2$ for energy density $u(\mathbf{r})$. The force law is $\mathbf{F} = m\mathbf{g} + m\mathbf{v} \times \mathbf{b}$. At this point, it would be reasonable to use $\nabla \cdot \mathbf{b} = 0$ to introduce a gravitomagnetic potential $\mathbf{a}$ whose curl is $\mathbf{b}$, and reproduce all of electro- and magneto-statics, and we leave that as an exercise for the reader. Instead, we will forge ahead to the dynamics of this extended theory, and we will spend even less time there, since ultimately, if any of this were the end of the story, then there would be no need for the remaining chapters of this book.

1.6.1 Dynamics

To finish exploring the parallels between the modified Newtonian gravity and E&M, we will sketch two arguments. First, we'll use mass conservation to establish that the gravitomagnetic curl source should include the time derivative of the gravitational field, precisely the same line of reasoning that led Maxwell to combine the current $\mathbf{J}$ with the temporal derivative of the electric field in his correction to Ampere's law. Then, to get the gravitomagnetic time derivative as a source for the curl of the gravitational field, we'll use a particular configuration of $\mathbf{b}$ and rely on relativistic consistency.

Mass conservation can be expressed in differential form as

$$\frac{\partial \rho(\mathbf{r}, t)}{\partial t} = -\nabla \cdot \mathbf{J}(\mathbf{r}, t),$$
(1.122)

where $\rho(\mathbf{r}, t)$ is a time-varying mass density, and we use $\mathbf{J}(\mathbf{r}, t)$ as shorthand for $\rho(\mathbf{r}, t)\mathbf{v}(\mathbf{r}, t)$, where $\mathbf{v}(\mathbf{r}, t)$ is a velocity field that tells us how the pieces of mass in $\rho(\mathbf{r}, t)$ are moving. For a volume Ω with boundary $\partial\Omega$, integrating both sides of (1.122) over Ω and using the divergence theorem, we can write an integral form of the conservation law

$$\frac{d}{dt} \int_\Omega \rho(\mathbf{r}, t)\, d\tau = -\oint_{\partial\Omega} \mathbf{J}(\mathbf{r}, t) \cdot d\mathbf{a},$$
(1.123)

where the integral on the left represents the total amount of mass enclosed by the volume Ω; on the right, we have a surface integral that measures how much $\mathbf{J}$ flows out of the surface. The minus sign makes sure that for outward pointing $\mathbf{J}$ and outward normal $d\mathbf{a}$, the product $\mathbf{J} \cdot d\mathbf{a}$ registers as a *loss* of mass in the volume Ω. A picture of the integral form of the conservation law is shown in Figure 1.14.

Returning to the source equation for $\mathbf{b}$, $\nabla \times \mathbf{b} = -4\pi G/c^2 \mathbf{J}$, if we take the divergence of both sides and note that $\nabla \cdot (\nabla \times \mathbf{b}) = 0$, then we find $\nabla \cdot \mathbf{J} = 0$,

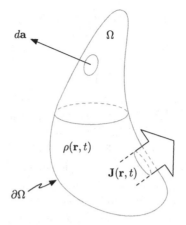

In a domain Ω with boundary $\partial\Omega$, the mass contained inside Ω changes because of mass flowing in and out of the boundary (in the form of the current $\mathbf{J}$). The unit normal to $\partial\Omega$ points outward, so outward pointing $\mathbf{J}$ corresponds to a loss of mass within Ω, that is the source of the minus sign in (1.123).

which isn't true unless ρ is time-independent. That's been fine for the primarily static configurations we've considered so far, but in general, ρ could depend on time in a nontrivial way. We need to add a term to the curl of $\mathbf{b}$ that reduces to the left-hand side of (1.122) when the divergence is taken. We know that $\nabla \cdot \mathbf{g} = -4\pi G\rho$, so the time derivative of $\mathbf{g}$ will have divergence proportional to the time derivative of ρ. To get the constant right, we'll add $\alpha\frac{\partial \mathbf{g}}{\partial t}$ to the current source and then set α to ensure (1.122) holds (we could, of course, peek at the relevant Maxwell equation and make appropriate substitutions):

$$\nabla \times \mathbf{b}(\mathbf{r}, t) = -\frac{4\pi G}{c^2}\mathbf{J}(\mathbf{r}, t) + \alpha\frac{\partial \mathbf{g}(\mathbf{r}, t)}{\partial t}. \tag{1.124}$$

The divergence of this equation is

$$0 = -\frac{4\pi G}{c^2}\nabla \cdot \mathbf{J}(\mathbf{r}, t) - 4\pi G\alpha\frac{\partial\rho(\mathbf{r}, t)}{\partial t}, \tag{1.125}$$

and we must have $\alpha = 1/c^2$ to recover a zero from the conservation law (1.122) on the right. The final form for the field equations governing $\mathbf{b}$ is

$$\nabla \times \mathbf{b}(\mathbf{r}, t) = -\frac{4\pi G}{c^2}\mathbf{J}(\mathbf{r}, t) + \frac{1}{c^2}\frac{\partial \mathbf{g}(\mathbf{r}, t)}{\partial t}, \tag{1.126}$$

and we still have $\nabla \cdot \mathbf{b}(\mathbf{r}, t) = 0$.

Next suppose that we take a mass m and shoot it with speed $v \ll c$ (working in the nonrelativistic limit) into a thin strip of nonzero gravitomagnetic field. To be concrete, let the field point into the page and extend from $x = 0$ to $x = \epsilon$,

$$\mathbf{b} = -b_0\big(\theta(x) - \theta(x - \epsilon)\big)\hat{\mathbf{z}}. \tag{1.127}$$

The massive particle comes in along the x axis, encounters the field region, and moves along a constant-speed circular arc of radius R through an angle

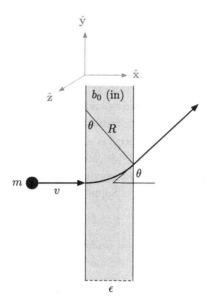

Fig. 1.15 A uniform gravitomagnetic field points into the page and extends from $x = 0 \to \epsilon$. A massive particle entering the region with speed v emerges at speed v making an angle θ with respect to the x axis.

θ, emerging with the same speed v but traveling along a line that makes the angle θ with respect to its original trajectory. The configuration and particle trajectory are shown in Figure 1.15. Since the particle undergoes uniform circular motion within the field region, we know that $R = v/b_0$, and from the geometry of the setup, $R \sin \theta = \epsilon$, so that $\sin \theta = \epsilon b_0 / v$.

While the particle is in the field region, the force acting on it can be written in terms of its x location as

$$\mathbf{F} = -mvb_0 \frac{x}{R} \hat{\mathbf{x}} + mvb_0 \sqrt{1 - \frac{x^2}{R^2}} \hat{\mathbf{y}}. \tag{1.128}$$

Taking $\epsilon/R \ll 1$, so that $x < \epsilon \ll R$, this force can be Taylor expanded to first order in x/R,

$$\mathbf{F} = -mvb_0 \frac{x}{R} \hat{\mathbf{x}} + mvb_0 \hat{\mathbf{y}} + O\big((x/R)^2\big). \tag{1.129}$$

If we imagine shrinking $\epsilon \to 0$, the only force that remains is in the $\hat{\mathbf{y}}$ direction.[13]

How does the story change if we instead take the gravitomagnetic field region and move it with speed v, past the mass m sitting at rest at $x = 0$? Now

---

[13]   This limit is a little tricky, since it implies that the speed of the particle is not constant. A force in the $\hat{\mathbf{y}}$ direction with no corresponding $\hat{\mathbf{x}}$ force tells us that acceleration will occur in $\hat{\mathbf{y}}$ but not in $\hat{\mathbf{x}}$, leading to a particle speed that is not the original v. It's part of the approximation here, and nothing to worry about.

the field is

$$\mathbf{b} = -b_0\big(\theta(x + vt) - \theta(x + vt - \epsilon)\big)\hat{\mathbf{z}}, \tag{1.130}$$

moving to the left, and encountering the particle from $t = 0$ to $t \approx \epsilon/v$. It must be the case that the same net force acts on the particle, only the *relative* motion of the particle and field can matter. Yet according to the Lorentz force law governing the motion of the mass m, there is no gravitomagnetic force on m since its speed is zero, $\mathbf{F} = m\mathbf{v} \times \mathbf{b} = 0$. We need *some other force* that points in the $\hat{\mathbf{y}}$ direction to accelerate the particle in the y direction to its final ϵb_0 velocity component. If it's not a gravitomagnetic force, it must be a gravitational force. Suppose we set

$$\mathbf{g} = b_0 v\big(\theta(x + vt) - \theta(x + vt - \epsilon)\big)\hat{\mathbf{y}}, \tag{1.131}$$

giving a force $\mathbf{F} = mvb_0\hat{\mathbf{y}}$ while the gravitomagnetic field passes overhead. That would have the same effect as the gravitomagnetic field, and over the same timeframe. A massive particle initially at rest would end up moving with velocity $\mathbf{v} = \epsilon b_0\hat{\mathbf{y}}$ after interacting with the field, matching the prediction from the configuration shown in Figure 1.15.

It was the temporal variation of the gravitomagnetic field that led to this new gravitational field, so in terms of generalizing the relation between $\mathbf{g}$ and $\mathbf{b}$ in (1.131), we expect $\frac{\partial \mathbf{b}}{\partial t}$ to play a role. Looking at the form of (1.131), we can introduce the time derivative of $\mathbf{b}$, which gives delta functions, with an integral in x to restore the step function dependence. In this case, we have

$$\frac{\partial g_y}{\partial x} = -\frac{\partial b_z}{\partial t}. \tag{1.132}$$

The derivative on the left looks like part of a curl, which is really the only place that $\frac{\partial \mathbf{b}}{\partial t}$ could go as a source. The units are already in good shape, so we expect the new coupling term to look like

$$\nabla \times \mathbf{g} = -\frac{\partial \mathbf{b}}{\partial t}. \tag{1.133}$$

We have gone as far as we can reasonably go, arriving at an extension of Newtonian gravity, which is, on its face, identical to Maxwell's E&M, but with the "wrong" signs for the sources and different units. The full set of equations is

$$\boxed{\begin{array}{ll} \nabla \cdot \mathbf{g}(\mathbf{r}, t) = -4\pi G\rho(\mathbf{r}, t), & \nabla \times \mathbf{g}(\mathbf{r}, t) = -\dfrac{\partial \mathbf{b}(\mathbf{r}, t)}{\partial t}, \\[2ex] \nabla \cdot \mathbf{b}(\mathbf{r}, t) = 0, & \nabla \times \mathbf{b}(\mathbf{r}, t) = -\dfrac{4\pi G}{c^2}\mathbf{J}(\mathbf{r}, t) + \dfrac{1}{c^2}\dfrac{\partial \mathbf{g}(\mathbf{r}, t)}{\partial t}, \end{array}} \tag{1.134}$$

with "Lorentz force law" $\mathbf{F} = m\mathbf{g} + m\mathbf{v} \times \mathbf{b}$. Almost any prediction you can make about the electromagnetic fields, and the motion of particles moving under their influence can be made about this system, and many gravitational experiments can be understood qualitatively from their electromagnetic analogs (see, for example, Problem 1.35). This intermediate theory is known as "gravito-electro-magnetism" (or "GEM").

Perhaps the most important new prediction we get from the above set is the existence of gravitational radiation. Working in vacuum, the curls of $\mathbf{g}$ and $\mathbf{b}$ become

$$\nabla \times \mathbf{g} = -\frac{\partial \mathbf{b}}{\partial t}, \qquad \nabla \times \mathbf{b} = \frac{1}{c^2}\frac{\partial \mathbf{g}}{\partial t}, \qquad (1.135)$$

and taking the curl of the first equation, we obtain

$$\nabla \times (\nabla \times \mathbf{g}) = -\nabla^2 \mathbf{g} = -\frac{\partial}{\partial t}(\nabla \times \mathbf{b}) = -\frac{1}{c^2}\frac{\partial^2 \mathbf{g}}{\partial t^2}, \qquad (1.136)$$

which can be written

$$-\frac{1}{c^2}\frac{\partial^2 \mathbf{g}}{\partial t^2} + \nabla^2 \mathbf{g} = 0. \qquad (1.137)$$

So the wave equation governs $\mathbf{g}$ in vacuum. You can similarly take the curl of the equation for $\mathbf{b}$ in vacuum and arrive at the wave equation for $\mathbf{b}$. The fields here are gravitational fields, and these waves are gravitational waves (not really, read on!). We offer (1.134) as a close to the chapter, one that allows some gravitational intuition to be built. You should think of $\mathbf{g}$ and $\mathbf{b}$ as qualitative descriptors but flawed quantitative predictors. In some cases, the predictions we make match those of general relativity (almost always purely by accident); in other cases, they are off by factors of two or four, as we will see later on. In all cases, the physical picture of fields and forces is, as I mentioned at the start of the chapter, not the story general relativity will tell.

As an epilogue to our first attempt to square Newtonian gravity with special relativity, I want to indicate two openings that we have glossed over and that give a hint about the target structure of a gravitational theory. Beyond that, a postscript (or entré, depending on your point of view) that gives some indication of why gravity is different from other interactions. In particular, the following two sections establish targets for a theory of gravity beyond (1.134).

1.7 Epilogue (What's Wrong with GEM?)

Although (1.134) looks just like Maxwell's equations, the sources are quite different. In particular, there are two different types of source we have considered here, "actual" mass and "effective" mass. We have allowed energy to play the role of mass both in sourcing fields and in responding to them. The former case is unfinished. We studied the gravitational field that is generated by the energy stored in electric and magnetic fields, but given the focus in Section 1.2, on the energy stored in the gravitational field, shouldn't we have started there? Prior to coupling the field to external energy sources, we can couple the field to itself using

$$u(\mathbf{r}) = -\frac{1}{8\pi G}\nabla\varphi(\mathbf{r}) \cdot \nabla\varphi(\mathbf{r}), \qquad (1.138)$$

so that our starting point (1.82) should make explicit reference to the energy density of the gravitational field. Taking $\rho(\mathbf{r})$ to be material sources and $u(\mathbf{r})$ to be energetic sources other than gravitational, we'll begin with

$$\nabla \cdot \mathbf{g}(\mathbf{r}) = -4\pi G\left(\rho(\mathbf{r}) + \frac{u(\mathbf{r})}{c^2} - \frac{1}{8\pi G c^2}\mathbf{g}(\mathbf{r}) \cdot \mathbf{g}(\mathbf{r})\right) \quad (1.139)$$

or, in terms of the potential,

$$\nabla^2\varphi(\mathbf{r}) = 4\pi G\left(\rho(\mathbf{r}) + \frac{u(\mathbf{r})}{c^2}\right) - \frac{1}{2c^2}\nabla\varphi(\mathbf{r}) \cdot \nabla\varphi(\mathbf{r}). \quad (1.140)$$

The problem here is one of consistency. Let's focus just on the self-coupling and matter sources by setting $u = 0$. Recall how we got the energy density from the expression for the work required to build the mass density ρ: Starting from

$$W = \frac{1}{2}\int \rho(\mathbf{r})\varphi(\mathbf{r})\,d\tau, \quad (1.141)$$

we replaced ρ in the integrand using $\nabla^2\varphi = 4\pi G\rho$, but that is no longer the field equation; now we have $\nabla^2\varphi = 4\pi G\rho - \nabla\varphi \cdot \nabla\varphi/(2c^2)$, and if we use *that* to replace ρ in the expression for W,

$$W = \int \frac{1}{8\pi G}\left(\varphi(\mathbf{r})\nabla^2\varphi(\mathbf{r}) + \frac{1}{2c^2}\varphi(\mathbf{r})\nabla\varphi(\mathbf{r}) \cdot \nabla\varphi(\mathbf{r})\right)d\tau, \quad (1.142)$$

and use integration by parts as we did before, then we can turn the first term into $-\nabla\varphi \cdot \nabla\varphi$:

$$W = \int \frac{1}{8\pi G}\nabla\varphi(\mathbf{r}) \cdot \nabla\varphi(\mathbf{r})\left(-1 + \frac{\varphi(\mathbf{r})}{2c^2}\right)d\tau. \quad (1.143)$$

Now we get a new expression for the energy density of the field,

$$u(\mathbf{r}) = -\frac{1}{8\pi G}\nabla\varphi(\mathbf{r}) \cdot \nabla\varphi(\mathbf{r})\left(1 + \frac{\varphi(\mathbf{r})}{2c^2}\right), \quad (1.144)$$

and it is *this* that we should use as the self-coupled field energy. Again ignoring external energies, we will now start with

$$\nabla^2\varphi(\mathbf{r}) = 4\pi G\rho(\mathbf{r}) - \frac{1}{2c^2}\left(1 - \frac{\varphi(\mathbf{r})}{2c^2}\right)\nabla\varphi(\mathbf{r}) \cdot \varphi(\mathbf{r}). \quad (1.145)$$

Of course, we'll run into the same problem again (and again) – this field equation has a different solution for ρ that we should use in (1.141), which will lead to a new energy density starting point. The process continues ad infinitum (even, seemingly, in a summable manner), but we won't pursue this further, since our next job is to show that this energy density is not the right one to begin with. Still, there is a message here, noting that the form of the field equation (and its infinite regress) starts off quadratic in φ and its derivatives (and gets worse), we expect that:

> A gravitational field theory couples gravity to all forms of energy, including that stored in the gravitational field itself, and that self-coupling will *necessarily* render the field equation nonlinear.

The first problem with GEM is that it is linear. The nonlinearity of any self-consistent theory of gravity tells us that we will not have superposition as a tool for solving problems incrementally. That is, indeed, true in general relativity.

Now for the form of the source. When we think of sources like ρ and $\mathbf{J}$ that appear in (1.134), our inclination is to assume that they come from a four-vector J^μ as they do in E&M, with $J^0 = \rho c$, and spatial components $\mathbf{J}$. If we started with a mass distribution at rest, like the infinite line of mass from Section 1.6, call it $\bar{J}^\mu$ with $\bar{J}^0 = \bar{\rho} c$ and $\bar{\mathbf{J}} = 0$, then a Lorentz boost in the x direction has

$$J^\mu \doteq \begin{pmatrix} \gamma & \frac{v}{c}\gamma & 0 & 0 \\ \frac{v}{c}\gamma & \gamma & 0 & 0 \\ 0 & 0 & 1 & 0 \\ 0 & 0 & 0 & 1 \end{pmatrix} \begin{pmatrix} \bar{\rho} c \\ 0 \\ 0 \\ 0 \end{pmatrix} = \begin{pmatrix} \gamma \bar{\rho} c \\ v \gamma \bar{\rho} \\ 0 \\ 0 \end{pmatrix}, \qquad (1.146)$$

so that $J^0 = \rho c = \gamma \bar{\rho} c$, which tells us that $\rho = \gamma \bar{\rho}$, precisely what we used to find the gravitational field in (1.109) (there we wrote $\lambda = \gamma_v \lambda_0$).

But the infinite line of mass is part of a broader source structure. If we take the energy as the source for the gravitational field, then for an infinite line of mass at rest with mass density $\bar{\lambda}$, the energy density is $\bar{u} = \bar{\lambda} c^2$. When the line of mass is moving, there are two effects: (1) length contraction $\Delta x = \gamma^{-1} \Delta \bar{x}$, causing $\lambda = \gamma \bar{\lambda}$ as we have seen already, but also (2) the kinetic energy of the moving line of mass causes a change in the energy itself. For a particle at rest, $\bar{E} = mc^2$, whereas a particle moving with speed v has energy given by (1.81), $E = \gamma mc^2$. Then the relation between the energy density for the line of mass at rest, $\bar{u} = \bar{E}/\Delta \bar{x}$, and the energy density when it is moving, $u = E/\Delta x$, is

$$u = \frac{E}{\Delta x} = \frac{\gamma \bar{E}}{\gamma^{-1} \Delta \bar{x}} = \gamma^2 \bar{u}. \qquad (1.147)$$

Whereas the mass density picks up one factor of γ when going from the rest frame of the infinite line of mass to a frame in which it is moving, the actual gravitational source picks up *two* factors of γ. As we shall see in detail in Chapter 2, the energy density is part of a second-rank tensor. Since each side of an equation must have the same tensorial transformation structure, if the source for gravity is a second-rank tensor, then

> The gravitational field is a second-rank tensor.

So problem two with GEM is that its field does not have the correct tensorial character. Putting the two targets together, we expect the full theory of gravity to be governed by a second-rank tensor with nonlinear field equations owing to its self-coupling.

1.8 Postscript

What makes gravity so different from E&M? One important difference, as we have seen, is the nature of the source. In E&M, charge and moving charge are the sources. Some particles have charge, others don't, so E&M is a limited phenomenon. Gravity, in light of special relativity, must have energy as its source, and everything has energy (rest energy for particles, field energy for light, for example), including gravity itself. This "universal coupling" means nothing is immune from both generating and reacting to gravity, there are no "neutral" particles here. That makes the theory unique, among the four "forces."

Another special piece of gravity is the role of mass both in the force $m\mathbf{g}$ and in the acceleration side of Newton's second law, $m\ddot{\mathbf{r}}(t)$. That's the same mass appearing in both places, and $m\ddot{\mathbf{r}}(t) = m\mathbf{g}$ allows us to cancel the mass from both sides, $\ddot{\mathbf{r}}(t) = \mathbf{g}$; *all* masses respond with the same acceleration. That doesn't happen for electric forces, where Newton's second law reads $m\ddot{\mathbf{r}}(t) = q\mathbf{E}$, and you can't cancel m and q. Note that it doesn't *have* to be the case that the "inertial" mass, m_i in $m_i\ddot{\mathbf{r}}(t)$ is the same as the gravitational mass m_g appearing in $m_g\mathbf{g}$. They *could* be different. To modern experimental accuracy, these two masses appear to be the same, $m_i = m_g$, and the exact equality is what allows gravity to have its geometric interpretation. That equality is known as the "equivalence principle," and it leads to predictions that are very different from all the other forces (see Problem 1.37).

Because mass cancels in the equation of motion involving gravity, all objects respond with the same acceleration. For gravity near the surface of the Earth, that acceleration is universal. Einstein postulated a strict equivalence between gravity and uniform acceleration so that anything that happens in a uniformly accelerated frame should also occur in a frame that is at rest in a gravitational field. As an example of the type of prediction this allows, think of a uniformly accelerating box of height h that is moving upwards with constant acceleration a as shown in Figure 1.16.

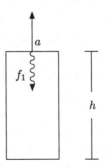

Fig. 1.16 A box of height h travels with constant acceleration a. Light is emitted from the ceiling with frequency f_1.

At time t the box is traveling with speed v_1 and light of frequency f_1 is emitted from the top, to be received at the bottom. It takes the light a time $\Delta t = h/c$ to go from the top to the bottom (neglecting the distance traveled by the accelerating elevator), and during that time, the speed of the elevator increases by $a\,\Delta t$, so that $v_2 = v_1 + ah/c$. The Doppler shift associated with the change in speed between emission and reception means that the light has frequency f_2 when it is received, with

$$f_2 = \frac{\sqrt{1 + (ah/c)/c}}{\sqrt{1 - (ah/c)/c}} f_1 \approx \left(1 + \frac{ah}{c^2}\right) f_1. \qquad (1.148)$$

Now using Einstein's equivalence principle, the same shift in frequency should occur in a box at rest in a gravitational field. We just replace a with g, the acceleration associated with gravity, to get the relation between the frequency f_1 of light emitted from a height h and the frequency f_2 of light received on the ground (at the bottom of the box),

$$f_2 \approx \left(1 + \frac{gh}{c^2}\right) f_1. \qquad (1.149)$$

Here nothing is moving, so the frequency shift comes "from" the gravitational field. One way to interpret the result is to imagine that just as time intervals are different for observers in relative motion, they are also different for observers at different heights in a gravitational field (this effect was observed experimentally in the Pound–Rebka experiment [38]). Instead of frequencies, we can focus on the time interval over which one cycle of light passes by source and observer, $T = 1/f$,

$$T_2 \approx \left(1 - \frac{gh}{c^2}\right) T_1. \qquad (1.150)$$

This must be true of time intervals in general, not just for the period of light. The implication is that if you have a clock at height h, and time Δt_h elapses on that clock, then for a clock on the ground, time Δt_0 elapses with

$$\Delta t_0 \approx \left(1 - \frac{gh}{c^2}\right) \Delta t_h, \qquad (1.151)$$

so that *less* time passes on the ground. You can work out the expression for gravitational potentials other than gh in Problem 1.38.

Exercises

1.1 From the definition of the Dirac delta "function" (more properly, a "distribution") with $\delta(x) = 0$ for $x \neq 0$, $\delta(x) \to \infty$ at $x = 0$, and

$$\int_{-\infty}^{\infty} \delta(x)\,dx = 1, \qquad (1.152)$$

show that for a function $f(x)$ (assume finite everywhere),

$$\int_{-\infty}^{\infty} \delta(x) f(x) dx = f(0) \qquad (1.153)$$

and

$$\int_{-\infty}^{\infty} \delta(x - a) f(x) dx = f(a). \qquad (1.154)$$

1.2   Using the definition of the Dirac delta function, from Problem 1.1 show that for constant k,

$$\int_{-\infty}^{\infty} \delta(kx) f(x) dx = \frac{f(0)}{|k|}. \qquad (1.155)$$

1.3   Find the gravitational field above and below an infinite sheet of mass with uniform mass density σ_0 lying in the xy plane.

1.4   What is the gravitational field inside and outside an infinite cylinder of radius R that has a uniform mass density ρ_0?

1.5   Find the gravitational field inside and outside a ball of mass with uniform mass density ρ_0 and radius R. What is the field inside and outside if the ball has mass density $\rho(r) = \alpha r$ for constant α? What are the units of α?

1.6   A pendulum of length ℓ has a "bob" of mass m that moves under the influence of the Earth's gravitational field. Assuming that motion occurs near the surface of the Earth (so that the distance to the center of the Earth is always R, the Earth's radius) and making the usual small angle assumptions (but do not assume that the gravitational force points straight down, that is, use the correct force direction), what is the period of the resulting oscillatory motion? What happens to that period as $\ell \rightarrow \infty$?

1.7   What is the energy density of the gravitational field associated with a point mass m placed at the origin? What is the total energy stored in that field?

1.8   You observe a gravitational field $\mathbf{g} = g_0 \hat{\mathbf{r}}$ for constant g_0. What is the mass density that generates this field?

1.9   In special relativity, the constant energy of a particle of mass m moving under the influence of a force with potential energy U is

$$E = \frac{mc^2}{\sqrt{1 - v^2/c^2}} + U(\mathbf{r}), \qquad (1.156)$$

where $\mathbf{v} \equiv \frac{d\mathbf{r}}{dt}$. Find the speed of a particle that starts from rest a distance r_0 from the center of a spherical source mass M centered at the origin. Sketch the speed and establish that it does not exceed c (even for a point source). This is the (special) relativistic resolution of the problem posed in Section 1.3.

1.10   Starting from (1.45), find the total energy E for a particle of mass m that starts at the Earth's surface R and ends at spatial infinity at rest. Using that value, find the minimum speed that you must launch the particle so as to reach spatial infinity, the escape speed. What is this special speed for the Earth?

1.11 What is the Schwarzschild radius $R_M \equiv 2GM/c^2$ for the Earth? How about the Sun?

1.12 Find the four acceleration vectors shown in Figure 1.6. Taylor expand your expressions assuming that d and h are small compared to r. Using these approximate expressions, find the relative accelerations $\mathbf{a}_t - \mathbf{a}_b$ and $\mathbf{a}_\ell - \mathbf{a}_r$.

1.13 In Figure 1.7, where does the minimum effective potential energy occur (write $r_\circ$ in terms of G, M, m, and L_z)? What is the value of the effective potential at the minimum?

1.14 Referring to Figure 1.7, given some $E_s > 0$ and central mass M, find the location of the point of closest approach to the central body.

1.15 For a point mass M sitting at the origin, show that the circular orbits of test masses in Newtonian gravity are stable and can occur at any radius.

1.16 In Newtonian gravity, there are stable circular orbits at any radius. Given a target radius R, what is the speed associated with the stable circular orbit at R? What values of R lead to a speed greater than c? In the relativistic setting, Newton's second law reads

$$\frac{d}{dt}\left(\frac{m\mathbf{v}}{\sqrt{1-v^2/c^2}}\right) = \mathbf{F}. \tag{1.157}$$

Work out the condition for circular orbits given $\mathbf{F} = -mGM/r^2\hat{\mathbf{r}}$ (i.e., the condition analogous to $mv^2/R = F$) and find the speed of the circular orbit in the relativistic setting. Does this limit the available radii, values of R, for which stable circular orbits exist?

1.17 Solving (1.66) for $\frac{d\rho(\phi)}{d\phi}$, you should get a solution of the form $\frac{d\rho(\phi)}{d\phi} = \pm\sqrt{f(\rho)}$ for some function $f(\rho)$. Take the ϕ derivative of this equation taking each sign in turn and show that you get the same expression for $\frac{d^2\rho(\phi)}{d\phi^2}$ in either case.

1.18 We saw that the equation governing $\rho(\phi)$ for Newtonian gravity (1.67) is of the form $\rho''(\phi) = -\rho(\phi) + C$ for a constant C. In some situations, this equation gets modified to read $\rho''(\phi) = -(1+\alpha)^2\rho(\phi) + C$, where $(1+\alpha)^2$ is a constant. Motivated by this form, and adding some specific names for the constants involved, solve

$$\rho''(\phi) = -(1+\alpha)^2\left[\rho(\phi) - \frac{1}{p}\right], \tag{1.158}$$

with "initial" conditions $\rho(0) = (1+e)/p$, $\rho'(0) = 0$, where e and p are both constants. What is $r(\phi) = 1/\rho(\phi)$? This form looks similar to the elliptical solution in (1.72), but the α makes it so that the point of closest approach moves. Find expressions for the closest and furthest approach radii (relative to the implicit spherical central mass at the origin) and the angles ϕ_j at which the closest approach occurs – j indexes the orbit number here. As an example, for a closed ellipse, $\phi_j = 2\pi j$. Using Mathematica (or other graphing software), make a plot of the trajectory of a particle moving according to your $r(\phi)$ using $x(\phi) = r(\phi)\cos(\phi)$, $y(\phi) = r(\phi)\sin(\phi)$ (the command `ParametricPlot` in `Mathematica` will do it) for $p = 2$, $e = 0.7$, and $\alpha = 0.1$, and taking $\phi = 0 \rightarrow 10\pi$. Superimpose on that plot a circle of radius equal to the closest approach distance. Characterize the motion that you see. If possible, include in your plot the points associated with the ϕ_j for $j = 1, 2, 3, 4, 5$.

1.19 Find expressions for the semimajor and semiminor axes a and b of an ellipse in terms of the closest and furthest approach points r_p and r_a.

1.20 For the special case of a circular orbit (around a spherical central body of mass M), find the period of the motion of an orbiting mass m in terms of the radius of the circle R and the source mass M.

1.21 For a volume domain Ω, integrate (1.82) over it to get an extended form of the gravitational "Gauss's law." Using this integral, find the gravitational field outside of a sphere of radius R with uniformly distributed mass M and charge Q.

1.22 Starting from the field equation for Newtonian gravity, $\nabla^2\varphi = 4\pi G\rho$, given a mass density ρ, use the "work required to build ρ" starting point,

$$W = \frac{1}{2}\int \rho(\mathbf{r})\varphi(\mathbf{r})\,d\tau \qquad (1.159)$$

(where the integral is over all space), eliminate ρ in favor of φ (and its derivative(s)), and use integration by parts (assuming that $\varphi \rightarrow 0$ at spatial infinity) to rewrite W (it will end up, at this stage, with an integrand that is proportional to $\nabla\varphi \cdot \nabla\varphi$). The integrand can now be interpreted as an energy density u, and we can update the field equation to include it, $\nabla^2\varphi = 4\pi G(\rho + u/c^2)$. Repeat the process using this updated form – infuriatingly, the substitution for ρ has changed. Iterate . . .

1.23 Find the Schwarzschild radius of a proton. What is its "charge" radius $R_Q \equiv \sqrt{GQ^2/(4\pi\epsilon_0 c^4)}$? Compare these with the actual (average) radius of the proton. Do these special lengths exist inside or outside the proton?

1.24 For a central sphere of mass M with charge Q and radius R, find the time it takes for a neutral particle of mass m starting from rest at $r_0 > R$ to run into the sphere (i.e., how long does it take the particle to reach R?). Use the gravitational field that includes both mass and charge (energy density of the field) as sources, from (1.88).

1.25 For a central sphere of mass M with charge Q and radius R, find the orbital period for a neutral particle of mass m moving in a circle of radius r_0 around the central body. Use the nonrelativistic equation of motion with gravitational field given by (1.88). Under what condition (relation between R_M, R_Q, and r_0) will this uniform circular motion exist?

1.26 Starting from the potential for a charged massive "point source,"

$$\varphi = \frac{c^2}{2}\left(-\frac{R_M}{r} + \frac{R_Q^2}{2r^2}\right), \qquad (1.160)$$

use the expressions for angular momentum $L_z = mr(t)^2\dot{\phi}(t)$ and energy

$$E = \frac{1}{2}m\dot{r}(t)^2 + \frac{1}{2}mr(t)^2\dot{\phi}(t)^2 + m\varphi(r(t)) \qquad (1.161)$$

and carry out the same steps we did in the Newtonian gravity case in Section 1.4.2: switch to $\rho(t) = 1/r(t)$, change the parameter from t to ϕ, and write out the second-order differential equation for $\rho(\phi)$. From its form, what type of motion will you get for a "bound" trajectory (i.e., not one of the scattering trajectories). Characterize it as completely as you can (the result from Problem 1.18 may prove useful).

1.27 For the magnetic dipole field (outside of the source) in (1.96), find the scalar potential W such that $\mathbf{B} = -\nabla W$.

1.28 A laser puts out 1 W of power in its beam. Treating the beam as an infinite line source, what is the gravitational force on a (neutral) one kilogram mass a centimeter away from the beam?

1.29 The expression for $\mathbf{g}$ in (1.94) refers directly to the electric potential V, which means that the (detectable) gravitational force could depend on the choice of zero for the potential. For a point charge at the origin, take $V = q/(4\pi\epsilon_0)(1/r - 1/r_0)$ with the zero at r_0 and construct the field $\mathbf{g}$. What happens to the constant term? The issue here is similar to the one resolved in Problem 1.21.

1.30 Referring to the "standard setup" for a Lorentz boost from Section A.1, we have two labs L and $\bar{L}$ with $\bar{L}$ moving to the right at speed v through L along a shared x axis. A mass at rest in $\bar{L}$ is acted on by a force of magnitude $\bar{F}_\perp$ in the $\hat{\bar{\mathbf{y}}}$ direction (perpendicular to the boost direction). What is the force $F_\perp$ measured in the L lab? (One way to do this is to think about Newton's second law in "difference" form: $F = \frac{\Delta p}{\Delta t}$). Next, consider a mass at rest in $\bar{L}$ that is acted on by a force $\bar{F}_\parallel$ in the $\hat{\bar{\mathbf{x}}}$ direction. Again, find the magnitude $F_\parallel$ of this force in L.

1.31 For

$$\gamma_v \equiv \frac{1}{\sqrt{1 - v^2/c^2}}, \qquad \gamma_u \equiv \frac{1}{\sqrt{1 - u^2/c^2}}, \qquad \gamma_{\bar{w}} \equiv \frac{1}{\sqrt{1 - \bar{w}^2/c^2}}, \qquad (1.162)$$

with

$$\bar{w} = \frac{v - u}{1 - uv/c^2}, \qquad (1.163)$$

show that $\gamma_{\bar{w}}/(\gamma_v\gamma_u) = (1 - uv/c^2)$, as suggested in (1.113).

1.32 What is the gravitational field inside and outside a massive, infinite cylindrical shell of radius R with uniform mass per unit area on the surface, σ_0, lying along the $\hat{\mathbf{z}}$ axis and moving with constant velocity $\mathbf{v} = v\hat{\mathbf{z}}$? How about the gravitomagnetic field $\mathbf{b}$ for this configuration? Use the gravitational analogue of Ampere's law from (1.116).

1.33 What are the gravitational and gravitomagnetic fields inside and outside a massive cylindrical shell of radius R with uniform mass per unit area on the surface, σ_0, lying along the $\hat{\mathbf{z}}$ axis and spinning with constant angular velocity $\boldsymbol{\omega} = \omega\hat{\mathbf{z}}$?

1.34 Find the form of the gravitational and gravitomagnetic field(s) (both **g** and **b**) outside a spinning massive sphere with mass M and radius R, spinning with constant angular velocity $\boldsymbol{\omega} = \omega\hat{\mathbf{z}}$ (you can look up the electromagnetic analogue and use the E&M to gravity mapping).

1.35 A ring of radius R with mass m spread uniformly around it spins with constant angular velocity $\boldsymbol{\omega} = \omega\hat{\mathbf{z}}$. What is the gravitomagnetic dipole moment of the ring (work in analogy with the magnetic dipole moment of a ring of charge)? If you put this ring in a uniform gravitomagnetic field $\mathbf{b} = b_0\hat{\mathbf{z}}$ such that the dipole moment makes an angle θ with respect to the $\hat{\mathbf{z}}$ axis, find the torque on the dipole. Using Newton's second law, solve for the angular momentum **L** of the ring and find the motion of the gravitational dipole moment as a function of time. This motion is what allows the Gravity Probe B experiment [15] to measure the gravitomagnetic field of the Earth using spinning spheres.

1.36 Find the gravitational versions of the electromagnetic potentials V and **A** (electric and magnetic, respectively) and write the field equations (1.134) in terms of them. From the relation between the potentials and fields, identify the gauge freedom here and find a gauge in which the vacuum form of the equations for the potentials becomes the wave equation.

1.37 Two masses m_1 and m_2 lie along the x axis separated by a distance d. What is the force that each exerts on the other? What is the associated acceleration of each mass? Taking $m_1 = m$ and $m_2 = -m$, what are the force and acceleration now? Assuming that the masses start from rest, find the solution to the equation of motion in this case and comment on conservation of momentum and energy.

1.38 Generalize expression (1.151) to give the relation between Δt_{r_1} and Δt_{r_2}, where $r_1 < r_2$, and both are arbitrary (although above the Earth's surface). Express your result in terms of the gravitational potential of the Earth, $\varphi(r) = -GM/r$.

1.39 In Problem 1.38, we learned that

$$\left(1 - \frac{1}{c^2}\varphi(r_1)\right)\Delta t_{r_1} = \left(1 - \frac{1}{c^2}\varphi(r_2)\right)\Delta t_{r_2}, \qquad (1.164)$$

where $\varphi(r)$ was the gravitational potential associated with the Earth, $\varphi(r) = -\frac{GM}{r}$ above its surface. If you take this relation seriously (as we will see is warranted), then you can work out the time-dilation associated with any gravitational potential $\varphi(r)$. Assuming that the Earth is a uniformly massed sphere of radius R and mass M, find the potential $\varphi(r)$ inside the Earth. Using realistic values for M and R and the age of the Earth, find the amount of time elapsed at the Earth's center as compared with the Earth's crust – that is, compute the difference $\Delta t_R - \Delta t_0$.

2 Transformation and Tensors

General relativity is a theory that is "generally covariant," meaning that its field equations look the same in *any* coordinate system. Tensors provide a language for ensuring that this is the case for any physical theory. We will go through the usual list of tensorial objects in a moment, but first, I want to highlight the defining feature of tensors, their response to a change of coordinates (or "coordinate transformation"). To talk about how a function or collection of functions responds to a transformation, it helps to have some models in mind. There are two general types: ones that have physical significance, like Lorentz transformations (both rotations and boosts), and ones that provide structural utility, changing to spherical coordinates for physical configurations involving spherical symmetry, for example. Let's review each of these; we'll refer back to them periodically when explicit examples are needed.

2.0.1 Lorentz Transformations

The set of Lorentz transformations is defined as the linear transformations that leave the Minkowski line element unchanged. In four-dimensional spacetime, the line element tells us how to measure distance, updating the familiar Pythagorean distance to include time. In Cartesian coordinates, the line element is

$$ds^2 = -c^2 dt^2 + dx^2 + dy^2 + dz^2. \tag{2.1}$$

To describe transformations in general, I'll use unbarred letters to refer to the "original" coordinate system, and then barred letters will denote the "new" coordinates. As an example, a coordinate system that is rotated about the z axis through an angle θ would be expressed as

$$c\bar{t} = ct, \qquad \bar{x} = x \cos\theta - y \sin\theta, \qquad \bar{y} = x \sin\theta + y \cos\theta, \qquad \bar{z} = z \tag{2.2}$$

(using ct for the temporal coordinate, so that all coordinates have dimension of length). It is easy to see that this transformation preserves the line element, just take $d\bar{x} = dx \cos\theta - dy \sin\theta, d\bar{y} = dx \sin\theta + dy \cos\theta$, with $cd\bar{t} = cdt$, $d\bar{z} = dz$, and construct $d\bar{s}^2$,

$$d\bar{s}^2 = -c^2 d\bar{t}^2 + d\bar{x}^2 + d\bar{y}^2 + d\bar{z}^2 = -c^2 dt^2 + dx^2 + dy^2 + dz^2. \tag{2.3}$$

Another type of transformation, a Lorentz boost, mixes time and space together. As an example, take a boost along the x axis with "parameter" v,

$$c\bar{t} = \gamma\left(ct - \frac{v}{c}x\right), \qquad \bar{x} = \gamma(-vt + x), \qquad \bar{y} = y,$$

$$\bar{z} = z \text{ with } \gamma \equiv \frac{1}{\sqrt{1 - v^2/c^2}}. \tag{2.4}$$

Here the barred coordinates represent a frame moving at speed v in the x direction. You can check that the infinitesimal lengths satisfy $d\bar{s}^2 = ds^2$ in Problem 2.2, a defining property of the Lorentz boost. A brief review of special relativity conventions and results can be found in Appendix A.

The relations between new and old coordinates found in (2.2) and (2.4) are linear in the original (unbarred) coordinates and can be written as a matrix multiplying the vector (using the infinitesimals themselves to define the transformation),

$$d\mathbf{x} \doteq \begin{pmatrix} cdt \\ dx \\ dy \\ dz \end{pmatrix}. \tag{2.5}$$

The infinitesimal form of (2.2) can then be written as

$$d\bar{\mathbf{x}} \doteq \begin{pmatrix} 1 & 0 & 0 & 0 \\ 0 & \cos\theta & -\sin\theta & 0 \\ 0 & \sin\theta & \cos\theta & 0 \\ 0 & 0 & 0 & 1 \end{pmatrix} d\mathbf{x}, \tag{2.6}$$

and you can generate the matrix associated with the boost transformation (2.4) in Problem 2.3.

We will use this opportunity to define and begin to use Einstein's "index notation." Instead of writing $d\mathbf{x}$ to refer to a vector, we'll define an index that picks out the entries directly. Let dx^μ have $\mu = 0$ corresponding to cdt, $\mu = 1$ gives dx, and so on. For the matrix appearing in (2.6), we'll also address entries by indices, let $L^\mu_{\ \nu}$ be associated with the $\mu\nu$ (row-column) entry of the matrix, so that $L^0_{\ 0} = 1$, $L^1_{\ 0} = 0$, $L^1_{\ 1} = \cos\theta$, $L^1_{\ 2} = -\sin\theta$, etc. Then the matrix-vector multiplication in (2.6) can be captured by the sum,

$$d\bar{x}^\mu = \sum_{\nu=0}^{3} L^\mu_{\ \nu} dx^\nu \text{ for } \mu = 0, 1, 2, 3. \tag{2.7}$$

The first rule of the index notation is that any index appearing once in an expression, like μ above, must appear in *every* term of the equation, in the same place (up or down), with the implicit instruction "for $\mu = 0, 1, 2, 3$" tacked on (in (2.7), that instruction is explicit, but we'll drop it from here on). So (2.7) is really four equations, as is clear from its predecessor (2.6) and *its* predecessor (2.2).

The second rule of Einstein's notation comes from the observation that whenever an index appears twice in a term, like the ν on the right in (2.7), there is always an associated sum over the repeated index, so there is no need to write the summation sign at all. Thus

$$d\bar{x}^\mu = L^\mu_{\ \nu} dx^\nu, \tag{2.8}$$

with summation on ν implicit is our shorthand for (2.7). To reiterate, an "open" index appears once in each term, it must have the same letter and appear in the same up or down position, and it literally "indexes" the equation. A "closed" or "dummy" index appears twice in a term, once up and once down, and it is summed over from $0 \rightarrow 3$. Try your hand at turning explicitly summed statements into index notation ones, and vice versa, in Problem 2.4.

The matrix representing the transformation in these cases shows up because the transformation is linear, but its entries turn out to be useful in defining the response of tensors to transformations even when that transformation is nonlinear. From $d\bar{x}^\mu = L^\mu_{\ \nu} dx^\nu$ it is clear that $L^\mu_{\ \nu}$ has entries that are the derivatives of the new coordinates with respect to the old:

$$L^\mu_{\ \nu} = \frac{\partial \bar{x}^\mu}{\partial x^\nu}. \tag{2.9}$$

The linearity of rotations and boosts means that $\frac{\partial \bar{x}^\mu}{\partial x^\nu}$ does not depend on the coordinates.

For a more general change of coordinates, the relation

$$d\bar{x}^\mu = \frac{\partial \bar{x}^\mu}{\partial x^\nu} dx^\nu \tag{2.10}$$

is an expression of the chain rule, and this is true even when $\frac{\partial \bar{x}^\mu}{\partial x^\nu}$ has coordinate dependence (a nonlinear relation between the barred coordinates and original set).

2.0.2 Spherical Coordinates

In spherical coordinates, we replace the Cartesian x, y, and z with the distance to the origin coordinate, r, the polar angle θ, and the azimuthal angle ϕ, as shown in Figure 2.1. From the figure, we can relate the spherical coordinates

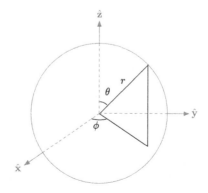

Fig. 2.1   The three coordinates r, θ, and ϕ that define the spherical coordinate system.

to the original Cartesian set, again using bars over the spherical coordinates to distinguish them as the "new" ones:

$$c\bar{t} = ct, \qquad x = \bar{r}\sin\bar{\theta}\cos\bar{\phi}, \qquad y = \bar{r}\sin\bar{\theta}\cos\bar{\phi}, \qquad z = \bar{r}\cos\bar{\theta}, \tag{2.11}$$

where for consistency, we have included the trivial $c\bar{t} = ct$ update to the temporal coordinate. We can invert this set, writing the new coordinates in terms of the old:

$$c\bar{t} = ct, \qquad \bar{r} = \sqrt{x^2 + y^2 + z^2},$$
$$\bar{\theta} = \tan^{-1}\left(\frac{\sqrt{x^2 + y^2}}{z}\right), \qquad \bar{\phi} = \tan^{-1}\left(\frac{y}{x}\right). \tag{2.12}$$

Take the ordering for the indexed dx^{μ} to be $dx^0 = cdt$, $dx^1 = dx$, $dx^2 = dy$, and $dx^3 = dz$ as before, and for the new coordinates, $d\bar{x}^0 = cd\bar{t}$, $d\bar{x}^1 = d\bar{r}$, $d\bar{x}^2 = d\bar{\theta}$, and $d\bar{x}^3 = d\bar{\phi}$. Then the matrix with entries $\frac{\partial \bar{x}^{\mu}}{\partial x^{\nu}}$ can be obtained by taking the derivatives of the equations in (2.12):

$$\frac{\partial \bar{x}^{\mu}}{\partial x^{\nu}} \doteq \begin{pmatrix} 1 & 0 & 0 & 0 \\ 0 & \frac{x}{\sqrt{x^2+y^2+z^2}} & \frac{y}{\sqrt{x^2+y^2+z^2}} & \frac{z}{\sqrt{x^2+y^2+z^2}} \\ 0 & \frac{x/z}{\sqrt{x^2+y^2}(1+(x^2+y^2)/z^2)} & \frac{y/z}{\sqrt{x^2+y^2}(1+(x^2+y^2)/z^2)} & -\frac{\sqrt{x^2+y^2}}{(1+(x^2+y^2)/z^2)z^2} \\ 0 & -\frac{y/x^2}{1+y^2/x^2} & \frac{1/x}{1+y^2/x^2} & 0 \end{pmatrix}. \tag{2.13}$$

In this case, the matrix entries depend on the coordinates, and we can express that dependence either in the old coordinate system, as has been done above, or in the new, replacing x, y, and z with their expressions in terms of $\bar{r}$, $\bar{\theta}$, and $\bar{\phi}$ from (2.11):

$$\frac{\partial \bar{x}^{\mu}}{\partial x^{\nu}} \doteq \begin{pmatrix} 1 & 0 & 0 & 0 \\ 0 & \sin\bar{\theta}\cos\bar{\phi} & \sin\bar{\theta}\sin\bar{\phi} & \cos\bar{\theta} \\ 0 & \frac{1}{\bar{r}}\cos\bar{\theta}\cos\bar{\phi} & \frac{1}{\bar{r}}\cos\bar{\theta}\sin\bar{\phi} & -\frac{1}{\bar{r}}\sin\bar{\theta} \\ 0 & -\frac{\sin\bar{\phi}}{\bar{r}\sin\bar{\theta}} & \frac{\cos\bar{\phi}}{\bar{r}\sin\bar{\theta}} & 0 \end{pmatrix}. \tag{2.14}$$

Of equal importance is the matrix that comes from taking the spherical coordinates to be the "original" set, with Cartesian coordinates as the "new" ones. To contrast with the matrix we have already constructed, we leave the spherical coordinates as the barred set; then the new matrix of interest has entries $\frac{\partial x^{\alpha}}{\partial \bar{x}^{\beta}}$

and can be expressed in either set of coordinates,

$$\frac{\partial x^\alpha}{\partial \bar{x}^\beta} \doteq \begin{pmatrix} 1 & 0 & 0 & 0 \\ 0 & \sin\bar{\theta}\cos\bar{\phi} & \bar{r}\cos\bar{\theta}\cos\bar{\phi} & -\bar{r}\sin\bar{\theta}\sin\bar{\phi} \\ 0 & \sin\bar{\theta}\sin\bar{\phi} & \bar{r}\cos\bar{\theta}\sin\bar{\phi} & \bar{r}\sin\bar{\theta}\cos\bar{\phi} \\ 0 & \cos\bar{\theta} & -\bar{r}\sin\bar{\theta} & 0 \end{pmatrix}$$

$$\doteq \begin{pmatrix} 1 & 0 & 0 & 0 \\ 0 & \frac{x}{\sqrt{x^2+y^2+z^2}} & \frac{xz}{\sqrt{x^2+y^2}} & -y \\ 0 & \frac{y}{\sqrt{x^2+y^2+z^2}} & \frac{yz}{\sqrt{x^2+y^2}} & x \\ 0 & \frac{z}{\sqrt{x^2+y^2+z^2}} & -\sqrt{x^2+y^2} & 0 \end{pmatrix}. \tag{2.15}$$

This matrix is the inverse of $\frac{\partial \bar{x}^\mu}{\partial x^\nu}$, which you can verify explicitly in either Cartesian or spherical coordinates (although you should express both matrices in the same coordinate system prior to multiplying them). It makes sense that $\frac{\partial x^\alpha}{\partial \bar{x}^\beta}$ is the inverse of $\frac{\partial \bar{x}^\mu}{\partial x^\nu}$. Think of the identity matrix with entries expressed as $\frac{\partial x^\alpha}{\partial x^\beta}$. If you view x^α as a function of the new coordinates, the barred set, then the chain rule gives

$$\frac{\partial x^\alpha}{\partial x^\beta} = \frac{\partial x^\alpha}{\partial \bar{x}^\nu}\frac{\partial \bar{x}^\nu}{\partial x^\beta}. \tag{2.16}$$

On the left, we have the identity with ones along the diagonal where $\alpha = \beta$ and zero for $\alpha \neq \beta$. On the right is the index notation form of the multiplication of the matrix (with entries) $\frac{\partial x^\alpha}{\partial \bar{x}^\nu}$ by $\frac{\partial \bar{x}^\nu}{\partial x^\beta}$. In the context of the Einstein notation, the index ν is a closed summation index, it appears "up" on the $\bar{x}$ derivative, but since one of those is in the denominator, that one counts as "down."[1]

We will see the utility of this seemingly esoteric notation soon enough. You should pause and try creating the matrices from this section for a Cartesian to cylindrical coordinate transformation in Problem 2.5. Cylindrical coordinates are defined by $\{c\bar{t}, \bar{s}, \bar{\phi}, \bar{z}\}$ with

$$c\bar{t} = ct, \qquad \bar{s} = \sqrt{x^2 + y^2}, \qquad \bar{\phi} = \tan^{-1}\left(\frac{y}{x}\right) \qquad \bar{z} = z, \tag{2.17}$$

so the temporal and z coordinate remain as before, $\bar{s}$ is the distance to the $\bar{z}$ axis, and $\bar{\phi}$ is the azimuthal angle, defined with respect to the x axis as for spherical coordinates.

---

[1] The other direction is true, too, if a lower index appears in a denominator, it counts as an upper index for the purposes of summation.

2.1 Tensor Transformations

With our example transformations in mind, we are ready to define tensors in terms of their response to a coordinate transformation. A tensor has indices dangling off of it, the placement of these tells us about the transformation response, and the number of indices is called the "rank." A tensor of rank 0, with no indices, is a "scalar," and we'll start with it.

2.1.1 Scalars

Scalars do not transform under a coordinate transformation. For a scalar function $\psi(t, x, y, z)$, if you move to a new coordinate system, you just take the Cartesian coordinates and express them in terms of the new targets $\{\bar{x}^0, \bar{x}^1, \bar{x}^2, \bar{x}^3\}$. The rule is

$$
\bar{\psi}\left(\bar{x}^0, \bar{x}^1, \bar{x}^2, \bar{x}^3\right) = \psi\left(t\left(\bar{x}^0, \bar{x}^1, \bar{x}^2, \bar{x}^3\right), x\left(\bar{x}^0, \bar{x}^1, \bar{x}^2, \bar{x}^3\right), \\ y\left(\bar{x}^0, \bar{x}^1, \bar{x}^2, \bar{x}^3\right), z\left(\bar{x}^0, \bar{x}^1, \bar{x}^2, \bar{x}^3\right)\right).
$$

(2.18)

As an example, take the scalar function, written initially in Cartesian coordinates, $\psi(t, x, y, z) = x^2 - yz$, and move to spherical coordinates. The "transformed" function is

$$
\bar{\psi}(\bar{t}, \bar{r}, \bar{\theta}, \bar{\phi}) = \bar{r}^2 \sin^2 \bar{\theta} \cos^2 \bar{\phi} - \bar{r}^2 \sin \bar{\theta} \sin \bar{\phi} \cos \bar{\theta},
$$

(2.19)

just using (2.11) to write the Cartesian components in terms of spherical ones. There's nothing else to it, this is the simplest object there is. We will sometimes write $\psi(x)$ with argument "x" as a stand-in for some original set of four coordinates; then the transformation rule (or lack thereof) can be expressed succinctly as $\bar{\psi}(\bar{x}) = \psi(x(\bar{x}))$.

2.1.2 Vectors

A vector has one index, it is a "first-rank" tensor. There are two places that index can appear, up or down, and these correspond to the manner in which the vector transforms. A "contravariant" first-rank tensor transforms with one factor of $\frac{\partial \bar{x}^\mu}{\partial x^\nu}$ and is written with its index up. For contravariant $A^\mu(x)$, we have

$$
\bar{A}^\mu(\bar{x}) = \frac{\partial \bar{x}^\mu}{\partial x^\nu} A^\nu\left(x(\bar{x})\right).
$$

(2.20)

Notice that we write $\bar{A}^\mu(\bar{x})$ in terms of the new coordinates using the old coordinates' expression in terms of the new, $x(\bar{x})$. That is separate from the transformation of the vector components, which is really a sum (or matrix-vector product, if you prefer).

The archetype of a contravariant first-rank tensor is the coordinate differential itself, dx^μ. Focusing on the zeroth component to be concrete, when we move to the new coordinates, $\bar{x}^0(x)$, that is, the new coordinate $\bar{x}^0$ depends on the old coordinates (like $\bar{r} = \sqrt{x^2 + y^2 + z^2}$, for example). Then the differential $d\bar{x}^0(x)$ is related to the differentials of the old coordinate system:

$$d\bar{x}^0(\bar{x}) = d\bar{x}^0\big(x(\bar{x})\big) = \frac{\partial \bar{x}^0}{\partial x^0}dx^0 + \frac{\partial \bar{x}^0}{\partial x^1}dx^1 + \frac{\partial \bar{x}^0}{\partial x^2}dx^2 + \frac{\partial \bar{x}^0}{\partial x^3}dx^3 = \frac{\partial \bar{x}^0}{\partial x^\nu}dx^\nu, \tag{2.21}$$

and this is true for each new coordinate, so we have

$$d\bar{x}^\mu = \frac{\partial \bar{x}^\mu}{\partial x^\nu}dx^\nu. \tag{2.22}$$

We have used the chain rule here, and (2.22) establishes that dx^α is a contravariant first-rank tensor. Contravariant tensors are probably the most familiar in physics, since when we have vector objects like the magnetic vector potential, $\mathbf{A} = A^x\hat{\mathbf{x}} + A^y\hat{\mathbf{y}} + A^z\hat{\mathbf{z}}$, the components $\{A^x, A^y, A^z\}$ are pieces of a contravariant tensor (for reasons that we will explain in a moment).

The other way that a first-rank tensor can respond to a transformation is with the *inverse* of $\frac{\partial \bar{x}^\mu}{\partial x^\nu}$, namely $\frac{\partial x^\alpha}{\partial \bar{x}^\beta}$. Such vectors are "covariant," and their index is placed down. For covariant $B_\mu(x)$, the transformation rule is defined to be

$$\boxed{\bar{B}_\beta(\bar{x}) = \frac{\partial x^\alpha}{\partial \bar{x}^\beta}B_\alpha\big(x(\bar{x})\big),} \tag{2.23}$$

where again, in addition to introducing the factor of $\frac{\partial x^\alpha}{\partial \bar{x}^\beta}$, we also have to write all expressions in terms of the new coordinates ("have to" is a little strong, we sometimes work with the old coordinates, whichever is easier). Referring to Footnote 1, the open β index appears up: $\partial \bar{x}^\beta$, but since that term is in the denominator, the overall effect is down, hence the lower placement of β on the left.

Here the example to keep in mind is the gradient of a scalar function $\psi(x)$. If you take the derivative of $\psi(x)$ with respect to the old coordinates, then you get $\frac{\partial \psi(x)}{\partial x^\mu}$, which we sometimes write as $\psi_{,\mu} \equiv \frac{\partial \psi}{\partial x^\mu}$ to highlight the index location. Under the coordinate transformation, we have a new gradient:

$$\frac{\partial \bar{\psi}(\bar{x})}{\partial \bar{x}^\beta} = \frac{\partial \psi(x(\bar{x}))}{\partial \bar{x}^\beta} = \frac{\partial \psi(x)}{\partial x^\alpha}\frac{\partial x^\alpha}{\partial \bar{x}^\beta}, \tag{2.24}$$

where to get the first equality, we used the scalar transformation rule from (2.18) (with shorthand $\bar{\psi}(\bar{x}) = \psi(x(\bar{x}))$) and then used the chain rule to get the second equality. Writing this expression in terms of the "comma" form of the derivative, we have

$$\bar{\psi}_{,\beta} = \frac{\partial x^\alpha}{\partial \bar{x}^\beta}\psi_{,\alpha}, \tag{2.25}$$

which allows easy visual confirmation of the covariant transformation defined in (2.23). It is also common to see the derivative operator written as a covariant first-rank tensor itself, $\partial_\mu \equiv \frac{\partial}{\partial x^\mu}$, so the above is sometimes written as

$$\bar{\partial}_\beta \bar{\psi} = \frac{\partial x^\alpha}{\partial \bar{x}^\beta} \partial_\alpha \psi. \tag{2.26}$$

It is important to note that for familiar vectors from E&M, like the current $\mathbf{J}$, what we are really looking at is the projection of components onto basis vectors. If we think of the current as having four components, with $J^0 = \rho c$, and we call the ct direction's basis vector $\hat{\mathbf{t}}$, then we really have

$$\mathbf{J} = J^0 \hat{\mathbf{t}} + J^x \hat{\mathbf{x}} + J^y \hat{\mathbf{y}} + J^z \hat{\mathbf{z}}, \tag{2.27}$$

and both the components, $\{J^0, J^x, J^y, J^z\}$ and basis vectors $\{\hat{\mathbf{t}}, \hat{\mathbf{x}}, \hat{\mathbf{y}}, \hat{\mathbf{z}}\}$, change under a coordinate transformation.

What happens to the basis vectors? Let's look at how they change under the Cartesian-to-spherical transformation, which provides a concrete setting. Basis vectors point in the direction of increasing coordinate value. That's familiar from the usual $\hat{\mathbf{x}}$, $\hat{\mathbf{y}}$. and $\hat{\mathbf{z}}$, which are constants. To write the basis vectors in the new coordinates, we need to find their "increasing" direction. Taking the $\bar{r} = \sqrt{x^2 + y^2 + z^2}$ coordinate as an example, the increasing direction is given by the gradient,

$$\bar{\mathbf{r}} = \nabla \bar{r} = \frac{\partial \bar{r}}{\partial x} \hat{\mathbf{x}} + \frac{\partial \bar{r}}{\partial y} \hat{\mathbf{y}} + \frac{\partial \bar{r}}{\partial z} \hat{\mathbf{z}} = \frac{x \hat{\mathbf{x}} + y \hat{\mathbf{y}} + z \hat{\mathbf{z}}}{\sqrt{x^2 + y^2 + z^2}}. \tag{2.28}$$

The other two coordinates have, similarly,

$$\bar{\boldsymbol{\theta}} = \frac{xz}{\sqrt{x^2 + y^2}(x^2 + y^2 + z^2)} \hat{\mathbf{x}} + \frac{yz}{\sqrt{x^2 + y^2}(x^2 + y^2 + z^2)} \hat{\mathbf{y}} - \frac{\sqrt{x^2 + y^2}}{x^2 + y^2 + z^2} \hat{\mathbf{z}} \tag{2.29}$$

and

$$\bar{\boldsymbol{\phi}} = -\frac{y}{x^2 + y^2} \hat{\mathbf{x}} + \frac{x}{x^2 + y^2} \hat{\mathbf{y}}. \tag{2.30}$$

These vectors do not come to us normalized in any obvious way, although we can always form unit vectors from them by dividing each by their length. Written in terms of the spherical coordinates themselves (but using the constant Cartesian basis vectors for clarity), we have

$$\bar{\mathbf{r}} = \sin \bar{\theta} \cos \bar{\phi} \hat{\mathbf{x}} + \sin \bar{\theta} \sin \bar{\phi} \hat{\mathbf{y}} + \cos \bar{\theta} \hat{\mathbf{z}},$$

$$\bar{\boldsymbol{\theta}} = \frac{1}{\bar{r}} (\cos \bar{\theta} \cos \bar{\phi} \hat{\mathbf{x}} + \cos \bar{\theta} \sin \bar{\phi} \hat{\mathbf{y}} - \sin \bar{\theta} \hat{\mathbf{z}}), \tag{2.31}$$

$$\bar{\boldsymbol{\phi}} = \frac{1}{\bar{r} \sin \bar{\theta}} (-\sin \bar{\phi} \hat{\mathbf{x}} + \cos \bar{\phi} \hat{\mathbf{y}}).$$

If we take $\mathbf{e}_\alpha$ to have the elements $\mathbf{e}_0 = \hat{\mathbf{t}}$, $\mathbf{e}_1 = \hat{\mathbf{x}}$, $\mathbf{e}_2 = \hat{\mathbf{y}}$, and $\mathbf{e}_3 = \hat{\mathbf{z}}$, then the transformation rule implied by (2.31) is

$$\bar{\mathbf{e}}_\beta = \frac{\partial x^\alpha}{\partial \bar{x}^\beta} \mathbf{e}_\alpha, \tag{2.32}$$

which gives, written in spherical coordinates,

$$\bar{\mathbf{e}}_1 = \sin\bar\theta\cos\bar\phi\hat{\mathbf{x}} + \sin\bar\theta\sin\bar\phi\hat{\mathbf{y}} + \cos\bar\theta\hat{\mathbf{z}},$$
$$\bar{\mathbf{e}}_2 = \bar{r}(\cos\bar\theta\cos\bar\phi\hat{\mathbf{x}} + \cos\bar\theta\sin\bar\phi\hat{\mathbf{y}} - \sin\bar\theta\hat{\mathbf{z}}), \qquad (2.33)$$
$$\bar{\mathbf{e}}_3 = \bar{r}\sin\bar\theta(-\sin\bar\phi\hat{\mathbf{x}} + \cos\bar\phi\hat{\mathbf{y}}).$$

While the normalizations for these three are different from the three vectors in (2.31), it is clear that the associated unit vectors in each case point in the same direction. We will account for the additional scaling factors out front in a bit. For now, it is enough to note that while the *components* of $\mathbf{J}$ in (2.27) transform as a contravariant first-rank tensor, the *basis* transforms covariantly. This is good since it means that the geometrically relevant $\mathbf{J}$ is *the same* in all coordinate systems. We think of an object like $\mathbf{J}$ as pointing in a particular direction with a particular magnitude at every point, and this should not depend on the coordinate system we use to express that magnitude and direction. To see that this is the case, write $\mathbf{J} = J^\alpha\mathbf{e}_\alpha$. Under the Cartesian $\rightarrow$ spherical transformation, we have

$$\bar{\mathbf{J}} = \bar{J}^\alpha\bar{\mathbf{e}}_\alpha = \frac{\partial\bar{x}^\alpha}{\partial x^\gamma}J^\gamma\frac{\partial x^\beta}{\partial\bar{x}^\alpha}\mathbf{e}_\beta = \frac{\partial x^\beta}{\partial x^\gamma}J^\gamma\mathbf{e}_\beta, \qquad (2.34)$$

and we know that $\frac{\partial x^\beta}{\partial x^\gamma}$ is one if $\beta = \gamma$ and zero otherwise, so that

$$\bar{\mathbf{J}} = \frac{\partial x^\beta}{\partial x^\gamma}J^\gamma\mathbf{e}_\beta = J^\beta\mathbf{e}_\beta = \mathbf{J}, \qquad (2.35)$$

as desired. There is more to say about this components-versus-basis transformation story, but we will hold off until we have a little more machinery to finish the discussion.

2.1.3 Tensors of Higher Rank

Once we have the contravariant and covariant first-rank forms, tensors with more indices can be defined in terms of products of these simpler constructs. As an example, if we have a pair of contravariant tensors, A^μ and B^ν, then we can form a second-rank object out of the product, $S^{\mu\nu} = A^\mu B^\nu$, with

$$\bar{S}^{\mu\nu} = \bar{A}^\mu\bar{B}^\nu = \frac{\partial\bar{x}^\mu}{\partial x^\alpha}A^\alpha\frac{\partial\bar{x}^\nu}{\partial x^\beta}B^\beta = \frac{\partial\bar{x}^\mu}{\partial x^\alpha}\frac{\partial\bar{x}^\nu}{\partial x^\beta}S^{\alpha\beta}. \qquad (2.36)$$

While only a model second-rank (two index) tensor, this does serve to define the second-rank contravariant transformation rule. For a generic $F^{\mu\nu}$, we have

$$\boxed{\bar{F}^{\mu\nu}(\bar{x}) = \frac{\partial\bar{x}^\mu}{\partial x^\alpha}\frac{\partial\bar{x}^\nu}{\partial x^\beta}F^{\alpha\beta}\big(x(\bar{x})\big)} \qquad (2.37)$$

even when $F^{\alpha\beta}$ can't necessarily be expressed as a product of first-rank tensors.

Similarly, we can motivate the transformation for second-rank covariant tensors by thinking about the product of first-rank covariant A_μ and B_ν tensors (see Problem 2.6). For a general $G_{\mu\nu}$, the transformation is

$$\bar{G}_{\mu\nu}(\bar{x}) = \frac{\partial x^\alpha}{\partial \bar{x}^\mu} \frac{\partial x^\beta}{\partial \bar{x}^\nu} G_{\alpha\beta}\big(x(\bar{x})\big). \tag{2.38}$$

Finally, there is a last option for second-rank tensors: "mixed," with one contravariant and one covariant index. Our model is a contravariant A^μ multiplying a covariant B_ν, then for $M^\mu_{\ \nu} \equiv A^\mu B_\nu$, the transformation is

$$\bar{M}^\mu_{\ \nu} = \bar{A}^\mu \bar{B}_\nu = \frac{\partial \bar{x}^\mu}{\partial x^\alpha} \frac{\partial x^\beta}{\partial \bar{x}^\nu} A^\alpha B_\beta = \frac{\partial \bar{x}^\mu}{\partial x^\alpha} \frac{\partial x^\beta}{\partial \bar{x}^\nu} M^\alpha_{\ \nu}, \tag{2.39}$$

and the general rule that defines the mixed rank tensor $H^\mu_{\ \nu}$ is

$$\bar{H}^\mu_{\ \nu}(\bar{x}) = \frac{\partial \bar{x}^\mu}{\partial x^\alpha} \frac{\partial x^\beta}{\partial \bar{x}^\nu} H^\alpha_{\ \beta}\big(x(\bar{x})\big). \tag{2.40}$$

Example 2.1 (Kronecker Delta). All tensors come with a natural index placement, and we will highlight those natural placements as we go. For example, the "Kronecker delta" tensor $\delta^\mu_{\ \nu}$ takes on the same numerical values in all coordinate systems:

$$\delta^\mu_{\ \nu} = \begin{cases} 1 & \text{if } \mu = \nu, \\ 0 & \mu \neq \nu. \end{cases} \tag{2.41}$$

You will show in Problem 2.8 that if $\bar{\delta}^\mu_{\ \nu} = \delta^\mu_{\ \nu}$ for any coordinate transformation, then the Kronecker delta is a mixed second-rank tensor as its index placement indicates.

Example 2.2 (Matrix Inverse). Given a second-rank covariant tensor $h_{\mu\nu}$, we can make a contravariant second-rank tensor by taking the matrix inverse (when it exists). Let $h^{\mu\nu}$ be the matrix inverse of $h_{\mu\nu}$, so that

$$h^{\mu\alpha} h_{\alpha\nu} = \delta^\mu_{\ \nu}. \tag{2.42}$$

Then for $h_{\mu\nu}$ a second-rank covariant tensor, $h^{\mu\nu}$ is contravariant, as I will now confirm. Transforming each term in (2.42), we get

$$\bar{h}^{\mu\alpha} \bar{h}_{\alpha\nu} = \bar{\delta}^\mu_{\ \nu} = \delta^\mu_{\ \nu}, \tag{2.43}$$

using the property of the Kronecker delta that you established in Problem 2.8. This equation establishes that $\bar{h}^{\mu\alpha}$ is the matrix inverse of $\bar{h}_{\rho\sigma}$. By assumption,

$$\bar{h}_{\alpha\nu} = \frac{\partial x^\beta}{\partial \bar{x}^\alpha} \frac{\partial x^\sigma}{\partial \bar{x}^\nu} h_{\beta\sigma}, \tag{2.44}$$

and putting this in to (2.43), we can begin to isolate $\bar{h}^{\mu\alpha}$:

$$\bar{h}^{\mu\alpha} \frac{\partial x^\beta}{\partial \bar{x}^\alpha} \frac{\partial x^\sigma}{\partial \bar{x}^\nu} h_{\beta\sigma} = \delta^\mu_{\ \nu}. \tag{2.45}$$

We'll start peeling off factors of $\frac{\partial x}{\partial \bar{x}}$ by looking for an open index that will accept matrix multiplication. Since the β, σ, and α indices on the left are locked up in sums, we have only the open ν to work with. Multiply both sides of the equation by $\frac{\partial \bar{x}^\nu}{\partial x^\rho}$,

$$\bar{h}^{\mu\alpha} \frac{\partial x^\beta}{\partial \bar{x}^\alpha} \underbrace{\frac{\partial x^\sigma}{\partial \bar{x}^\nu} \frac{\partial \bar{x}^\nu}{\partial x^\rho}}_{=\delta^\sigma_\rho} h_{\beta\sigma} = \delta^\mu_\nu \frac{\partial \bar{x}^\nu}{\partial x^\rho},$$

$$\bar{h}^{\mu\alpha} \frac{\partial x^\beta}{\partial \bar{x}^\alpha} h_{\beta\rho} = \frac{\partial \bar{x}^\mu}{\partial x^\rho}. \tag{2.46}$$

Continuing to move terms from left to right we have an open ρ index, so multiply both sides by $h^{\rho\gamma}$ and use $h_{\beta\rho} h^{\rho\gamma} = \delta^\gamma_\beta$ on the left,

$$\bar{h}^{\mu\alpha} \frac{\partial x^\gamma}{\partial \bar{x}^\alpha} = \frac{\partial \bar{x}^\mu}{\partial x^\rho} h^{\rho\gamma}. \tag{2.47}$$

Finally, we can multiply by $\frac{\partial \bar{x}^\nu}{\partial x^\gamma}$ and simplify using $\frac{\partial x^\gamma}{\partial \bar{x}^\alpha} \frac{\partial \bar{x}^\nu}{\partial x^\gamma} = \delta^\nu_\alpha$ to arrive at the transformation rule for $\bar{h}^{\mu\nu}$:

$$\bar{h}^{\mu\nu} = \frac{\partial \bar{x}^\mu}{\partial x^\rho} \frac{\partial \bar{x}^\nu}{\partial x^\gamma} h^{\rho\gamma}, \tag{2.48}$$

establishing, by comparison with (2.37), that $h^{\mu\nu}$ is, indeed, a contravariant second-rank tensor. The moral of the story is that given a second-rank covariant tensor, there is a mathematically available associated (by matrix inverse, assuming it exists) second-rank contravariant tensor.

We can keep going, defining tensors of third rank, fourth rank, etc. In each case, there is a proliferation of index position possibilities. But I hope the general pattern is clear: Every upper index transforms with a factor of $\frac{\partial \bar{x}}{\partial x}$, appropriately contracted with the index, and every lower index picks up a factor of $\frac{\partial x}{\partial \bar{x}}$ upon coordinate transformation. The general transformation rule, for an $(m + n)$-rank tensor with m up indices, n down, and coordinates $x \to \bar{x}$, is

$$\bar{T}^{\mu_1 \mu_2 \ldots \mu_m}_{\nu_1 \nu_2 \ldots \nu_n} = \frac{\partial \bar{x}^{\mu_1}}{\partial x^{\alpha_1}} \frac{\partial \bar{x}^{\mu_2}}{\partial x^{\alpha_2}} \cdots \frac{\partial \bar{x}^{\mu_m}}{\partial x^{\alpha_m}} \frac{\partial x^{\beta_1}}{\partial \bar{x}^{\nu_1}} \frac{\partial x^{\beta_2}}{\partial \bar{x}^{\nu_2}} \cdots \frac{\partial x^{\beta_n}}{\partial \bar{x}^{\nu_n}} T^{\alpha_1 \alpha_2 \ldots \alpha_m}_{\beta_1 \beta_2 \ldots \beta_n}. \tag{2.49}$$

2.2 The Metric

There is a special second-rank covariant tensor that plays a central role in general relativity. We have already seen how a "line element" tells us how to measure distances. For Minkowski spacetime, in Cartesian coordinates, we had

$$ds^2 = -c^2 dt^2 + dx^2 + dy^2 + dz^2, \tag{2.50}$$

but what happens if we switch to spherical coordinates? The line element is a scalar, it cannot matter what coordinate system we use to describe the infinitesimal lengths, the total length (squared) must always come out the same. Using that notion, we could just take the transformation to spherical coordinates from (2.11), compute dt, dx, dy, dz in terms of the new differentials $\{d\bar{t}, d\bar{r}, d\bar{\theta}, d\bar{\phi}\}$, and put those in to (2.50). But given that the temporal piece is unchanged and that the spatial piece still represents the Pythagorean combination, we can develop the correct expression by thinking about the orthogonal directions at a point in spherical coordinates. If you moved a distance $d\bar{r}$ in the radial direction, then $ds^2 = d\bar{r}^2$. Moving through an angle in the θ direction, $d\bar{\theta}$, the distance traveled is $\bar{r}d\bar{\theta}$, depending on how far from the origin you are, so in that case, $ds^2 = \bar{r}^2 d\bar{\theta}^2$. Finally, for an infinitesimal displacement of $d\bar{\phi}$ in the ϕ direction, if we are at $\bar{r}$ with polar angle $\bar{\theta}$, then the distance is $\bar{r}\sin\bar{\theta}d\bar{\phi}$ (motion in the ϕ direction occurs on a circle of radius $\bar{r}\sin\bar{\theta}$; see Figure 2.1). We can combine these orthogonal infinitesimal displacements by adding the squares to get

$$ds^2 = -c^2 d\bar{t}^2 + d\bar{r}^2 + \bar{r}^2 d\bar{\theta}^2 + \bar{r}^2 \sin^2\bar{\theta}d\bar{\phi}^2. \qquad (2.51)$$

The pattern continues for other types of transformation (work out the line element for cylindrical coordinates in Problem 2.10), we will always end up with a function that is quadratic in the coordinate differentials, and the coefficients are themselves functions of the coordinates. Defining the coordinate differential to have elements dx^μ, in any coordinate system now, the distance squared associated with the individual differentials can be written

$$ds^2 = dx^\mu g_{\mu\nu} dx^\nu \qquad (2.52)$$

for a second-rank covariant tensor $g_{\mu\nu}$ that is called the "metric." This tensor is naturally covariant since dx^μ is naturally contravariant and we are making a scalar, so the transformation of the second-rank contravariant $dx^\mu dx^\nu$ must be "undone" by a second-rank covariant tensor. The metric is often displayed as a matrix, although it is not particularly useful to think of tensor entries as matrix entries for higher-rank tensors. For Cartesian coordinates with $dx^0 = cdt$, $dx^1 = dx$, $dx^2 = dy$, $dx^3 = dz$, the Minkowski metric can be read off from (2.50),

$$g_{\mu\nu} \doteq \begin{pmatrix} -1 & 0 & 0 & 0 \\ 0 & 1 & 0 & 0 \\ 0 & 0 & 1 & 0 \\ 0 & 0 & 0 & 1 \end{pmatrix}. \qquad (2.53)$$

If we transform to spherical coordinates with $d\bar{x}^0 = cd\bar{t}$, $d\bar{x}^1 = d\bar{r}$, $d\bar{x}^2 = d\bar{\theta}$, $d\bar{x}^3 = d\bar{\phi}$, then from (2.51) we have

$$\bar{g}_{\mu\nu} \doteq \begin{pmatrix} -1 & 0 & 0 & 0 \\ 0 & 1 & 0 & 0 \\ 0 & 0 & \bar{r}^2 & 0 \\ 0 & 0 & 0 & \bar{r}^2 \sin^2\bar{\theta} \end{pmatrix}. \qquad (2.54)$$

As defined here, the metric is a feature of the coordinate system you are using, and its general quadratic structure is informed by the underlying Cartesian definition (2.50). However, the metric can do more than tell you what coordinate system is expressing the Minkowski length definition; it can change the very meaning of length. This dual role of both defining length (beyond our familiar notion) and setting the coordinate system in which that length is measured makes the metric a little slippery. How do we know when we have moved beyond the Minkowski length definition, versus disguised it in an unfamiliar coordinate choice? That will be an important question later on, but I want to highlight it here. For now, the metric only has one job, to tell us how to measure Minkowski lengths in a particular coordinate system.

There is one more structural point to make about this new object $g_{\mu\nu}(x)$. We can take it to be symmetric with $g_{\nu\mu}(x) = g_{\mu\nu}(x)$. You will show in Problem 2.11 that any second-rank tensor can be split into symmetric and antisymmetric pieces, and then you will establish in Problem 2.12 that when you multiply a symmetric tensor, like $dx^\mu dx^\nu$ in (2.52) by a second-rank tensor like $g_{\mu\nu}$, summing over both indices, only the symmetric part of $g_{\mu\nu}$ contributes to the sum. If there was an antisymmetric piece to the metric, it wouldn't show up in the defining equation (2.52), so we then take the metric to be symmetric from the start.

2.2.1 Raising and Lowering Indices

In general relativity (and in all of the other physical theories discussed along the way), there will always be a metric. We work in particular coordinate systems, so there is always a form for $g_{\mu\nu}$ that expresses length in that coordinate system. For more exotic geometries, we will describe their deviation from the Minkowski length definition using a metric, so there, too, the metric will always be available. As a second-rank covariant tensor, we know from Example 2.2 that there is an associated contravariant form that can be obtained by taking the matrix inverse of $g_{\mu\nu}$, so not only do we always have a metric, we will always have its inverse[2] $g^{\alpha\beta}$ defined by

$$g_{\mu\nu}g^{\nu\beta} = \delta^\beta_{\ \mu}. \tag{2.55}$$

Suppose you had a naturally contravariant tensor A^μ. It is easy to see that $g_{\mu\nu}A^\mu$ is a covariant first-rank tensor. Consider its transformation

$$\bar{g}_{\mu\nu}\bar{A}^\mu = \frac{\partial x^\alpha}{\partial \bar{x}^\mu}\frac{\partial x^\beta}{\partial \bar{x}^\nu}g_{\alpha\beta}\frac{\partial \bar{x}^\mu}{\partial x^\rho}A^\rho, \tag{2.56}$$

---

[2]   We restrict our attention to metrics that are invertible almost everywhere. For a metric that is not invertible, there exists a coordinate system in which the metric is diagonal with at least one zero entry on its diagonal, so that one of the coordinates doesn't contribute to the length calculation. In that case, the dimension of our space is less than four.

and we know, from the chain rule, that

$$\delta^{\alpha}{}_{\rho} = \frac{\partial x^{\alpha}}{\partial x^{\rho}} = \frac{\partial x^{\alpha}}{\partial \bar{x}^{\mu}} \frac{\partial \bar{x}^{\mu}}{\partial x^{\rho}}, \tag{2.57}$$

which we can use to simplify the right-hand side of (2.56):

$$\bar{g}_{\mu\nu} \bar{A}^{\mu} = \frac{\partial x^{\alpha}}{\partial \bar{x}^{\mu}} \frac{\partial x^{\beta}}{\partial \bar{x}^{\nu}} g_{\alpha\beta} \frac{\partial \bar{x}^{\mu}}{\partial x^{\rho}} A^{\rho} = \frac{\partial x^{\beta}}{\partial \bar{x}^{\nu}} g_{\rho\beta} A^{\rho}. \tag{2.58}$$

This is the transformation rule for a first-rank covariant tensor. For convenience (and no other reason!), it is common to *define* this covariant object associated with the contravariant A^{μ} to be $A_{\nu} \equiv g_{\mu\nu} A^{\mu}$. Similarly, if we have a naturally covariant B_{μ}, then we define the contravariant form to be $B^{\mu} \equiv g^{\mu\nu} B_{\nu}$. Notice that in expressions with closed indices, like the scalar $A^{\mu} B_{\mu}$, we can swap the lower and upper indices, since we can always put an implicit metric in place, and it can act to lower either index:

$$A^{\mu} B_{\mu} = A^{\mu} B^{\nu} g_{\mu\nu} = A_{\nu} B^{\nu} = A_{\mu} B^{\mu}. \tag{2.59}$$

We could make these same definitions with any other covariant/contravariant pair of second-rank tensors. It just so happens that the metric and its inverse are ubiquitous, so they form a natural choice (and one with geometric implications). For this reason, we use the metric and its inverse to raise and lower tensor indices. An upper μ (say) index on any tensor that is contracted into $g_{\mu\nu}$ yields a covariant ν index, so the identification with a "lower" index is appropriate. Similarly, a lower ν index on a tensor, when contracted with $g^{\alpha\nu}$, gives the resulting tensor a contravariant α index, and we write it as an "upper" index. The raising and lowering of indices with the metric can be thought of as part of the Einstein index notation convention. But it is important to remember that most of the tensors we encounter have "natural" index placement based on how they transform. We can raise or lower as necessary to satisfy the conventions of the notation, while keeping in mind which form is actually associated with a physical quantity versus a "derived" form in which indices have been raised or lowered with a factor of the metric. The next two examples make the pitfalls clear.

Example 2.3 (Basis Vectors). In this and the next example, we will work in three-dimensional spherical coordinates from the start, so all implicit sums will be from $1 \rightarrow 3$, and we will get rid of the bars over the spherical coordinates $\{r, \theta, \phi\}$ since we are not really transforming from one coordinate system to another. Let's return to the (unnormalized) basis vectors we computed in spherical coordinates, written in terms of the constant Cartesian set from (2.31),

$$\mathbf{r} = \sin\theta \cos\phi \hat{\mathbf{x}} + \sin\theta \sin\phi \hat{\mathbf{y}} + \cos\theta \hat{\mathbf{z}},$$

$$\boldsymbol{\theta} = \frac{1}{r}(\cos\theta \cos\phi \hat{\mathbf{x}} + \cos\theta \sin\phi \hat{\mathbf{y}} - \sin\theta \hat{\mathbf{z}}), \tag{2.60}$$

$$\boldsymbol{\phi} = \frac{1}{r \sin\theta}(-\sin\phi \hat{\mathbf{x}} + \cos\phi \hat{\mathbf{y}}).$$

We computed these by finding the "direction of increase" of the coordinates themselves.

Then we showed that these vectors are proportional to the *covariant* $\mathbf{e}_\alpha$, computed using the transformation from Cartesian to spherical coordinates. The covariant form was

$$\mathbf{e}_1 = \sin\theta\cos\phi\hat{\mathbf{x}} + \sin\theta\sin\phi\hat{\mathbf{y}} + \cos\theta\hat{\mathbf{z}},$$
$$\mathbf{e}_2 = r(\cos\theta\cos\phi\hat{\mathbf{x}} + \cos\theta\sin\phi\hat{\mathbf{y}} - \sin\theta\hat{\mathbf{z}}), \qquad (2.61)$$
$$\mathbf{e}_3 = r\sin\theta(-\sin\phi\hat{\mathbf{x}} + \cos\phi\hat{\mathbf{y}}).$$

If the basis vectors are supposed to be covariant, why isn't $\mathbf{e}_1 = \mathbf{r}$, $\mathbf{e}_2 = \boldsymbol{\theta}$, and $\mathbf{e}_3 = \boldsymbol{\phi}$? What we have in (2.60) is really the *contravariant* components of the basis vectors, the derived form. In three dimensions, the metric and its inverse in spherical coordinates are (for ordering $dx^1 = dr$, $dx^2 = d\theta$, $dx^3 = d\phi$)

$$g_{\mu\nu} \doteq \begin{pmatrix} 1 & 0 & 0 \\ 0 & r^2 & 0 \\ 0 & 0 & r^2\sin^2\theta \end{pmatrix}, \qquad g^{\mu\nu} \doteq \begin{pmatrix} 1 & 0 & 0 \\ 0 & \frac{1}{r^2} & 0 \\ 0 & 0 & \frac{1}{r^2\sin^2\theta} \end{pmatrix}, \qquad (2.62)$$

so that the upper form of the $\mathbf{e}_\alpha$ from (2.61) is $\mathbf{e}^\beta \equiv \mathbf{e}_\alpha g^{\alpha\beta}$. The contravariant elements are

$$\mathbf{e}^1 = \sin\theta\cos\phi\hat{\mathbf{x}} + \sin\theta\sin\phi\hat{\mathbf{y}} + \cos\theta\hat{\mathbf{z}},$$
$$\mathbf{e}^2 = \frac{1}{r}(\cos\theta\cos\phi\hat{\mathbf{x}} + \cos\theta\sin\phi\hat{\mathbf{y}} - \sin\theta\hat{\mathbf{z}}), \qquad (2.63)$$
$$\mathbf{e}^3 = \frac{1}{r\sin\theta}(-\sin\phi\hat{\mathbf{x}} + \cos\phi\hat{\mathbf{y}}).$$

It is this contravariant form of the basis vectors that we have in (2.60). If we normalize the basis vectors, then the distinction goes away, so there is yet a third important form here, the $\hat{\mathbf{r}}$, $\hat{\boldsymbol{\theta}}$, and $\hat{\boldsymbol{\phi}}$ we find in, for example, [21]. We can relate these *unit* basis vectors to the $\mathbf{e}_\alpha$:

$$\hat{\mathbf{r}} = \mathbf{e}_1, \qquad \hat{\boldsymbol{\theta}} = \frac{\mathbf{e}_2}{r}, \qquad \hat{\boldsymbol{\phi}} = \frac{\mathbf{e}_3}{r\sin\theta}. \qquad (2.64)$$

The basis vectors are covariant, why did we end up with the contravariant form when we computed them earlier using the gradient operator? Read on.

Example 2.4 (The Gradient). We are familiar with the gradient operator from vector calculus (working again in three spatial dimensions):

$$\nabla = \hat{\mathbf{x}}\frac{\partial}{\partial x} + \hat{\mathbf{y}}\frac{\partial}{\partial y} + \hat{\mathbf{z}}\frac{\partial}{\partial z}, \qquad (2.65)$$

where the components of the operator are $\frac{\partial}{\partial x}$, $\frac{\partial}{\partial y}$, and $\frac{\partial}{\partial z}$, but are these the components of a contravariant or covariant tensor? In Cartesian coordinates, the metric has ones along the diagonal and zero elsewhere, as does its inverse, so we don't actually know whether we're working with the contravariant or covariant components here.

Moving to spherical coordinates, the situation is clear, since there is a numerical distinction between the metric and its inverse (shown in (2.62)). The gradient of a scalar $\psi(r, \theta, \phi)$ is, again from vector calculus,

$$\nabla \psi = \frac{\partial \psi}{\partial r} \hat{\mathbf{r}} + \frac{1}{r} \frac{\partial \psi}{\partial \theta} \hat{\boldsymbol{\theta}} + \frac{1}{r \sin \theta} \frac{\partial \psi}{\partial \phi} \hat{\boldsymbol{\phi}}. \tag{2.66}$$

We can replace the unit basis vectors here with the covariant basis using (2.64):

$$\nabla \psi = \frac{\partial \psi}{\partial r} \mathbf{e}_1 + \frac{1}{r^2} \frac{\partial \psi}{\partial \theta} \mathbf{e}_2 + \frac{1}{r^2 \sin^2 \theta} \frac{\partial \psi}{\partial \phi} \mathbf{e}_3. \tag{2.67}$$

Now we have components multiplying the natural basis vectors, so we can evaluate their transformation properties. Remember that the gradient of a scalar is a covariant tensor $\psi_{,\mu} = \frac{\partial \psi}{\partial x^\mu}$ with the upper form obtained by raising the index using the (inverse) metric: $\psi_{\cdot}^{\ \nu} \equiv \psi_{,\mu} g^{\mu\nu}$. The two versions are

$$\psi_{,\mu} \equiv \frac{\partial \psi}{\partial x^\mu} \doteq \begin{pmatrix} \frac{\partial \psi}{\partial r} \\ \frac{\partial \psi}{\partial \theta} \\ \frac{\partial \psi}{\partial \phi} \end{pmatrix}, \qquad \psi_{\cdot}^{\ \nu} \equiv \psi_{,\mu} g^{\mu\nu} \doteq \begin{pmatrix} \frac{\partial \psi}{\partial r} \\ \frac{1}{r^2} \frac{\partial \psi}{\partial \theta} \\ \frac{1}{r^2 \sin^2 \theta} \frac{\partial \psi}{\partial \phi} \end{pmatrix}. \tag{2.68}$$

Aha – the components of $\nabla \psi$ appearing in (2.67) are associated with the *contravariant* form of the gradient. This makes sense: going back to our "vector arrow" notation, we had $\mathbf{J} = J^\alpha \mathbf{e}_\alpha$, so that the components are generally contravariant. In this case, using ∇ gives us the contravariant elements of the gradient. We can write $\nabla \psi$ in indexed form to focus our attention on its tensorial transformation character,

$$\nabla \psi = g^{\alpha\beta} \psi_{,\beta} \mathbf{e}_\alpha = \psi_{\cdot}^{\ \alpha} \mathbf{e}_\alpha \tag{2.69}$$

is clearly an object with contravariant components. It was precisely ∇ that we applied to the coordinates to define their direction of increasing value, which is what gave us the set of basis vectors in (2.31). It is no surprise, then, that these represented the contravariant form of the basis vectors.

2.3 Tensorial Expressions

We needed to get to the metric, since it is an important geometric quantity, but what is the broader point of tensors? Their primary utility comes in expressing "generally covariant" equations. We would like to write physical laws in mathematical language that does not depend on the coordinates we choose. A physical principle cannot depend on us writing it in spherical coordinates instead of Cartesian. This idea is codified, in a limited way, in the special relativistic idea that "the laws of physics are the same in all inertial frames." That means that physical predictions must hold in two coordinate systems related by a Lorentz transformation. Properly formulated tensor equations automatically hold in all coordinate systems. As an example, if you have two contravariant tensor functions $A^\mu(x)$ and $B^\mu(x)$ and you know that, in some coordinate system, $A^\mu(x) = B^\mu(x)$, then multiplying both sides of this equation by $\frac{\partial \bar{x}^\nu}{\partial x^\mu}$ (keep in mind that because of the repeated μ index in the product, this is really a linear combination of the original equations weighted by the coordinate derivatives) gives

$$\frac{\partial \bar{x}^\nu}{\partial x^\mu} A^\mu = \frac{\partial \bar{x}^\nu}{\partial x^\mu} B^\mu \longrightarrow \bar{A}^\nu = \bar{B}^\nu, \tag{2.70}$$

so that if the equality holds in one coordinate system, then it holds in all of them.

As a physically relevant example, take the four-potential of E&M, A^μ, with $A^0 = V/c$ and the spatial components that are precisely the magnetic vector potential. Maxwell's equations can be written as (in Lorenz gauge)

$$g^{\alpha\beta} \frac{\partial}{\partial x^\alpha} \frac{\partial A^\mu}{\partial x^\beta} = -\mu_0 J^\mu \tag{2.71}$$

for four-current J^μ (the temporal component is $J^0 = \rho c$ with spatial components $\mathbf{J}$). The second derivative operator here is really just the D'Alembertian[3]

$$\Box = g^{\alpha\beta} \frac{\partial}{\partial x^\alpha} \frac{\partial}{\partial x^\beta} \equiv \partial_\mu \partial^\mu \tag{2.72}$$

(using $\frac{\partial}{\partial x^\mu} \equiv \partial_\mu$ in the last definition). and it is clear that this is a scalar. In Cartesian coordinates, it is the familiar wave operator $\Box = -\frac{1}{c^2} \frac{\partial^2}{\partial t^2} + \nabla^2$. If we take the equation $\Box A^\mu = -\mu_0 J^\mu$ and transform it to a new coordinate system, either with a Lorentz boost or with a more general transformation, then we have $\Box \to \bar{\Box}$, and multiplying by $\frac{\partial \bar{x}^\nu}{\partial x^\mu}$ we have

$$\bar{\Box} \bar{A}^\nu = -\mu_0 \bar{J}^\nu; \tag{2.73}$$

the equation holds for all inertial frames, for any choice of coordinates.[4] Writing physical laws in tensorial form ensures this general covariance. It also allows for easy error-checking in expressions. The complete list of rules for index notation, all of which we've seen, but reiterated and collected here, are:

1. Any index appearing once on a term in an equation is an "open" index and takes on values $0 \to 3$ (or more for higher dimension). The index must appear as an open index with the same placement (up or down) on every term in the equation, so that you have a new valid equation for each value the index takes on. You are allowed to change the index's name as long as you do it across the board.

---

[3] We will update this in Section 2.5, for now the D'Alembertian as written is only a scalar under Lorentz transformations.

[4] What about the Lorenz gauge condition $\nabla \cdot \mathbf{A} = -\frac{1}{c^2} \frac{\partial V}{\partial t}$ itself, is it generally covariant?

2. An index appearing twice on a term in an equation is called a "closed" index. It is implicitly summed over, from $0 \to 3$. The index must appear in both the up and down positions on the term. These closed indices can be re-labeled as necessary, just as in an explicit summation, but both occurrences must be renamed with the same letter.

If you find one of these has been violated, it means that either a typographical or algebraic error has crept in to a calculation, or that the starting point did not represent a valid tensorial expression. Both situations are easy to fix. Try looking at the expressions in Problem 2.17, see if you can spot the ones that are invalid, and suggest a fix (these are not unique in the setting of an artificially produced error, you can't just go back through the derivation to find out what went wrong).

2.4 Relativistic Length of Curves

For a coordinate differential dx^μ, we have seen how the metric defines the total length associated with the individual infinitesimal displacements via $ds^2 = dx^\mu g_{\mu\nu} dx^\nu$. Suppose we had a curve parameterized by σ, so that the vector $x^\mu(\sigma)$ points from the origin to the spacetime point along the curve at parameter value σ (a generalization of $\mathbf{r}(t)$ from Section 1.3, for example) as shown in Figure 2.2. Then the differential becomes $dx^\mu = \dot{x}^\mu(\sigma)d\sigma$, where the dot here denotes the derivative with respect to the argument of the curve. In a "time" $d\sigma$, then, the distance travelled along the curve is

$$ds = \sqrt{\dot{x}^\mu(\sigma)g_{\mu\nu}\dot{x}^\nu(\sigma)}d\sigma. \qquad (2.74)$$

We are imagining a particle traveling along this curve with σ some sort of temporal parameterization (in special relativity, both the coordinate time and proper time are available). Suppose the particle is at rest (traveling only through time), so that $\dot{x}^0(\sigma) = c\dot{t}(\sigma)$ with all other components zero, then the distance, squared, is $ds^2 = -c^2\dot{t}(\sigma)^2 d\sigma^2$, where we are assuming the metric takes the form shown in (2.53). To avoid an imaginary length for particles at rest, we typically work with $-ds^2 = (c\dot{t}(\sigma)d\sigma)^2$, so the square root is real. To avoid confusion and a proliferation of i's, we'll write $d\ell^2 = -ds^2$ and compute lengths using $d\ell$.

In general, the particle travels a total distance (in the Minkowski sense)

$$\ell = \int d\ell = \int_{\sigma_0}^{\sigma_f} \sqrt{-\dot{x}^\mu(\sigma)g_{\mu\nu}\dot{x}^\nu(\sigma)}d\sigma, \qquad (2.75)$$

given some initial and final parameter values σ_0 and σ_f with $x(\sigma_0)$ and $x(\sigma_f)$ associated with the initial and final positions of the particle. This expression is

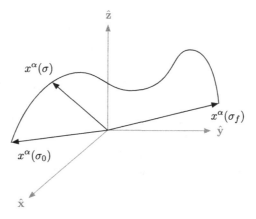

Fig. 2.2   A sample curve parameterized by σ with components $x^\alpha(\sigma)$. The curve starts at some initial parameter value σ_0, and ends at σ_f.

reparameterization invariant, as it should be, since the total length of a curve cannot depend on how we decide to put tick marks along it (sure, you could measure it in meters or miles, but the actual number must correspond to the same distance in each unit). Let $\rho(\sigma)$ be a new parameter. Then

$$\dot{x}^\mu(\sigma) = \frac{dx^\mu(\sigma)}{d\sigma} = \frac{dx^\mu(\rho(\sigma))}{d\rho}\frac{d\rho}{d\sigma} \tag{2.76}$$

using the chain rule. Putting this expression for $\dot{x}^\mu$ in (2.75), we get

$$\ell = \int_{\rho(\sigma_0)}^{\rho(\sigma_f)} \sqrt{-\frac{dx^\mu(\rho)}{d\rho}g_{\mu\nu}\frac{dx^\nu(\rho)}{d\rho}\frac{d\rho}{d\sigma}}d\sigma$$
$$= \int_{\rho_0}^{\rho_f} \sqrt{-\frac{dx^\mu(\rho)}{d\rho}g_{\mu\nu}\frac{dx^\nu(\rho)}{d\rho}}d\rho, \tag{2.77}$$

and the identical form, for either σ or ρ parameterization, is a manifestation of the reparameterization invariance. Try calculating ℓ for familiar particle motions in one spatial dimension in Problem 2.18.

In classical mechanics, we typically describe trajectories using time as the parameter. From the point of view of special relativity, that choice cannibalizes one of the coordinates themselves. You wouldn't do all of classical mechanics using the coordinate value of y as the parameter to describe motion in the x direction, an $x(y)$ curve.[5] The other obvious choice in special relativity is the arc length (proper time) parameterization,[6] where we parameterize with

---

[5] Of course, we did precisely this in Section 1.4.2, where we used the azimuthal angle ϕ to parameterize motion in r, the $r(\phi)$ of (1.72), for example.

[6] Arc length parameterization of a curve in mathematics is defined by a unit tangent vector. Given a curve $x^\alpha(\lambda)$, λ is the arc length parameter if $\dot{x}^\alpha(\lambda)g_{\alpha\beta}\dot{x}^\beta(\lambda) = 1$. See Appendix C for an expanded discussion.

respect to τ such that

$$- \frac{dx^\mu(\tau)}{d\tau} g_{\mu\nu} \frac{dx^\nu(\tau)}{d\tau} = c^2 \tag{2.78}$$

with c^2 the natural update for the Mathematician's "1" in arc length parameter-ization. This special parameter τ is the proper time associated with relativistic motion. At any point along the curve representing the motion of a particle, it is possible to make a Lorentz boost into the local rest frame of the particle, where it moves in time only, for an infinitesimal interval. That amounts to orienting the temporal axis along the tangent to the curve. Formally, we have

$$- c^2 d\tau^2 = dx^\mu g_{\mu\nu} dx^\nu \tag{2.79}$$

as the defining feature, where on the left, we have set $dx = dy = dz = 0$ in Cartesian coordinates. It is instructive to use both of these natural choices: in the case of coordinate time t, to make contact with nonrelativistic mechanics, and in the case of proper time τ, to simplify calculations.

Example 2.5 (Finding the Instantaneous Rest Frame). In proper time param-eterization, a curve has tangent $\dot{x}^\mu(\tau)$ with $\dot{x}^\mu \dot{x}_\mu = -c^2$ by definition, and $\dot{x}^\mu(\tau) \equiv \frac{dx^\mu}{d\tau}$ is a first-rank tensor under a Lorentz transformation, since the differential dx^μ is a tensor, and $d\tau$ is a scalar,[7] $-c^2 d\tau^2 = dx^\mu dx_\mu$. To set the physical scene, suppose you have a particle traveling along a curve with tangent $\dot{x}^\mu(\tau)$. At any point along the curve, we can perform a Lorentz boost into a coordinate system in which the particle is instantaneously at rest, so that at a particular value of τ, we have $\dot{\bar{x}}^\mu(\tau) = -c\bar{t}(\tau)\delta_0^\mu$. All we have to do is take the spatial direction of the tangent vector as the boost direction and use the instantaneous speed of the particle (in temporal parameterization) to define the boost parameter. The procedure for doing so is outlined in Appendix A.

As an example, consider the curve tracing out a circle in the xy plane,

$$x^\mu(\tau) \doteq \begin{pmatrix} c\tau\sqrt{1 + (\frac{\omega R}{c})^2} \\ R\sin(\omega\tau) \\ R\cos(\omega\tau) \\ 0 \end{pmatrix}, \quad \dot{x}^\mu(\tau) \doteq \begin{pmatrix} c\sqrt{1 + (\frac{\omega R}{c})^2} \\ \omega R\cos(\omega\tau) \\ -\omega R\sin(\omega\tau) \\ 0 \end{pmatrix}. \tag{2.80}$$

At $\tau = 0$ the tangent vector has spatial component ωR in the x direction, and the speed at this point can be obtained from the values of $\dot{t}(\tau)$ and $\dot{x}(\tau)$ at $\tau = 0$ (since the velocity is in the x direction, the speed only depends on this nonzero component of the spatial piece of the tangent vector),

$$v = \frac{dx}{dt}\bigg|_{\tau=0} = \frac{\dot{x}(\tau)}{\dot{t}(\tau)}\bigg|_{\tau=0} = \frac{\omega R}{\sqrt{1 + (\omega R/c)^2}}. \tag{2.81}$$

---

[7] We have scalar $d\bar{s}^2 = ds^2$, and for the instantaneous rest frame, $ds^2 = -c^2 d\tau^2$, so that we also have $d\bar{s}^2 = -c^2 d\tau^2$, and $d\tau$ is, therefore, a scalar.

Using a Lorentz boost in the x direction with this v, we get, for the tangent vector at this point,

$$\dot{\bar{x}}^\mu(0) = \frac{\partial \bar{x}^\mu}{\partial x^\alpha} \dot{x}^\alpha(0) \doteq \begin{pmatrix} c \\ 0 \\ 0 \\ 0 \end{pmatrix}. \tag{2.82}$$

Example 2.6 (The Tangent as the Temporal Basis Vector). We know there is always a Lorentz transformation that will bring $\dot{x}^\mu(\tau)$ at a particular τ to the simple form (2.82). We can use this observation to make scalars that have specific interpretation in the particle's rest frame. Given a vector V_μ, we can evaluate the scalar product $\dot{x}^\mu(\tau)V_\mu$ at proper time τ in the rest frame coordinates,

$$\dot{\bar{x}}^\mu(\tau)\bar{V}_\mu = c\bar{V}_0. \tag{2.83}$$

The tangent vector, expressed in its rest frame, picks out the zero component of $\bar{V}_\mu$, so that $\dot{\bar{x}}^\mu(\tau)/c$ acts as a temporal basis vector at τ. The product is a scalar, so in any coordinate system, at time τ,

$$\frac{1}{c}\dot{x}^\mu(\tau)V_\mu = \frac{1}{c}\dot{\bar{x}}^\mu(\tau)\bar{V}_\mu = \bar{V}_0. \tag{2.84}$$

2.4.1 Lagrangian and Geodesics

Think of the Lagrangians you know from classical mechanics[8] in three dimensions, $L = T - U$, where T is the kinetic energy, U the potential energy. In Cartesian coordinates, the kinetic energy is

$$T = \frac{1}{2}m\big(\dot{x}(t)^2 + \dot{y}(t)^2 + \dot{z}(t)^2\big), \tag{2.85}$$

which we could write as $T = m/2(\dot{x}^\mu(t)g_{\mu\nu}\dot{x}^\nu(t))$ for the metric in three spatial, Cartesian, coordinates. Or we could work in spherical coordinates,

$$T = \frac{1}{2}m\big(\dot{r}(t)^2 + r(t)^2\dot{\theta}(t)^2 + r(t)^2\sin^2\theta(t)\dot{\phi}(t)^2\big). \tag{2.86}$$

This can also be written in metric form as $T = m/2(\dot{x}^\mu(t)g_{\mu\nu}\dot{x}^\nu(t))$ using the metric from (2.62) and $\dot{x}^1(t) = \dot{r}(t)$, $\dot{x}^2(t) = \dot{\theta}(t)$, $\dot{x}^3(t) = \dot{\phi}(t)$.

Lagrangian mechanics starts from a functional (taking a function to a number), the "action"

$$S\big[x(t)\big] = \int_{t_0}^{t_f} L\big(x(t), \dot{x}(t)\big)dt, \tag{2.87}$$

---

[8] For a review of variational calculus in the context of classical mechanics, see [20] and [30].

and extremizing the action gives us the Euler–Lagrange equations of motion

$$-\frac{d}{dt}\left(\frac{\partial L}{\partial \dot{x}^\beta}\right) + \frac{\partial L}{\partial x^\beta} = 0. \tag{2.88}$$

These reduce to the familiar form of Newton's second law in Cartesian coordinates, but because the Lagrangian is a scalar, the equations of motion in (2.88) are appropriately tensorial (although not yet in the four-dimensional spacetime sense, that is coming up).

For a Lagrangian with no potential energy, when we minimize the action, what we are really doing is minimizing the distance (squared) between two boundary values. That will give us a line in whatever coordinate system we choose. The same idea holds when we make a "kinetic" relativistic action. Moving back to the four-dimensional spacetime with a Minkowski metric written in some coordinate system, we propose a free particle action proportional to the length along a curve,

$$S\big[x(\sigma)\big] = \alpha \int_{\sigma_0}^{\sigma_f} \sqrt{-\dot{x}^\mu(\sigma) g_{\mu\nu} \dot{x}^\nu(\sigma)}\, d\sigma \tag{2.89}$$

for a constant α, where again σ can be any parameter we like (proper time, time, some other coordinate), and we'll take it to be both t and τ as we go. The square root in the integrand is important in enforcing reparameterization invariance – if you use $-\dot{x}^\mu(\sigma) g_{\mu\nu} \dot{x}^\nu(\sigma)$, then you lose the invariance and cannot use the same expression when taking $\sigma = t$ versus $\sigma = \tau$.

The constant α out front will be used to set the units and make contact with the nonrelativistic limit (for coordinate time parameterization). As always, the integrand is the "Lagrangian" here, $L(x, \dot{x}) = \alpha\sqrt{-\dot{x}^\mu g_{\mu\nu} \dot{x}^\nu}$. This Lagrangian has explicit dependence on $\dot{x}^\mu$, but it could also contain coordinate dependence through the metric $g_{\mu\nu}$. In Cartesian coordinates, the spacetime metric is coordinate-independent, but this is not the case in general (think of the Minkowski metric in spherical coordinates). The Euler–Lagrange equations are just as in (2.88), but with the understanding that $\beta = 0 \rightarrow 3$ (unlike the nonrelativistic, three-dimensional case, in which $\beta = 1, 2, 3$).

In Cartesian coordinates with coordinate time parameterization $\sigma = t$, the Lagrangian reduces to

$$L = \alpha\sqrt{c^2\left(\frac{dt}{dt}\right)^2 - \left(\frac{dx(t)}{dt}\right)^2 - \left(\frac{dy(t)}{dt}\right)^2 - \left(\frac{dz(t)}{dt}\right)^2}. \tag{2.90}$$

Using $\mathbf{v} = \dot{x}(t)\hat{\mathbf{x}} + \dot{y}(t)\hat{\mathbf{y}} + \dot{z}(t)\hat{\mathbf{z}}$ with $v^2 = \mathbf{v}\cdot\mathbf{v}$, we can factor out c and write the Lagrangian as

$$L = \alpha c\sqrt{1 - \frac{v^2}{c^2}}. \tag{2.91}$$

Now we can think about the $v \ll c$ limit using Taylor expansion to get the leading order behavior in v/c,

$$L \approx \alpha c\left(1 - \frac{v^2}{2c^2}\right). \tag{2.92}$$

This Lagrangian should limit to $L = mv^2/2$, the usual kinetic free particle Lagrangian. We can enforce this by picking $\alpha = -mc$; then our small speed limit looks like

$$L \approx -mc^2 + \frac{1}{2}mv^2, \tag{2.93}$$

and we have picked up a constant offset. That's no problem, as it is unde-tectable in the equations of motion. It will contribute to the particle's energy, of course, when we form the Hamiltonian, but that is expected in the relativistic setting, where the constant is just the rest energy of the massive particle.

Now that we have set $\alpha = -mc$ so as to recover classical mechanics, we can think of the other natural parameterization, proper time. For $x^\mu(\tau)$, proper time means that $\dot{x}^\mu(\tau) g_{\mu\nu} \dot{x}^\nu(\tau) = -c^2$. The Lagrangian is

$$L = -mc\sqrt{c^2\left(\frac{dt(\tau)}{d\tau}\right)^2 - \left(\frac{dx(\tau)}{d\tau}\right)^2 - \left(\frac{dy(\tau)}{d\tau}\right)^2 - \left(\frac{dz(\tau)}{d\tau}\right)^2}, \tag{2.94}$$

which is, of course, equal to $-mc^2$ numerically by the very definition of proper time, but that is not a helpful observation prior to finding the equations of motion (it does not allow us to probe the velocity and position dependence of L, even though in proper time, $L = -mc^2$ in any coordinate system). The Euler–Lagrange equations of motion are still (2.88), but with the parameter $t \to \tau$. Working out these equations for $ct(\tau)$ and $x(\tau)$ to see the pattern,[9]

$$0 = -\frac{d}{d\tau}\left(\frac{-mc^2 \dot{t}(\tau)}{\sqrt{c^2\dot{t}(\tau)^2 - \dot{x}(\tau)^2 - \dot{y}(\tau)^2 - \dot{z}(\tau)^2}}\right),$$
$$0 = -\frac{d}{d\tau}\left(\frac{-mc\dot{x}(\tau)}{\sqrt{c^2\dot{t}(\tau)^2 - \dot{x}(\tau)^2 - \dot{y}(\tau)^2 - \dot{z}(\tau)^2}}\right), \tag{2.95}$$

and it is now, before we take the τ derivative of the term in parentheses, that we use the proper time definition to simplify the denominators, which become c, leading to highly simplified equations of motion,

$$mc\ddot{t}(\tau) = 0, \qquad m\ddot{x}(\tau) = 0, \qquad m\ddot{y}(\tau) = 0, \qquad m\ddot{z}(\tau) = 0. \tag{2.96}$$

While we work explicitly in Cartesian coordinates, we can easily use a generic metric $g_{\mu\nu}(x)$; the equations of motion just become a little more in-volved. In proper time parameterization, we again have $\sqrt{-\dot{x}^\mu(\tau) g_{\mu\nu} \dot{x}^\nu(\tau)} = c$. The Lagrangian is $L = -mc\sqrt{-\dot{x}^\mu(\tau) g_{\mu\nu} \dot{x}^\nu(\tau)}$, and the equation of motion

---

[9] The top equation in (2.95) is really the Euler–Lagrange equation with respect to $x^0 \equiv ct$, not t, if you were worried about a missing factor of c.

for x^α is

$$-\frac{d}{d\tau}\frac{\partial L}{\partial \dot{x}^\alpha} + \frac{\partial L}{\partial x^\alpha}$$

$$= 0$$

$$= -\frac{d}{d\tau}\left(\frac{mc}{2}\frac{(\partial \dot{x}^\mu/\partial \dot{x}^\alpha)g_{\mu\nu}\dot{x}^\nu + \dot{x}^\mu g_{\mu\nu}(\partial \dot{x}^\nu/\partial \dot{x}^\alpha)}{\sqrt{-\dot{x}^\mu g_{\mu\nu}\dot{x}^\nu}}\right) \tag{2.97}$$

$$+ \frac{mc}{2}\frac{\dot{x}^\mu \dot{x}^\nu(\partial g_{\mu\nu}/\partial x^\alpha)}{\sqrt{-\dot{x}^\mu g_{\mu\nu}\dot{x}^\nu}}.$$

Note that the expression $\frac{\partial \dot{x}^\mu}{\partial \dot{x}^\alpha} = \delta^\mu_\alpha$ is nonzero only when $\mu = \alpha$ (think of $\frac{\partial \dot{x}}{\partial \dot{y}} = 0$ versus $\frac{\partial \dot{x}}{\partial \dot{x}} = 1$), and similarly, $\frac{\partial \dot{x}^\nu}{\partial \dot{x}^\alpha} = \delta^\nu_\alpha$. Using the proper time parameterization to simplify,

$$\frac{d}{d\tau}\left(mg_{\alpha\nu}\dot{x}^\nu\right) - \frac{m}{2}\dot{x}^\mu \dot{x}^\nu \frac{\partial g_{\mu\nu}}{\partial x^\alpha} = 0. \tag{2.98}$$

Finally, we can take the τ derivative of the first term, noting that there is co-ordinate dependence in the metric, so that $\frac{dg_{\alpha\nu}}{d\tau} = \frac{\partial g_{\alpha\nu}}{\partial x^\beta}\dot{x}^\beta$ by the chain rule, leaving us with

$$g_{\alpha\nu}\ddot{x}^\nu + \frac{\partial g_{\alpha\nu}}{\partial x^\beta}\dot{x}^\beta \dot{x}^\nu - \frac{1}{2}\frac{\partial g_{\beta\nu}}{\partial x^\alpha}\dot{x}^\beta \dot{x}^\nu = 0. \tag{2.99}$$

The middle term has contracted β and ν indices, and we can use the result of Problem 2.12 to pull out just the symmetric part, combining with the last term to get

$$\boxed{g_{\alpha\nu}\ddot{x}^\nu + \frac{1}{2}\left(\frac{\partial g_{\alpha\nu}}{\partial x^\beta} + \frac{\partial g_{\alpha\beta}}{\partial x^\nu} - \frac{\partial g_{\beta\nu}}{\partial x^\alpha}\right)\dot{x}^\beta \dot{x}^\nu = 0,} \tag{2.100}$$

called the "geodesic equation of motion." A geodesic is a length-minimizing curve, and we obtained this equation by extremizing an action that was proportional to the Minkowski length, and so it represents the minimum length curve connecting the initial, $x^\alpha(\tau_0)$, and final, $x^\alpha(\tau_f)$, points.[10] As is evident from its form in Cartesian coordinates, $\ddot{x}^\mu = 0$, these length-minimizing curves are just straight lines, and the non-$\ddot{x}^\mu$ detritus in (2.100) comes from the coordinate dependence of the metric, a direct consequence of the coordinates in which you choose to do the length minimizing. Since we are, at this point, free to *choose* Cartesian coordinates, we may as well solve the equation of motion in those. After all, parameterizing a straight line in spherical coordinates is difficult. But later on, when our metric does not represent the geometry of Minkowski spacetime, the same equation and interpretation will hold.

[10]   Technically, the geodesic equation of motion in (2.100) comes from extremizing length, and so could represent either a minimum or a maximum. For the quadratic form of ds^2 and the metrics we will use, the length will be minimized.

Example 2.7 (Geodesics on a Sphere). As an example, to demonstrate the utility of the geodesic equation of motion in determining the length-minimizing curve between two points, let's find the geodesic curve taking us from the north to the south pole of a sphere. One of the advantages of (2.100) is that it makes no explicit reference to the dimension or type of geometry we are working in, that is all encapsulated in the metric. So our first job is to find the metric for the surface of the sphere using some set of coordinates. The fastest way to do this is to start in our four-dimensional spacetime written in spherical coordinates, where the metric is

$$g_{\mu\nu} \doteq \begin{pmatrix} -1 & 0 & 0 & 0 \\ 0 & 1 & 0 & 0 \\ 0 & 0 & r^2 & 0 \\ 0 & 0 & 0 & r^2 \sin^2\theta \end{pmatrix} \tag{2.101}$$

for spatial coordinates $\{r, \theta, \phi\}$ from Figure 2.1 and a temporal coordinate ct. To trap ourselves on the surface of a sphere of radius R, we just set $r = R$ and remove the piece of the coordinate differential pointing in the r direction (a technique available here because the vector $\hat{\mathbf{r}}$ is perpendicular to the surface of the sphere), so that

$$\begin{pmatrix} cdt \\ dr \\ d\theta \\ d\phi \end{pmatrix} \longrightarrow \begin{pmatrix} cdt \\ d\theta \\ d\phi \end{pmatrix} \tag{2.102}$$

with $dr = 0$. We are left in two spatial dimensions, with coordinates θ and ϕ, and the usual temporal dimension, ct. The coordinate differential dx^α has $dx^0 = cdt$, $dx^1 = d\theta$, $dx^2 = d\phi$, and the metric obtained by taking the lower 2×2 block of the three-dimensional metric in (2.101), together with the temporal top-left corner, has entries

$$g_{\mu\nu} \doteq \begin{pmatrix} -1 & 0 & 0 \\ 0 & R^2 & 0 \\ 0 & 0 & R^2 \sin^2\theta \end{pmatrix}. \tag{2.103}$$

Since we are in three-dimensional spacetime,[11] μ and ν take on values 0, 1, and 2, associated with the coordinates ct, θ, and ϕ respectively.

To use (2.100), we need the derivatives of the metric with respect to the coordinates. There is only one nonzero derivative term for the metric in (2.103),

$$\frac{\partial g_{22}}{\partial x^1} = 2R^2 \sin\theta \cos\theta. \tag{2.104}$$

Using proper time parameterization, the coordinates defining the curve are $ct(\tau)$, $\theta(\tau)$, and $\phi(\tau)$ with proper time constraint $\sqrt{-\dot{x}^\mu(\tau) g_{\mu\nu} \dot{x}^\nu(\tau)} = c$

---

[11] To highlight the number of temporal versus spatial dimensions, we sometimes denote the dimension of four-dimensional spacetime as $3 + 1$. In that notation, two spatial dimensions and time are referred to as $2 + 1$.

written out explicitly here,

$$\sqrt{c^2 \dot{t}(\tau)^2 - R^2 \dot{\theta}(\tau)^2 - R^2 \sin^2 \theta(\tau) \dot{\phi}(\tau)^2} = c. \qquad (2.105)$$

The geodesic equation of motion for the three coordinates comes from (2.100):

$$0 = -c\ddot{t}(\tau),$$
$$0 = R^2 \ddot{\theta}(\tau) - \sin \theta(\tau) \cos \theta(\tau) R^2 \dot{\phi}(\tau)^2, \qquad (2.106)$$
$$0 = R^2 \sin^2 \theta(\tau) \ddot{\phi}(\tau) + 2 \sin \theta(\tau) \cos \theta(\tau) R^2 \dot{\theta}(\tau) \dot{\phi}(\tau),$$

as you can verify in Problem 2.20. Remember that we want to go from the north pole to the south, so we'll have (taking $\tau_0 = 0$) $\theta(0) = 0$ and $\theta(\tau_f) = \pi$. It is not entirely clear how to set the boundary conditions for ϕ, since at the north pole, ϕ isn't even defined, but it hardly matters: $\dot{\phi}(\tau) = 0$ solves ϕ's equation of motion while also telling us that the value of $\phi(\tau) = \phi_0$ is a constant. We are left with $\ddot{\theta}(\tau) = 0$ and $\ddot{t}(\tau) = 0$. The solutions are $\theta(\tau) = A\tau + B$ with constants of integration A and B obtained by imposing the boundary conditions. Those leave us with $\theta(\tau) = \pi \tau / \tau_f$. For the temporal coordinate, we may as well take $t(0) = 0$, and we'll let $t(\tau_f) = T$ some final time. The solution to t's equation of motion is $t(\tau) = F\tau + G$, and the boundary conditions give us $t(\tau) = T\tau / \tau_f$. Finally, we promised that (2.105) would hold in order to write the equations of motion as (2.100), so we must enforce the proper time condition. With our current solution, this condition reads

$$\sqrt{\frac{c^2 T^2 - \pi^2 R^2}{\tau_f^2}} = c \longrightarrow T = \sqrt{\tau_f^2 + \frac{\pi^2 R^2}{c^2}} \qquad (2.107)$$

and serves to set the final coordinate time as a function of the final proper time.

We have a family of geodesic curves connecting the north pole to the south pole. The curves describe halves of great circles, and we can travel along them at any angle ϕ_0, while retaining the length-minimization requirement. There is nothing special about the north and south poles as points, so we conclude that arcs of great circles are the length-minimizing curves connecting any two points on the surface of a sphere.

The real point of this long proper time introduction is that when we use proper time to describe the geodesic Lagrangian, we can replace it with the equivalent, and more familiar, quadratic

$$L_\tau = \frac{1}{2} m \dot{x}^\mu(\tau) g_{\mu\nu} \dot{x}^\nu(\tau), \qquad (2.108)$$

since the Euler–Lagrange equations of motion for this Lagrangian are identical to (2.100) (as you should check in Problem 2.22). The Lagrangian itself is a constant, $L_\tau = -mc^2/2$ in proper time parameterization, and we can use its constant value to simplify the procedure for finding geodesics. In Minkowski spacetime, we know the geodesics already, so no simplification is necessary, but in the more interesting spacetimes that will show up later on, solving for geodesic curves is the name of the game.

Proper time is available for anything that has an instantaneous rest frame. Light travels with constant speed c for all inertial observers, so there is no rest frame, no way to make a Lorentz boost to a frame in which the light is at rest. What does this mean for the geodesics associated with light? Think about light traveling along the x axis with speed c. The location of the light front at time t is $x = ct$. Now if you used a different parameter to describe the motion of the light through space and time, say s, you could write $x(s) = ct(s)$ for functions $x(s)$ and $t(s)$. No matter what s is, we have

$$\left(-c^2\dot{t}(s)^2 + \dot{x}(s)^2\right)ds^2 = 0 \tag{2.109}$$

or, more generally,

$$\dot{x}^\mu(s)g_{\mu\nu}\dot{x}^\nu(s) = 0. \tag{2.110}$$

We can still use the form of (2.108) to get the geodesic equations of motion (although you should try it from the start in Problem 2.23). Only now, instead of $L_\tau = -mc^2/2$, we have $L_\tau = 0$, and τ is no longer the "proper time," it is just an arbitrary parameter.

We started with (2.89) to define an action proportional to length along a curve. Then the minimum of that action will naturally minimize length. To connect to nonrelativistic mechanics, we have a parameter α in front and set that to $-mc$ to recover the usual nonrelativistic, free-particle Lagrangian. The value of α is irrelevant to the free-particle equation of motion since it can be divided out at the end (effecting all terms equally), so we have the freedom to *choose* α in (2.89) without changing the form or interpretation of the resulting equations of motion. That is what we have done in this light example – there is no nonrelativistic equation of motion for light, so there is no natural value for α. Instead, we use (any) nonzero constant. The value cannot matter since, for light, $L = 0$ numerically.

2.5 Derivatives in Curvilinear Coordinates

So far we have focused on the derivative as it acts on scalars like ψ, or the Lagrangian L from the previous section. We know, by direct calculation, that the gradient ∂_μ of a scalar is a covariant first-rank tensor, and our notation highlights this, $\psi_{,\mu}$ and $\partial_\mu\psi$ are both explicit ways to remind ourselves of the natural transformation character of the gradient acting on a scalar. How about the derivative of a first-rank tensor; how does that transform?

Given a contravariant A^μ, the gradient is

$$\frac{\partial A^\mu}{\partial x^\nu} \equiv A^\mu{}_{,\nu} \equiv \partial_\nu A^\mu, \tag{2.111}$$

but does it actually transform as a second-rank tensor with one up, one down index? Let's see: the transformed version is

$$\bar{A}^{\alpha}{}_{,\nu}(\bar{x}) \equiv \frac{\partial \bar{A}^{\alpha}(\bar{x})}{\partial \bar{x}^{\nu}} = \frac{\partial}{\partial \bar{x}^{\nu}}\left(\frac{\partial \bar{x}^{\alpha}}{\partial x^{\mu}} A^{\mu}(x(\bar{x}))\right) = \frac{\partial \bar{x}^{\alpha}}{\partial x^{\mu}}\frac{\partial A^{\mu}}{\partial x^{\beta}}\frac{\partial x^{\beta}}{\partial \bar{x}^{\nu}} + \frac{\partial^2 \bar{x}^{\alpha}}{\partial \bar{x}^{\nu}\partial x^{\mu}} A^{\mu}.$$

(2.112)

The first term in the far right equality is fine; it expresses the desired transformation rule, but that is spoiled by the second term, which looks like nothing we've seen before. It is a tricky term, since at first glance, we imagine that cross-derivative equality holds, and then you could swap the order of the $\bar{x}$ and x derivatives, giving the x-derivative of a Kronecker delta, which is zero, problem solved. But, of course, it is not true. You can't necessarily change the order of the barred and unbarred derivatives. Let's do a quick example to see why. Suppose we've gone from Cartesian coordinates (unbarred) to spherical (barred). One term that might come up in the offending expression from (2.112) is $\frac{\partial^2 \bar{r}}{\partial \bar{\theta}\partial x}$. That would evaluate to

$$\frac{\partial^2 \bar{r}}{\partial \bar{\theta}\partial x} = \frac{\partial}{\partial \bar{\theta}}\frac{\partial \bar{r}}{\partial x} = \frac{\partial(x/\bar{r})}{\partial \bar{\theta}} = \cos\bar{\theta}\cos\bar{\phi}.$$

(2.113)

Meanwhile, the same expression evaluated with the derivatives in the other order gives

$$\frac{\partial^2 \bar{r}}{\partial x\partial\bar{\theta}} = \frac{\partial}{\partial x}\frac{\partial \bar{r}}{\partial \bar{\theta}} = 0.$$

(2.114)

The problem is that the cross-derivative equality doesn't hold when it comes to derivatives in different coordinate systems. We can clarify the situation a little by thinking of $\frac{\partial \bar{x}^{\alpha}}{\partial x^{\mu}}$ as a function of $x(\bar{x})$ and use the chain rule to get

$$\frac{\partial}{\partial \bar{x}^{\nu}}\left(\frac{\partial \bar{x}^{\alpha}}{\partial x^{\mu}}\right) = \frac{\partial^2 \bar{x}^{\alpha}}{\partial x^{\beta}\partial x^{\mu}}\frac{\partial x^{\beta}}{\partial \bar{x}^{\nu}} \neq \frac{\partial}{\partial x^{\mu}}\left(\frac{\partial \bar{x}^{\alpha}}{\partial \bar{x}^{\nu}}\right) = 0.$$

(2.115)

Remember the original problem, the transformation of the derivative of the contravariant tensor A^{μ}, which can be written as

$$\bar{A}^{\alpha}{}_{,\nu} = \frac{\partial \bar{x}^{\alpha}}{\partial x^{\mu}}\frac{\partial x^{\beta}}{\partial \bar{x}^{\nu}} A^{\mu}{}_{,\beta} + \frac{\partial^2 \bar{x}^{\alpha}}{\partial x^{\beta}\partial x^{\mu}}\frac{\partial x^{\beta}}{\partial \bar{x}^{\nu}} A^{\mu}.$$

(2.116)

Again, the first term is fine, but the second term is not. It doesn't even involve the derivative of A^{ν}, which provides us with a clue about how to fix the situation. What we want is a well-defined tensorial derivative. Remember the point of this chapter: making a tensor equation ensures that the equation holds in all coordinate systems. We can't have an "extra" term appearing in expressions involving the derivative of a tensor, but not in other equations.

At the heart of the problem is our focus on the components without the associated basis vectors. The derivative of a first-rank tensor is fundamentally a subtraction of the tensor elements at two different points. But that difference is ambiguous if we don't know the direction of the basis vector telling us in what direction the vector "points." As an example, the tensor component of the vector $\mathbf{r} = r\hat{\mathbf{r}}$ is "r." But the difference between $\mathbf{r}$ at location $\theta = 0, \phi = 0$ and $\mathbf{r}$ at $\theta = \pi/2, \phi = 0$ is $\mathbf{r}(0,0) - \mathbf{r}(\pi/2,0) = r(\hat{\mathbf{z}} - \hat{\mathbf{x}})$, which is not what you would get if you just subtracted the component value at either of these two points, that is, zero: $r - r = 0$, nor is it the same as subtracting the vector **r** at different choices of ϕ, for example.

If we consider the vector with both components and basis in place, $\mathbf{A} = A^\alpha \mathbf{e}_\alpha$, then the gradient is

$$\frac{\partial \mathbf{A}}{\partial x^\alpha} = A^\mu{}_{,\alpha} \mathbf{e}_\mu + A^\mu \frac{\partial \mathbf{e}_\mu}{\partial x^\alpha}, \tag{2.117}$$

and the changes in both components and basis vectors are accounted for. We are going to develop a tensorial version of (2.116), which refers only to components, but transforms correctly as a mixed second-rank tensor. Our model will be (2.117), where the first term represents the piece found in (2.116), and the second term is new. Our job is to capture the second term without referring directly to the basis vectors. Note that the second term is linear in A^μ and involves a vector expression with two indices, so that we need three indices to express $\frac{\partial \mathbf{e}_\mu}{\partial x^\alpha}$: μ, α, and the component of $\mathbf{e}_\mu$.

Starting from the definition of the gradient of the tensor A^μ, we have

$$A^\mu{}_{,\nu} \equiv \frac{\partial A^\mu(x)}{\partial x^\nu} = \lim_{dx^\nu \to 0} \frac{A^\mu(x + dx^\nu) - A^\mu(x)}{dx^\nu}, \tag{2.118}$$

where dx^ν is one of the coordinates (and there is no sum implied in its appearance in both the numerator and denominator on the right above). The numerator has the problematic (as discussed above) difference between A^μ at two different points, a specific locale x and the nearby $x + dx^\nu$ (here x represents the full set of coordinates of a point, and dx^ν indicates the distance and specific direction to the nearby point). What we'll do is add in a "fudge factor," motivated by the form of the second term in (2.117). Define a new derivative by (again, ν is a fixed index)

$$A^\mu{}_{;\nu} = \lim_{dx^\nu \to 0} \frac{A^\mu(x + dx^\nu) - A^\mu(x) + \delta A^\mu(x)}{dx^\nu}, \tag{2.119}$$

where δA^μ is a function whose properties are up to us. There are two special cases that will help us determine the form of δA^μ: (1) If $dx^\nu = 0$, then $\delta A^\mu = 0$, since then we are not subtracting the vector A^μ at two different points and therefore have no need to fix the numerator of (2.118); (2) If $A^\mu = 0$, then $\delta A^\mu = 0$, and the zero vector *can* be subtracted at different points unambiguously. Then δA^μ should be linear in both dx^ν and A^μ (the linearity in A^μ is to capture the second term in (2.117)), and in order to do this as generally as possible, let $\delta A^\mu \equiv C^\mu{}_{\nu\alpha} dx^\nu A^\alpha$, where $C^\mu{}_{\nu\alpha}$ is a collection of functions whose transformation properties will be defined so as to cancel the extra term in (2.116) when using the new, semicolon, derivative in (2.119) – since that additional term is not associated with a tensor transformation, we do not expect $C^\mu{}_{\nu\alpha}$ to be a tensor (despite its dangling indices). The new, "covariant derivative," now looks like

$$\boxed{A^\mu{}_{;\nu} = A^\mu{}_{,\nu} + C^\mu{}_{\nu\alpha} A^\alpha.} \tag{2.120}$$

Let's look at how the covariant derivative transforms; that will help us figure out how $C^{\mu}_{\ va}$ transforms. For $x \to \bar{x}$,

$$\bar{A}^{\alpha}_{\ ;v} = \bar{A}^{\alpha}_{\ ,v} + \bar{C}^{\alpha}_{\ v\beta}\bar{A}^{\beta} = \frac{\partial \bar{x}^{\alpha}}{\partial x^{\mu}}\frac{\partial x^{\beta}}{\partial \bar{x}^{v}}A^{\mu}_{\ ,\beta} + \frac{\partial^{2}\bar{x}^{\alpha}}{\partial x^{\beta}\partial x^{\mu}}\frac{\partial x^{\beta}}{\partial \bar{x}^{v}}A^{\mu} + \bar{C}^{\alpha}_{\ v\beta}\frac{\partial \bar{x}^{\beta}}{\partial x^{\sigma}}A^{\sigma}. \tag{2.121}$$

We want to get rid of all the terms linear in A, but first, we have to replace the $A^{\mu}_{\ ,\beta}$ "bad" derivative with the target tensorial $A^{\mu}_{\ ;\beta}$ using $A^{\mu}_{\ ,\beta} = A^{\mu}_{\ ;\beta} - C^{\mu}_{\ \beta\rho}A^{\rho}$ from (2.120),

$$\bar{A}^{\alpha}_{\ ;v} = \frac{\partial \bar{x}^{\alpha}}{\partial x^{\mu}}\frac{\partial x^{\beta}}{\partial \bar{x}^{v}}\left(A^{\mu}_{\ ;\beta} - C^{\mu}_{\ \beta\rho}A^{\rho}\right) + \frac{\partial^{2}\bar{x}^{\alpha}}{\partial x^{\beta}\partial x^{\mu}}\frac{\partial x^{\beta}}{\partial \bar{x}^{v}}A^{\mu} + \bar{C}^{\alpha}_{\ v\beta}\frac{\partial \bar{x}^{\beta}}{\partial x^{\sigma}}A^{\sigma}. \tag{2.122}$$

Isolating the very first term inside the parentheses on the right in this equation, we see that it defines the appropriate tensorial transformation, so that we want all the terms linear in A to cancel. Reindexing those terms so that they all refer to A^{σ}, we are demanding that

$$\left(-\frac{\partial \bar{x}^{\alpha}}{\partial x^{\mu}}\frac{\partial x^{\beta}}{\partial \bar{x}^{v}}C^{\mu}_{\ \beta\sigma} + \frac{\partial^{2}\bar{x}^{\alpha}}{\partial x^{\beta}\partial x^{\sigma}}\frac{\partial x^{\beta}}{\partial \bar{x}^{v}} + \bar{C}^{\alpha}_{\ v\beta}\frac{\partial \bar{x}^{\beta}}{\partial x^{\sigma}}\right)A^{\sigma} = 0. \tag{2.123}$$

This equation must hold for all A^{σ} (the covariant derivative shouldn't be a tensor only for *some* A^{σ}), so

$$\bar{C}^{\alpha}_{\ v\beta}\frac{\partial \bar{x}^{\beta}}{\partial x^{\sigma}} = \frac{\partial \bar{x}^{\alpha}}{\partial x^{\mu}}\frac{\partial x^{\beta}}{\partial \bar{x}^{v}}C^{\mu}_{\ \beta\sigma} - \frac{\partial^{2}\bar{x}^{\alpha}}{\partial x^{\beta}\partial x^{\sigma}}\frac{\partial x^{\beta}}{\partial \bar{x}^{v}}. \tag{2.124}$$

Finally, multiplying by $\frac{\partial x^{\sigma}}{\partial \bar{x}^{\mu}}$ introduces a δ^{β}_{μ} on the left, and we have

$$\boxed{\bar{C}^{\alpha}_{\ v\mu} = \frac{\partial \bar{x}^{\alpha}}{\partial x^{\mu}}\frac{\partial x^{\beta}}{\partial \bar{x}^{v}}\frac{\partial x^{\sigma}}{\partial \bar{x}^{\mu}}C^{\mu}_{\ \beta\sigma} - \frac{\partial^{2}\bar{x}^{\alpha}}{\partial x^{\beta}\partial x^{\sigma}}\frac{\partial x^{\beta}}{\partial \bar{x}^{v}}\frac{\partial x^{\sigma}}{\partial \bar{x}^{\mu}}.} \tag{2.125}$$

The first term here looks like the tensor transformation rule for a mixed third-rank tensor. The second term looks a lot like the bad term from the regular partial derivative in (2.116), and indeed, appears here to cancel precisely that term. So $C^{\alpha}_{\ v\mu}$, defined by its response to a coordinate transformation in (2.125), is not a tensor, but it is an important object, called a "connection."

You can think of the form in (2.120) as "correcting" the bad transformation behavior of the derivative acting on a tensor by soaking up the nontensorial detritus associated with the tensor index μ and derivative index v (as in (2.120)) with a compensating nontensorial connection having those indices, the $C^{\mu}_{\ va}$ appearing in (2.120). This is just a mnemonic, but it tells us how to generate the correct covariant derivative of a second-rank tensor $S^{\alpha\beta}$. The covariant derivative of $S^{\alpha\beta}$ needs two connection terms added together, one to "correct" the derivative for the α index, one for the β. So $S^{\alpha\beta}_{\ \ ;v}$ needs a $C^{\alpha}_{\ v\rho}$ term and a $C^{\beta}_{\ v\rho}$ term, with ρ contracted into the index we are correcting with the connection:

$$S^{\alpha\beta}_{\ \ ;v} \equiv S^{\alpha\beta}_{\ \ ,v} + C^{\alpha}_{\ v\rho}S^{\rho\beta} + C^{\beta}_{\ v\rho}S^{\alpha\rho}, \tag{2.126}$$

where the first term "fixes" the α index, and the second fixes the β index. For an expression with no open indices, like a scalar, there are no nontensorial terms to worry about, the covariant derivative is equal to the partial. If ψ is a scalar, then $\psi_{;\nu} = \psi_{,\nu}$.

2.6 Properties of the Covariant Derivative

This new derivative in (2.120) is more complicated than the usual partial derivative. The advantage is that we have a tensorial building block that we can use in constructing theories involving derivatives of contravariant tensors. How about the covariant derivative of naturally covariant tensors? We generate the correct form by appealing to scalar combinations of tensors and requiring that the covariant derivative, like the familiar partial derivative, supports the product rule, a convenient feature that allows familiar manipulations of partial derivatives to carry over to the covariant derivative.

Take two tensors, one contravariant A^μ, one covariant B_ν, and form the scalar $A^\mu B_\mu$. Since this is a scalar, we know that the partial derivative by itself is a covariant vector, and the covariant derivative is equal to the partial derivative, $(A^\mu B_\mu)_{;\nu} = (A^\mu B_\mu)_{,\nu}$. If we assume that the product rule holds for the covariant derivative as it does for partial derivatives, then

$$A^\mu_{;\nu} B_\mu + A^\mu B_{\mu;\nu} = A^\mu_{,\nu} B_\mu + A^\mu B_{\mu,\nu}, \tag{2.127}$$

and we can use the covariant derivative definition (2.120) to expand: $A^\mu_{;\nu} = A^\mu_{,\nu} + C^\mu_{\nu\rho} A^\rho$. Putting this in on the left and collecting terms with some reindexing gives

$$A^\mu \left[B_{\mu;\nu} - B_{\mu,\nu} + C^\sigma_{\nu\mu} B_\sigma \right] = 0. \tag{2.128}$$

For this equation to hold for all contravariant tensors A^μ, we must have

$$\boxed{B_{\mu;\nu} = B_{\mu,\nu} - C^\sigma_{\nu\mu} B_\sigma.} \tag{2.129}$$

We can again think of the role of the connection $C^\sigma_{\nu\mu}$, here as "correcting" the transformation of the partials, so we want the connection to have both the covariant tensor index and the derivative index (then the only place to put the closed index is in the up position to match the covariant B_σ, as is clear from (2.128), again, a useful mnemonic).

For a second-rank covariant tensor $Q_{\alpha\beta}$, we require two connections, just as for the second-rank contravariant case (2.126), one for each index. The covariant derivative of a second-rank covariant tensor is

$$Q_{\alpha\beta;\nu} = Q_{\alpha\beta,\nu} - C^\sigma_{\nu\alpha} Q_{\sigma\beta} - C^\sigma_{\nu\beta} Q_{\alpha\sigma}. \tag{2.130}$$

For a mixed rank tensor $M^\alpha_{\ \beta}$, we have, predictably,

$$M^\alpha_{\ \beta;\nu} = M^\alpha_{\ \beta,\nu} + C^\alpha_{\ \nu\rho}M^\rho_{\ \beta} - C^\rho_{\ \nu\beta}M^\alpha_{\ \rho}, \tag{2.131}$$

and the pattern continues to higher-rank tensors.

We have seen three different ways to refer to partial derivatives using indices. For the derivative with respect to x^μ, we have

$$\frac{\partial}{\partial x^\mu}, \qquad \partial_\mu, \qquad \underline{\quad}_{,\mu} \tag{2.132}$$

with the latter two highlighting the covariant nature of the gradient. For the covariant derivative, we have already seen the use of the semicolon in place of the comma, but another common way to denote the covariant derivative is D_μ, an update of ∂_μ. The notations we will use for the covariant derivative are

$$\frac{D}{Dx^\mu}, \qquad D_\mu, \qquad \underline{\quad}_{;\mu}. \tag{2.133}$$

Remember in the case of the comma and semicolon notation that any index coming after the comma/semicolon indicates additional derivatives.

2.6.1 Covariant Derivative of the Metric

In the tensor calculus we have developed so far there are currently two inputs. The first is a metric, required to define lengths and then used as the raising and lowering operator for tensor indices;[12] the second is the connection $C^\alpha_{\ \mu\nu}$, which we need to define the covariant derivative. That connection could, in theory, be provided separately. But given the role of the metric and its inverse, we can connect the connection (no pun intended) to derivatives of the metric.

Consider the covariant derivative of the Kronecker delta,

$$\delta^\alpha_{\ \beta;\nu} = \delta^\alpha_{\ \beta,\nu} + C^\alpha_{\ \nu\rho}\delta^\rho_{\ \beta} - C^\rho_{\ \nu\beta}\delta^\alpha_{\ \rho}. \tag{2.134}$$

The first term on the right vanishes since the Kronecker delta is either one or zero, and neither of those depend on the coordinates, so $\delta^\alpha_{\ \beta,\nu} = 0$. The covariant derivative is just the combination of connections,

$$\delta^\alpha_{\ \beta;\nu} = C^\alpha_{\ \nu\beta} - C^\alpha_{\ \nu\beta} = 0. \tag{2.135}$$

Good, that is reasonable, both the regular and covariant derivatives of the Kronecker delta are zero.

---

[12] That's important, the metric only raises and lowers indices for *tensors*; it can do nothing with nontensorial expressions since those do not have well-defined tensor behavior to begin with. For example, we cannot raise and lower the indices of the partial derivative acting on a tensor: $A_{\mu,\nu}$ does *not* have $A^\mu_{\ ,\nu} = g^{\mu\nu}A_{\mu,\nu}$, so we leave those expressions alone when multiplying terms in an equation by a metric. That is, we will always write $g^{\mu\nu}A_{\mu,\nu}$ and never "simplify" by raising either the μ or the ν index.

Now think of the metric and its inverse. We know that $g^{\alpha\mu} g_{\mu\beta} = \delta^{\alpha}{}_{\beta}$, and then $(g^{\alpha\mu} g_{\mu\beta})_{;\nu} = 0$, but by the product rule,

$$\left(g^{\alpha\mu} g_{\mu\beta}\right)_{;\nu} = g^{\alpha\mu}{}_{;\nu} g_{\mu\beta} + g^{\alpha\mu} g_{\mu\beta;\nu} = 0, \tag{2.136}$$

or, using the metric in each term to raise or lower an index on the *tensorial* covariant derivative,

$$g^{\alpha}{}_{\beta;\nu} + g^{\alpha}{}_{\beta;\nu} = 2g^{\alpha}{}_{\beta;\nu} = 0. \tag{2.137}$$

Since this is a tensor statement, we can lower the α index to get, finally, $g_{\alpha\beta;\nu} = 0$. The covariant derivative of the metric is zero, and this is true in all coordinate systems.

Write out what we mean by this covariant derivative,

$$g_{\alpha\beta;\nu} = g_{\alpha\beta,\nu} - C^{\rho}{}_{\nu\alpha} g_{\rho\beta} - C^{\rho}{}_{\nu\beta} g_{\alpha\rho} = 0, \tag{2.138}$$

and it is clear that we can relate the connection to derivatives of the metric. As you will show in Problem 2.27, the relation is

$$\boxed{C^{\alpha}{}_{\mu\nu} = \frac{1}{2} g^{\alpha\rho} (g_{\rho\mu,\nu} + g_{\rho\nu,\mu} - g_{\mu\nu,\rho}) \equiv \Gamma^{\alpha}{}_{\mu\nu}.} \tag{2.139}$$

Of course, don't take my word for it, you can show that this combination of derivatives transforms as a connection, as defined in (2.125). This type of connection, related to the derivatives of the metric, is called a "metric connection," and $\Gamma^{\alpha}{}_{\mu\nu}$ is the "Christoffel symbol of the second kind." An alternate form, the "Christoffel symbol of the first kind" is related: just multiply the above by $g_{\alpha\beta}$,

$$\boxed{\Gamma_{\beta\mu\nu} \equiv g_{\alpha\beta} \Gamma^{\alpha}{}_{\mu\nu} = \frac{1}{2} (g_{\beta\mu,\nu} + g_{\beta\nu,\mu} - g_{\mu\nu,\beta}).} \tag{2.140}$$

2.6.2 The Connection as Derivative of Basis Vectors

We motivated the introduction of the connection as a way of accounting for the change of basis vectors while comparing vector components at different points. If we take the partial derivative of $\mathbf{A} = A^{\mu} \mathbf{e}_{\mu}$, then we should be able to track the changes in basis and relate those to the connection coefficients. Referring to (2.117), we expect

$$\frac{\partial \mathbf{A}}{\partial x^{\alpha}} = A^{\mu}{}_{;\alpha} \mathbf{e}_{\mu}, \tag{2.141}$$

and equating the right-hand sides of (2.117) and (2.141) sets up the association:

$$A^{\mu}{}_{,\alpha} \mathbf{e}_{\mu} + A^{\mu} \frac{\partial \mathbf{e}_{\mu}}{\partial x^{\alpha}} = \left(A^{\mu}{}_{,\alpha} + \Gamma^{\mu}{}_{\alpha\rho} A^{\rho}\right) \mathbf{e}_{\mu} \longrightarrow \Gamma^{\mu}{}_{\alpha\rho} \mathbf{e}_{\mu} = \frac{\partial \mathbf{e}_{\rho}}{\partial x^{\alpha}}. \tag{2.142}$$

So given a set of coordinates, we should be able to compute the connection coefficients by taking derivatives of the basis vectors.

Example 2.8 (Spherical Coordinates). Let's evaluate the derivative $\frac{\partial \mathbf{e}_\rho}{\partial x^\alpha}$ for spherical coordinates and compare with the expression from (2.139). To take the derivative of the spherical basis vectors, it is easiest to write them in terms of the Cartesian ones (which are constant) as in (2.61), and then we only have to worry about the derivatives of the components. The nonzero derivatives are

$$\frac{\partial \mathbf{e}_1}{\partial \theta} = \cos\theta \cos\phi \hat{\mathbf{x}} + \cos\theta \sin\phi \hat{\mathbf{y}} - \sin\theta \hat{\mathbf{z}} = \frac{\mathbf{e}_2}{r},$$

$$\frac{\partial \mathbf{e}_1}{\partial \phi} = -\sin\theta \sin\phi \hat{\mathbf{x}} + \sin\theta \cos\phi \hat{\mathbf{y}} = \frac{\mathbf{e}_3}{r},$$

$$\frac{\partial \mathbf{e}_2}{\partial r} = \frac{\mathbf{e}_2}{r}, \qquad \frac{\partial \mathbf{e}_2}{\partial \theta} = -r\mathbf{e}_1, \qquad \frac{\partial \mathbf{e}_2}{\partial \phi} = \frac{\cos\theta}{\sin\theta}\mathbf{e}_3,$$

$$\frac{\partial \mathbf{e}_3}{\partial r} = \frac{\mathbf{e}_3}{r}, \qquad \frac{\partial \mathbf{e}_3}{\partial \theta} = \frac{\cos\theta}{\sin\theta}\mathbf{e}_3, \qquad \frac{\partial \mathbf{e}_3}{\partial \phi} = -r\sin^2\theta \mathbf{e}_1 - \sin\theta\cos\theta \mathbf{e}_2.$$

$$(2.143)$$

Using $\Gamma^\mu_{\alpha\rho}\mathbf{e}_\mu = \frac{\partial \mathbf{e}_\rho}{\partial x^\alpha}$, we can pick off the nonzero entries:

$$\Gamma^2_{12} = \frac{1}{r}, \qquad \Gamma^3_{13} = \frac{1}{r}, \qquad \Gamma^2_{21} = \frac{1}{r},$$

$$\Gamma^1_{22} = -r, \qquad \Gamma^3_{23} = \cot\theta, \qquad \Gamma^3_{31} = \frac{1}{r}, \qquad (2.144)$$

$$\Gamma^3_{32} = \cot\theta \qquad \Gamma^1_{33} = -r\sin^2\theta, \qquad \Gamma^2_{33} = -\sin\theta\cos\theta.$$

These match the values you got computing the connection using (2.139) in Problem 2.30.

2.6.3 Divergence and Laplacian

For a contravariant tensor A^μ, the divergence is defined to be $A^\mu_{;\mu}$, employing the covariant derivative to get a true scalar quantity. Writing out the semicolon derivative,

$$A^\mu_{;\mu} = A^\mu_{,\mu} + \Gamma^\mu_{\mu\rho}A^\rho. \tag{2.145}$$

We expect this scalar to match familiar expressions like (for time-independent A^μ),

$$\nabla \cdot \mathbf{A} = \frac{1}{r^2}\frac{\partial}{\partial r}\left(r^2 A^r\right) + \frac{1}{r\sin\theta}\frac{\partial}{\partial\theta}\left(A^\theta \sin\theta\right) + \frac{1}{r\sin\theta}\frac{\partial A^\phi}{\partial\phi} \tag{2.146}$$

for $\mathbf{A} = A^r\hat{\mathbf{r}} + A^\theta\hat{\boldsymbol{\theta}} + A^\phi\hat{\boldsymbol{\phi}}$. It does, of course, but there are a few steps to the verification that need to be taken carefully (see Problem 2.31).

We should get the same expression by raising the first index of the covariant derivative of A_μ: $A_{\mu;\nu}g^{\mu\nu}$, although at first glance the expressions look quite different,

$$g^{\mu\nu}A_{\mu;\nu} = g^{\mu\nu}A_{\mu,\nu} - g^{\mu\nu}\Gamma^\rho_{\nu\mu}A_\rho. \tag{2.147}$$

You can show that this expression does in fact equal (2.145), really a lesson in why $A^{\mu}{}_{,\nu} \neq g^{\mu\nu}A_{\mu,\nu}$ (you can't use the metric to raise and lower indices on nontensors).

From this second form in (2.147) we can write out the Laplacian of a scalar, $\psi_{;\mu\nu}g^{\mu\nu}$, just take $A_{\mu} \equiv \psi_{,\mu}$ in (2.147),

$$\psi^{\mu}{}_{;\;\mu} = g^{\mu\nu}\psi_{,\mu\nu} - g^{\mu\nu}\Gamma^{\rho}{}_{\nu\mu}\psi_{,\rho}. \tag{2.148}$$

This is really what we mean by $\Box\psi$. If you work out the expression in spherical coordinates, just to check it against a known expression, then you quickly recover

$$\psi^{\mu}{}_{;\;\mu} = -\frac{1}{c^2}\frac{\partial^2\psi}{\partial t^2} + \frac{1}{r^2}\frac{\partial}{\partial r}\left(r^2\frac{\partial\psi}{\partial r}\right) + \frac{1}{r^2\sin\theta}\frac{\partial}{\partial\theta}\left(\sin\theta\frac{\partial\psi}{\partial\theta}\right) + \frac{1}{r^2\sin^2\theta}\frac{\partial^2\psi}{\partial\phi^2}. \tag{2.149}$$

Notationally, we can use D_{μ} to express either the divergence or the Laplacian. The covariant divergence of A^{μ} is $D_{\mu}A^{\mu} \equiv A^{\mu}{}_{;\mu}$. The covariant Laplacian can be written $D_{\mu}D^{\mu}\psi = \psi^{\mu}{}_{;\;\mu}$.

2.7 Parallel Transport

Enough of the definitions for a while, let's get back to some geometrically (and physically) relevant calculations. Suppose you have a curve parameterized by τ, an $x^{\mu}(\tau)$ like the one in Figure 2.2, but we have chosen concrete proper time parameterization, $\dot{x}^{\mu}(\tau)g_{\mu\nu}\dot{x}^{\nu}(\tau) = -c^2$. We also have a scalar field $\psi(x)$, and we want to know how the scalar changes as we move along the trajectory. Using the chain rule, the change in $\psi(x(\tau))$ as a function of τ is

$$\frac{d\psi(x(\tau))}{d\tau} = \frac{\partial\psi}{\partial x^{\alpha}}\dot{x}^{\alpha} = \psi_{,\alpha}\dot{x}^{\alpha}. \tag{2.150}$$

Good, that's easy, and both $\psi_{,\alpha}$ and $\dot{x}^{\alpha}$ are tensors, so the result $\frac{d\psi}{d\tau}$ is a scalar.

Now suppose we have a contravariant tensor $A^{\mu}(x)$. How does *it* change along the curve? If we take the τ derivative as before, then

$$\frac{dA^{\mu}}{d\tau} = \frac{\partial A^{\mu}}{\partial x^{\alpha}}\dot{x}^{\alpha} = A^{\mu}{}_{,\alpha}\dot{x}^{\alpha}. \tag{2.151}$$

That's no good, though, since $A^{\mu}{}_{,\alpha}$ is *not* a tensor, so neither is $\frac{dA^{\mu}}{d\tau}$. We've already seen precisely this sort of expression, the $\ddot{x}^{\nu} \equiv \frac{d\dot{x}^{\nu}(\tau)}{d\tau}$ appearing in (2.100), for example. That geodesic had better look the same in all coordinate systems, since we already promised that the equation of motion coming from a scalar Lagrangian is generally covariant.

The problem, as with the partial derivatives of A^{μ}, is with the lack of basis vector information. The components of A^{μ} are changing as we move along the curve, but so, too, are the basis vectors to which those components refer. The fix is the same, just replace the comma with a semicolon in (2.151) and define the new τ derivative

$$\frac{DA^\mu}{D\tau} \equiv A^\mu{}_{;\alpha}\dot{x}^\alpha. \tag{2.152}$$

We can relate the two using the covariant derivative definition,

$$\frac{DA^\mu}{D\tau} \equiv \left(A^\mu{}_{,\alpha} + \Gamma^\mu{}_{\alpha\rho}A^\rho\right)\dot{x}^\alpha = \frac{dA^\mu}{d\tau} + \Gamma^\mu{}_{\alpha\rho}A^\rho\dot{x}^\alpha. \tag{2.153}$$

Working in Cartesian coordinates, the connection from (2.139) vanishes, and therefore $\frac{DA^\mu}{D\tau} = \frac{dA^\mu}{d\tau}$. Consider a constant vector f^α; it is clear that

$$\frac{Df^\alpha}{D\tau} = \frac{df^\alpha}{d\tau} = f^\alpha{}_{,\rho}\dot{x}^\rho = 0. \tag{2.154}$$

In general, we say that any vector with $\frac{DA^\alpha}{D\tau} = 0$ is "parallel transported along the curve with tangent specified by $\dot{x}^\mu(\tau)$." The name comes from the fact that in this Cartesian example, the constant vector is clearly parallel to itself at all locations along the path.

With the tensorial $D/D\tau$, the definition holds in any coordinate system, so that the general equation for parallel transport of a tensor A^μ is

$$\boxed{\frac{DA^\mu}{D\tau} = A^\mu{}_{;\rho}\dot{x}^\rho = 0.} \tag{2.155}$$

Example 2.9 (Parallel Transport in Spherical Coordinates). Let's take a vector that is clearly constant, say $\hat{\mathbf{z}}$. We can express the components of this vector in spherical coordinates by algebraically inverting the equations in (2.61),

$$\hat{\mathbf{z}} = \cos\theta\,\mathbf{e}_1 - \frac{\sin\theta}{r}\mathbf{e}_2, \tag{2.156}$$

so our constant vector's components are (written in the four-dimensional spacetime of interest)

$$A^\mu \doteq \begin{pmatrix} 0 \\ \cos\theta \\ -\frac{\sin\theta}{r} \\ 0 \end{pmatrix}. \tag{2.157}$$

As our curve, we'll go straight out the $\hat{\mathbf{x}}$ axis with $x^\mu(\tau) = \alpha\tau\delta^\mu_1$ (the δ^μ_1 here selects the $\hat{\mathbf{r}}$ component) at $\theta = \pi/2$, $\phi = 0$. Let's start by using the nontensorial $\frac{dA^\mu}{d\tau}$:

$$\frac{dA^\mu}{d\tau} = A^\mu{}_{,\rho}\dot{x}^\rho = (\alpha\tau)^{-2}\alpha\delta^\mu_2 \neq 0, \tag{2.158}$$

where we have set $\theta = \pi/2$ and $r = \alpha\tau$.

We would naively conclude that the constant vector is not transported parallel to itself along the curve. But since $\hat{\mathbf{z}}$ is a constant, you can pick it up and put it *anywhere*, and it is always parallel to itself. Now let's use the correct covariant $D/D\tau$. First, we need the connection coefficients for spherical coordinates. The nonzero entries associated with the spatial entries of the connection are

$$\Gamma^1_{22} = -r, \qquad \Gamma^1_{33} = -r \sin^2 \theta,$$

$$\Gamma^2_{12} = \Gamma^2_{21} = \frac{1}{r}, \qquad \Gamma^2_{33} = -\sin\theta \cos\theta, \qquad (2.159)$$

$$\Gamma^3_{13} = \Gamma^3_{31} = \frac{1}{r}, \qquad \Gamma^3_{23} = \Gamma^3_{32} = \cot\theta,$$

and then, referring to (2.153), we need the connection evaluated along the path,

$$\Gamma^\mu_{\alpha\rho} A^\rho \dot{x}^\alpha = -\alpha(\alpha\tau)^{-2} \delta^\mu_2, \qquad (2.160)$$

so the sum in (2.153) will give zero, with the pieces from (2.158) and the above canceling. Conclusion: for the $\hat{\mathbf{z}}$ vector, clearly parallel transported along the $\hat{\mathbf{x}}$ axis, we found that $\frac{dA^\mu}{d\tau} \neq 0$, but $\frac{DA^\mu}{D\tau} = 0$.

There are a number of physically relevant observations we can make about vectors parallel transported along a common path. All of these should seem obvious if you think of parallel transport as picking up a vector in Cartesian coordinates and moving it along the prescribed path. All we are doing, for now, is accounting for the change in the description of that process that comes from changing coordinates to some general curvilinear set where the connection doesn't vanish, and hence there is a difference between $\frac{dA^\mu}{d\tau} = 0$ and the correct $\frac{DA^\mu}{D\tau} = 0$. But it is important to reiterate again (to rereiterate) that all of this machinery holds in spacetimes that do not have the underlying Minkowski geometry, a fact that we will return to and exploit in the next chapter.

Two immediate consequences of our general parallel transport definition are that (1) the length of a vector is preserved under parallel transport and (2) the angle between two vectors parallel transported along the same curve is constant. Referring to the cartoon in Figure 2.3, this is obviously true in Cartesian coordinates, and of course that picture is coordinate independent, so it is also true in all other coordinate systems, all we are doing is establishing that the accounting is correct.

The length of a vector A^μ is given by $A = \sqrt{A^\mu g_{\mu\nu} A^\nu}$, which is the natural successor to $\sqrt{\mathbf{A} \cdot \mathbf{A}}$, to which it reduces in the Cartesian case. Suppose A^μ is parallel transported along a curve $x^\alpha(\tau)$, so that it satisfies (2.155). Let's take the derivative of A with respect to τ. Notice that since A is a scalar, we could take either $d/d\tau$ or $D/D\tau$ to find the change in A along the curve. We'll use the covariant version since that's the point of this section,

$$\frac{D}{D\tau} \sqrt{A^\mu g_{\mu\nu} A^\nu} = \frac{1}{2A} \frac{D}{D\tau} (A^\mu g_{\mu\nu} A^\nu) = \frac{1}{2A} (A^\mu g_{\mu\nu} A^\nu)_{;\alpha} \dot{x}^\alpha$$

$$= \frac{1}{2A} \left(\frac{DA^\mu}{D\tau} g_{\mu\nu} A^\nu + A^\mu g_{\mu\nu;\alpha} A^\nu \dot{x}^\alpha + A^\mu g_{\mu\nu} \frac{DA^\nu}{D\tau} \right) = 0, \qquad (2.161)$$

where we have used $g_{\mu\nu;\alpha} = 0$, our new best friend.

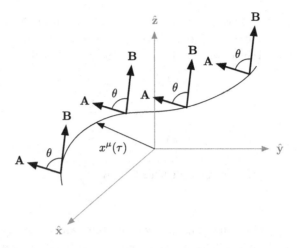

Fig. 2.3 The vectors **A** and **B** are parallel transported along the curve specified by $x^\mu(\tau)$. We are literally moving each one by picking it up and placing it along the curve. In Cartesian coordinates, the components of each vector satisfy $\frac{dA^\mu}{d\tau} = \frac{dB^\mu}{d\tau} = 0$. In any coordinate system, the components satisfy $\frac{DA^\mu}{D\tau} = \frac{DB^\mu}{D\tau} = 0$. Neither the length of the vectors nor the angle between them changes as they are moved along the curve.

For a pair of vectors A^μ and B^μ, the angle between is defined by

$$\cos\theta = \frac{A^\mu g_{\mu\nu} B^\nu}{AB}. \tag{2.162}$$

If each is parallel transported along the curve $x^\mu(\tau)$, then

$$\frac{D}{D\tau}(\cos\theta) = \frac{D(AB)^{-1}}{D\tau} A^\mu g_{\mu\nu} B^\nu + \frac{1}{AB}\frac{D}{D\tau}\left(A^\mu g_{\mu\nu} B^\nu\right), \tag{2.163}$$

and we have already established that the lengths A and B are constant, so the first term goes away, leaving us with

$$\frac{D}{D\tau}(\cos\theta) = \frac{1}{AB}\frac{D}{D\tau}\left(A^\mu g_{\mu\nu} B^\nu\right) = \frac{1}{AB}\left(A^\mu g_{\mu\nu} B^\nu\right)_{;\alpha} \dot{x}^\alpha$$
$$= \frac{1}{AB}\left(\frac{DA^\mu}{D\tau} g_{\mu\nu} B^\nu + A^\mu g_{\mu\nu;\alpha} B^\nu \dot{x}^\alpha + A^\mu g_{\mu\nu}\frac{DB^\nu}{D\tau}\right) = 0, \tag{2.164}$$

with the same simplifications as in (2.161).

Example 2.10 (Parallel Transport on a Sphere). We can use the equation that defines parallel transport to *construct* a parallel transported vector along some given curve. That is uninteresting in our usual Minkowski spacetime, where parallel transport just looks like the examples shown in Figure 2.3. But in spaces that are not Minkowski spacetime, parallel transport has significant and physically relevant implications.

As a toy example, let's work on the surface of a sphere of radius R as in Example 2.7, but we'll omit the temporal coordinate and just focus on the two-dimensional $\theta\phi$ subspace. This example of parallel transport will be instructive as we think about applications to non-Minkowski spacetimes in Chapter 3. From the metric on the surface of the sphere in (2.103), the nonzero connection coefficients are

$$\Gamma^1{}_{22} = -\sin\theta\cos\theta, \qquad \Gamma^2{}_{12} = \Gamma^2{}_{21} = \cot\theta. \tag{2.165}$$

For the curve, let's take a trip around the sphere at constant angle θ_0, with $\phi(\tau) = 2\pi\tau/\tau_f$, and define the constant $\alpha \equiv 2\pi/\tau_f$. Then the tangent to the curve is $\dot{x}^\mu(\tau) = \alpha\delta^\mu_2$. Our goal is to find the components of the transported $A^\mu(\phi(\tau))$ from the ODE that defines parallel transport (2.155) and some given initial condition. The ODEs are, using $\phi \equiv \alpha\tau$ as the parameter,

$$\frac{DA^\mu}{D\tau} = 0 = A^\mu{}_{;\alpha}\dot{x}^\alpha \doteq \begin{pmatrix} \alpha(\frac{dA^1}{d\phi} - A^2\sin\theta_0\cos\theta_0) \\ \alpha(\frac{dA^2}{d\phi} + A^1\cot\theta_0) \end{pmatrix}. \tag{2.166}$$

We can decouple the equations by taking the ϕ-derivative of each,

$$\begin{aligned} \frac{d^2A^1}{d\phi^2} &= \frac{dA^2}{d\phi}\sin\theta_0\cos\theta_0, \\ \frac{d^2A^2}{d\phi^2} &= -\frac{dA^1}{d\phi}\cot\theta_0, \end{aligned} \tag{2.167}$$

and then using the first derivatives on the right, we have, for A^1,

$$\frac{d^2A^1}{d\phi^2} = -\cos^2\theta_0 A^1 \longrightarrow A^1 = F\cos(\phi\cos\theta_0) + G\sin(\phi\cos\theta_0) \tag{2.168}$$

for constants F and G, and similarly we get $A^2 = J\cos(\phi\cos\theta_0) + K \times \sin(\phi\cos\theta_0)$ for constants J and K. We can set two of the constants by returning with these solutions to the first-order ODEs in (2.166), giving $G = J\sin\theta_0$ and $F = -K\sin\theta_0$. Our set now reads

$$\begin{aligned} A^1(\phi) &= \sin\theta_0\big(-K\cos(\phi\cos\theta_0) - J\sin(\phi\cos\theta_0)\big), \\ A^2(\phi) &= \big(J\cos(\phi\cos\theta_0) + K\sin(\phi\cos\theta_0)\big). \end{aligned} \tag{2.169}$$

We can fix J and K once the initial conditions have been given. If we take $A^1(0) = A^\theta$ and $A^2(0) = A^\phi$, then the vector parallel transported around our curve has the components

$$A^\mu(\phi) \doteq \begin{pmatrix} A^\theta\cos(\phi\cos\theta_0) + A^\phi\sin\theta_0\sin(\phi\cos\theta_0) \\ A^\phi\cos(\phi\cos\theta_0) - \frac{A^\theta}{\sin\theta_0}\sin(\phi\cos\theta_0) \end{pmatrix}. \tag{2.170}$$

We can check the length of this vector to verify that it does not change as we go along the curve. The length, squared, is

$$A^\mu g_{\mu\nu} A^\nu = R^2\big[(A^\theta)^2 + (A^\phi\sin\theta_0)^2\big], \tag{2.171}$$

which is constant.

If we have two vectors, A^μ from above and B^μ (with $B^1(0) = B^\theta$, $B^2(0) = B^\phi$) parallel transported along this curve, then the angle η between them is given by

$$\cos\eta = \frac{A^\mu g_{\mu\nu} B^\nu}{AB} = \frac{A^\theta B^\theta + A^\phi B^\phi \sin^2\theta_0}{\sqrt{(A^\theta)^2 + (A^\phi \sin\theta_0)^2}\sqrt{(B^\theta)^2 + (B^\phi \sin\theta_0)^2}}, \tag{2.172}$$

again, clearly a constant.

One interesting observation we can make right off the bat – take $A^\phi = 0$ and $A^\theta = 1$ as the initial condition. Then at $\phi = 0$, the vector is $A^\mu(0) = \delta_1^\mu$, but at $\phi = 2\pi$,

$$A^\mu(2\pi) \doteq \begin{pmatrix} \cos(2\pi\cos\theta_0) \\ -\frac{1}{\sin\theta_0}\sin(2\pi\cos\theta_0) \end{pmatrix}, \tag{2.173}$$

so that the initial and final vectors are *not* parallel; they make an angle ψ with

$$\cos\psi = \cos(2\pi\cos\theta_0). \tag{2.174}$$

Along the journey, the vector undergoes an overall rotation through an angle $\psi = 2\pi\cos\theta_0$.

A few example trajectories and associated parallel transported vectors on the sphere are shown in Figure 2.4. The path goes from $0 \to 2\pi$, so you can see the relative rotation of the initial and final vector components. Try Problem 2.33

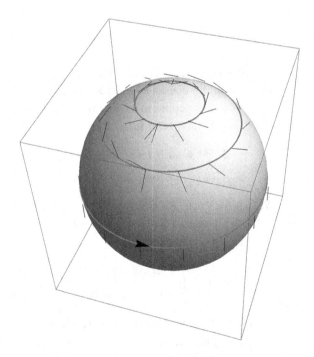

Fig. 2.4   Parallel transport of a vector that has initial value $A^\theta = 1$, $A^\phi = 0$ for $\theta_0 = \pi/2, \pi/4, \pi/9$, with gray circles showing the path for each. The arrow on the circle at $\theta_0 = \pi/2$ shows the starting point and direction of travel for the curves.

to see what happens to the same parallel transport story if you were to embed the problem here in three dimensions.

2.7.1 Geodesics Redux

There is a special class of curve that is parallel transported along its own tangent. It sounds circular (pun intended), but if we simply write down the ODE we must solve to find such a curve, then the meaning becomes clear. For a curve $\dot{x}^\sigma(\tau)$ to be parallel transported along its tangent, we have $\frac{D\dot{x}^\sigma}{D\tau} = 0$ or, in the form of (2.153),

$$\dot{x}^\sigma_{;\nu}\dot{x}^\nu = \frac{d\dot{x}^\sigma(\tau)}{d\tau} + \Gamma^\sigma_{\beta\nu}\dot{x}^\beta(\tau)\dot{x}^\nu(\tau) = 0. \tag{2.175}$$

Fine, but write this out

$$\frac{d\dot{x}^\sigma(\tau)}{d\tau} + \frac{1}{2}g^{\sigma\rho}(g_{\rho\beta,\nu} + g_{\rho\nu,\beta} - g_{\beta\nu,\rho})\dot{x}^\beta(\tau)\dot{x}^\nu(\tau) = 0; \tag{2.176}$$

multiplying by $g_{\alpha\sigma}$ gives precisely (2.100). A curve parallel-transported along its own tangent is a geodesic.

We can also see how the proper time definition, $\dot{x}^\mu\dot{x}_\mu = -c^2$, is consistent with the geodesic equation. Remember that we derived (2.100) in a particular parameterization; the form changes if we are not using τ with $\dot{x}^\mu(\tau)g_{\mu\nu} \times \dot{x}^\nu(\tau) = -c^2$. But is the second-order ODE we got supported by the proper time constraint itself? Yes, take the $D/D\tau$ derivative of the proper time definition, which, since it is a scalar equation, is the same as taking $d/d\tau$. On the right, we get zero. The left-hand side can be expanded using the product rule:

$$\frac{D}{D\tau}\big(\dot{x}^\mu(\tau)g_{\mu\nu}\dot{x}^\nu(\tau)\big)$$
$$= \frac{D\dot{x}^\mu(\tau)}{D\tau}g_{\mu\nu}\dot{x}^\nu(\tau) + \dot{x}^\mu(\tau)g_{\mu\nu;\alpha}\dot{x}^\alpha(\tau)\dot{x}^\nu(\tau) + \dot{x}^\mu(\tau)g_{\mu\nu}\frac{D\dot{x}^\nu(\tau)}{D\tau}, \tag{2.177}$$

and $g_{\mu\nu;\alpha} = 0$, the first and third terms are the same up to index relabeling and the symmetry of the metric, so the derivative of the proper time definition gives

$$2\frac{D\dot{x}^\mu(\tau)}{D\tau}\dot{x}_\mu(\tau) = 0, \tag{2.178}$$

which is true for geodesics, as we have just seen. Note that the particular value of c^2 in the proper time parameterization constraint is irrelevant; any constant value will do.

2.8 Tensor Field Equations

We'll close this chapter by thinking about how to build field theories that are generally covariant using tensors. This type of model building is popular in physics and lends a certain inevitability to physical theories like electricity and magnetism and general relativity. Don't be fooled, though; the "rules" by which we generate valid theories are themselves informed by the successful ones.

A field equation relates a field to its sources. Typically, in physics, that relation involves the field's derivatives, which is one of the reasons we have spent so much time working out the implications of the covariant derivative. There are some simple rules we follow when generating a field equation. The first, and most important, is that the field equation be tensorial, for then we are guaranteed that it holds in any coordinate system. The second rule is that the field equation contain at most second derivatives. This is a matter of observation, the Laplacian and D'Alembertian are, at least for fundamental theories, the end of the line. They also have the advantage of requiring boundary conditions where observation occurs. We are used to inferring what happens in the bulk of a domain from observations made on the boundary of the domain. Laplacians and D'Alembertians can be used to enforce those boundary observables mathematically.

Our third and fourth rules are more like guiding principles, they are a matter of simplicity and choice. In order of importance, the third rule will be that a field theory governing a particular tensor rank should not have contributions from tensors of any other rank. In other words, a first-rank tensor field theory won't have scalar field contributions, and a second-rank tensor field theory won't have vector or scalar field contributions, etc. This allows us to keep field theories for different objects separate and leads to the gauge freedom we see in E&M, for example. Finally, for our fourth rule, we will construct linear field equations. Most isolated field theories, when coupled with each other, become nonlinear, and we want to cut down on the amount of nonlinearity we will allow from the start, only introducing it when necessary. As we have already established, general relativity will be a nonlinear second-rank tensor field theory, but we will start by thinking about a generic linear second-rank theory that offers both a weak-field "linearized" limit and a natural starting point.

A zero-rank tensor is a scalar and presents the simplest field theory. All we get is the field ψ with some sort of scalar source j. The field equation itself must be a scalar to accommodate the source, and that means we can use ψ itself and the covariant Laplacian (or D'Alembertian if you prefer, since we are in a four-dimensional spacetime) $D_\mu D^\mu \psi$. The most general field equation consistent with our building guidelines is

$$\alpha D_\mu D^\mu \psi + \beta \psi = j \tag{2.179}$$

for constants α and β.

2.8.1 First-Rank Tensor Field

Let's construct the most general first-rank tensor field equation consistent with our model building rules from above. The goal of this section and the next is to demonstrate that once you have some simple rules in place, the form of the field equation is highly constrained and practically unique with our set of guidelines.

For a field A_μ, with first-rank tensor source J_μ, each term in the field equation has to be a first-rank covariant tensor to match the right-hand side J_μ. The available terms involving A_μ are

$$A_\mu, \qquad D_\nu D^\nu A_\mu, \qquad D^\nu D_\mu A_\nu, \tag{2.180}$$

and so the most general field equation is

$$\alpha D_\nu D^\nu A_\mu + \beta D^\nu D_\mu A_\nu + \sigma A_\mu = J_\mu. \tag{2.181}$$

According to our third guideline, we want to remove any scalar field from this equation. How could a scalar creep in? You could disguise a scalar as a vector by taking its derivative. Set $A_\mu = \psi_{,\mu}$ and put this into (2.181):

$$\alpha D_\nu D^\nu A_\mu + \beta D^\nu D_\mu A_\nu + \sigma A_\mu = \alpha D_\nu D^\nu D_\mu \psi + \beta D^\nu D_\mu D_\nu \psi + \sigma D_\mu \psi. \tag{2.182}$$

Assuming that $D_\mu D_\nu \psi = D_\nu D_\mu \psi$ (it does here since ψ is a scalar, as we will see in Chapter 3),[13] we can gather terms as follows:

$$\alpha D_\nu D^\nu A_\mu + \beta D^\nu D_\mu A_\nu + \sigma A_\mu = (\alpha + \beta) D_\nu D^\nu D_\mu \psi + \sigma D_\mu \psi, \tag{2.183}$$

from which we learn that if we want to remove any reference to ψ, then we must set $\alpha = -\beta$ and $\sigma = 0$. We have now projected out any possible reference to a scalar field ψ in our field equation, decoupling A_μ from potential scalars disguised as vectors. Again, the goal is to start with a pure vector field theory; if you want to study scalar fields, then you should be working with (2.179).

The field equation (2.181) is now

$$\beta D^\nu (D_\mu A_\nu - D_\nu A_\mu) = J_\mu. \tag{2.184}$$

In an interesting twist, you will show in Problem 2.34 that $D_\mu A_\nu - D_\nu A_\mu = \partial_\mu A_\nu - \partial_\nu A_\mu$, the combination of covariant derivatives is equal to the combination of partial derivatives here, and that combination of derivatives of the four-potential is what we call the field strength tensor in E&M, $F_{\mu\nu} \equiv \partial_\mu A_\nu - \partial_\nu A_\mu$.

We have built in the independence of A_μ on scalars; taking $A_\mu \to A_\mu + \psi_{,\mu}$ doesn't affect the field equation by design. That is the gauge freedom of E&M, you can add $\psi_{,\mu}$ without changing the physical predictions that are really based on the fields found in $F_{\mu\nu}$. We can use that freedom to pick a particular value

---

[13]   More generally $D_\mu D_\nu = D_\nu D_\mu$ as long as the underlying metric is the Minkowski metric written in any coordinate system.

for $\psi_{,\mu}$. If you had $\tilde{A}_\mu$ with $D^\mu \tilde{A}_\mu = H$, some scalar function, then $A_\mu = \tilde{A}_\mu + \psi_{,\mu}$ has the divergence

$$D^\mu A_\mu = H + D^\mu D_\mu \psi. \tag{2.185}$$

Take $D^\mu D_\mu \psi = -H$ by solving Poisson's equation with H as the source. Then the divergence of A_μ is zero: $D^\mu A_\mu = 0$. This is our gauge choice, known as "Lorenz gauge." There are other choices one might make in other settings, but this particular one can be used to further simplify the field equation. With $D^\mu A_\mu = 0$ in (2.184), we have

$$\beta D^\nu D_\nu A_\mu = -J_\mu, \tag{2.186}$$

a simple Poisson problem to solve for A_μ given the source vector J_μ.

Conservation of the source is built in to the field equation here. In curvilinear coordinates, conservation reads $D^\mu J_\mu = 0$, which is just our current way of expressing (1.122). Taking the divergence of both sides of (2.186) gives zero on the left when A_μ is in Lorenz gauge.

2.8.2 Second-Rank Tensor Field

For a second-rank tensor field $h_{\mu\nu}$ with a second-rank tensor source $T_{\mu\nu}$, each term in the field equation must be a second-rank covariant tensor. It suffices to consider only symmetric second-rank tensors, $h_{\nu\mu} = h_{\mu\nu}$ (as you will see in the next section), so in addition to being second-rank tensors, we must have symmetry under $\mu \leftrightarrow \nu$ interchange.

The building blocks proliferate with indices,

$$h_{\mu\nu}, \qquad D_\mu D_\nu g^{\alpha\beta} h_{\alpha\beta}, \qquad D_\alpha D^\alpha h_{\mu\nu}, \qquad D^\alpha D_\mu h_{\alpha\nu} + D^\alpha D_\nu h_{\alpha\mu},$$
$$g_{\mu\nu} g^{\alpha\beta} h_{\alpha\beta}, \qquad g_{\mu\nu} D^\rho D_\rho \big(g^{\alpha\beta} h_{\alpha\beta}\big), \qquad g_{\mu\nu} D^\alpha D^\beta h_{\alpha\beta}, \tag{2.187}$$

and each term is symmetric in $\mu \leftrightarrow \nu$. The most general field equation is

$$T_{\mu\nu} = \alpha D_\alpha D^\alpha h_{\mu\nu} + \beta D_\mu D_\nu g^{\alpha\beta} h_{\alpha\beta} + \sigma\big(D^\alpha D_\mu h_{\alpha\nu} + D^\alpha D_\nu h_{\alpha\mu}\big) + \rho h_{\mu\nu}$$
$$+ \mu g_{\mu\nu} g^{\alpha\beta} h_{\alpha\beta} + \gamma g_{\mu\nu} D^\rho D_\rho \big(g^{\alpha\beta} h_{\alpha\beta}\big) + \lambda g_{\mu\nu} D^\alpha D^\beta h_{\alpha\beta} \tag{2.188}$$

for constants $\{\alpha, \beta, \sigma, \rho, \mu, \gamma, \lambda\}$. You can "project out" the scalar and vector field contributions in Problem 2.35 to arrive at

$$T_{\mu\nu} = \alpha\big(D_\alpha D^\alpha h_{\mu\nu} + D_\mu D_\nu g^{\alpha\beta} h_{\alpha\beta} - \big(D^\alpha D_\mu h_{\alpha\nu} + D^\alpha D_\nu h_{\alpha\mu}\big)\big)$$
$$+ \gamma g_{\mu\nu}\big(D^\alpha D_\alpha \big(g^{\rho\sigma} h_{\rho\sigma}\big) - D^\alpha D^\beta h_{\alpha\beta}\big). \tag{2.189}$$

Gauge freedom for E&M came from the requirement that the field equation was unchanged by $A_\mu \rightarrow A_\mu + \psi_{,\mu}$. Since we have, similarly, removed scalar and vector field dependence in this second-rank tensor field equation, we expect to have similar gauge freedom here. As with the vector case, suppose you take $\tilde{h}_{\mu\nu}$ with some divergence $D^\mu \tilde{h}_{\mu\nu} = H_\nu$. Then let $h_{\mu\nu} =$

$\tilde{h}_{\mu\nu} + (A_{\mu;\nu} + A_{\nu;\mu}) + \psi_{;\mu\nu}$ for vector A_μ and scalar ψ with divergence

$$D^\mu h_{\mu\nu} = H_\nu + D_\nu\big(D^\mu A_\mu + D^\mu D_\mu \psi\big) + D^\mu D_\mu A_\nu. \qquad (2.190)$$

We can pick A_ν such that $D^\mu D_\mu A_\nu = -H_\nu$, and let ψ satisfy $D^\mu D_\mu \psi = -D^\mu A_\mu$. Then the divergence of $h_{\mu\nu}$ is zero. This is analogous to the Lorenz gauge for vectors. The field equation reduces to

$$\alpha\big(D_\alpha D^\alpha h_{\mu\nu} + D_\mu D_\nu\big(g^{\alpha\beta} h_{\alpha\beta}\big)\big) + \gamma g_{\mu\nu} D^\alpha D_\alpha\big(g^{\rho\sigma} h_{\rho\sigma}\big) = T_{\mu\nu}. \qquad (2.191)$$

We can keep going to higher-rank theories, but there is no obvious physical reason to do so. There is a scalar field in nature, the Higgs field, the lone vector field is E&M, and the second-rank symmetric tensor field is responsible for gravity. Field theories involving spinors can be generated using model building ideas similar to the ones described here, but we will not discuss those in this book.

2.8.3 Coupling

In thinking about the relativistic Lagrangian, we only considered the "kinetic" term with no potential energy. How should we incorporate forces into the relativistic Lagrangian, either in its reparameterization-invariant form or in its simplified, proper time L_τ, form? Well, first, we need an external source with which to couple. Since a Lagrangian is a scalar, we must ensure that whatever we add to it is itself a scalar, so that the overall coupled Lagrangian retains its scalar character. If there was a field theory that produced a real scalar field ψ, we could use that and start from an augmented Lagrangian of the form

$$L_\tau = \frac{1}{2}m\dot{x}^\mu g_{\mu\nu}\dot{x}^\nu + \alpha\psi, \qquad (2.192)$$

where α is some coupling constant that governs the strength of the particle's interaction with ψ.

Of the four forces of nature, one of them, E&M, is a "vector" theory. The four-potential of E&M is naturally covariant, A_μ is really the field in the theory.[14] But because of our preference to interpret four-vectors in terms of their contravariant elements, again due to their projection onto a covariant basis to form $\mathbf{A} = A^\mu \mathbf{e}_\mu$, the electric potential and magnetic vector potential are really the components of the upper form,

$$A^\nu \equiv g^{\mu\nu} A_\mu \doteq \begin{pmatrix} \frac{V}{c} \\ \mathbf{A} \end{pmatrix}. \qquad (2.193)$$

Regardless, we now have to make a scalar out of A_μ and the only other *a priori* available tensor lying around, $\dot{x}^\nu$ (x^ν is not a tensor, and the scalar $\partial_\mu A^\mu$

---

[14] A note on the overloaded use of the word "field." In E&M, the vectors $\mathbf{E}$ and $\mathbf{B}$ are the "fields," whereas the V and $\mathbf{A}$ are "potentials." In the context of "field theories," a "field" is any function of position and time, so that the potentials themselves are fields of the theory.

lacks coupling to the particle and is zero in Lorenz gauge). The relevant scalar is clear, we will use $\dot{x}^\mu A_\mu$. That combination is actually reparameterization invariant under the integral that we would use to construct the action. Suppose the motion is parameterized by an arbitrary σ. Then the action, with electromagnetic contribution, is

$$S[x(\sigma)] = \int_{\sigma_0}^{\sigma_f} \left(-mc\sqrt{-\dot{x}^\mu(\sigma)g_{\mu\nu}\dot{x}^\nu(\sigma)} + \alpha\dot{x}^\mu(\sigma)A_\mu \right)d\sigma, \qquad (2.194)$$

and you can quickly verify that switching to a parameter $\rho(\sigma)$ leaves the integral form unchanged. The coupling constant α can be found, as usual, from a comparison with the nonrelativistic limit, requiring us once again to choose temporal parameterization. In that case, the Lagrangian, defined to be the integrand of the action in (2.194), is

$$L = -mc^2\sqrt{1 - \frac{v^2}{c^2}} + \alpha(cA_0 + \mathbf{v} \cdot \mathbf{A}), \qquad (2.195)$$

where we are working in Cartesian coordinates and have used the usual Minkowski metric from (2.53). The lower A_0 is just the negative of $A^0 = V/c$, the electric potential, so picking $\alpha = q$ gives

$$L = -mc^2\sqrt{1 - \frac{v(t)^2}{c^2}} + q(-V + \mathbf{v} \cdot \mathbf{A}). \qquad (2.196)$$

This reduces to the familiar Lagrangian from electrodynamics and in its current form allows us to find the equations of motion for t-parameterized electromagnetic forcing (see Problem 2.36).

I suggested, at the end of the last chapter, that based on the source of a relativistic theory of gravity, we expect the field to be a second-rank tensor, call it $h_{\mu\nu}$. How might that couple to particle motion? Working in proper time parameterization with the free-particle starting point $L_\tau = (m/2)\dot{x}^\mu g_{\mu\nu}\dot{x}^\nu$, we must form a scalar to add in, and that scalar can be made out of $h_{\mu\nu}$ and the already present $\dot{x}^\mu$. The simplest scalar we can form, by analogy with the vector case, is $\dot{x}^\mu h_{\mu\nu}\dot{x}^\nu$. This additional term tells us that only the symmetric piece of the tensor $h_{\mu\nu}$ contributes to the motion of particles, since any antisymmetric contribution would vanish when contracted with the symmetric $\dot{x}^\mu\dot{x}^\nu$ (from Problem 2.12). So our gravitational field is a symmetric second-rank tensor. If we introduce a coupling constant α as in the scalar and vector examples, then the particle Lagrangian with interaction reads

$$L = L_\tau + \alpha\dot{x}^\mu h_{\mu\nu}\dot{x}^\nu = \frac{1}{2}m\dot{x}^\mu\left(g_{\mu\nu} + \frac{2\alpha}{m}h_{\mu\nu}\right)\dot{x}^\nu. \qquad (2.197)$$

This is an interesting structure right off the bat, since it suggests that what we really have is a geodesic Lagrangian for a new metric $g_{\mu\nu} + 2(\alpha/m)h_{\mu\nu}$.[15]

---

[15] The proper time parameterization with respect to the original $g_{\mu\nu}$ complicates the story here, but as we will see, the conclusion ends up being the same.

The interaction, unlike the electromagnetic one, can be reinterpreted as defining a new geometry, with a metric that is, at least potentially, different from Minkowski (in any coordinate system). This is one of the motivations for thinking about geodesics. There is no "force" in the equation of motion, just length-minimizing curves in an exotic setting.

Combined with the comments at the end of the last chapter, we are honing in on the correct form for a relativistic theory of gravity: A second-rank symmetric tensor field, which can be understood as a metric governing geodesic motion of all particles (so that everything responds to it) and which is sourced by all forms of energy, rendering its field equation necessarily nonlinear. All that, and we still haven't really done much beyond squint!

Exercises

2.1 For the transformation given in (2.2), sketch the $\bar{x}$ and $\bar{y}$ coordinate axes in the xy plane (the $\bar{x}$ axis is defined by the set of points with $\bar{y} = 0$, and vice versa). Are the new coordinate axes a clockwise or counterclockwise rotation of the original?

2.2 Show that the transformation defined by (2.4) preserves the Minkowski line element (i.e., show that $d\bar{s}^2 = ds^2$).

2.3 Find the matrix expressing the boost from (2.4) as in (2.6).

2.4 Write the following expressions in index notation:

$$A^\mu = B^\mu \sum_{\alpha=0}^{3} C^\alpha D_\alpha \text{ for } \mu = 0, 1, 2, 3,$$

$$F^\mu{}_\nu = \sum_{\sigma=0}^{3} A^{\mu\sigma} B_{\sigma\nu} \text{ for } \mu = 0, 1, 2, 3, \nu = 0, 1, 2, 3, \quad (2.198)$$

$$\sum_{\mu=0}^{3} A^\mu B_{\mu\rho} = \sum_{\beta=0}^{3} F^\beta G_{\beta\rho} \text{ for } \rho = 0, 1, 2, 3.$$

Take the following statements written in index notation and add back in the summations and open index magic words:

$$g_{\mu\nu} h^\nu = f_\mu, \qquad h^{\alpha\beta} j_\beta = k^{\alpha\sigma} m_\sigma, \qquad h^{\rho\sigma} m_{\sigma\delta} = a^\rho b_\delta c^\alpha{}_\alpha. \quad (2.199)$$

2.5 Form the matrices (with entries) $\frac{\partial \bar{x}^\mu}{\partial x^\nu}$ and $\frac{\partial x^\alpha}{\partial \bar{x}^\beta}$ for cylindrical coordinates defined in (2.17). For the new coordinates, use the ordering $d\bar{x}^0 = cd\bar{t}$, $d\bar{x}^1 = d\bar{s}$, $d\bar{x}^2 = d\bar{\phi}$, and $d\bar{x}^3 = d\bar{z}$. Express the matrix entries in both Cartesian and cylindrical coordinates. Verify, explicitly by matrix multiplication, that the two matrices are indeed inverses of each other.

2.6 For first-rank covariant tensors A_μ and B_ν, what is the transformation of the second-rank $A_\mu B_\nu$? What if you had an upper F^μ and lower G_ν, how does the "mixed" second-rank tensor $F^\mu G_\nu$ transform?

2.7 Given a matrix $\mathbb{A} \in \mathbb{R}^{4\times4}$ with entries $A^{\mu\nu}$ and a matrix $\mathbb{B} \in \mathbb{R}^{4\times4}$ with entries $B_{\alpha\beta}$, express the matrix product (entries) $\mathbb{AB}$, $\mathbb{A}^T\mathbb{B}$, and $\mathbb{AB}^T$ using index notation (e.g., start with $(\mathbb{AB})^\rho{}_\sigma$ and continue).

2.8 Show that if the Kronecker delta transforms as a mixed rank tensor,

$$\bar\delta^\mu{}_\nu = \frac{\partial \bar x^\mu}{\partial x^\alpha} \frac{\partial x^\beta}{\partial \bar x^\nu} \delta^\alpha{}_\beta, \tag{2.200}$$

and takes on its usual value in the original coordinate system ($\delta^\alpha{}_\beta = 1$ if $\alpha = \beta$, zero otherwise), then it takes on those same numerical values in all coordinate systems. You can run the argument backwards, showing that if the Kronecker delta takes on the same numerical value in all coordinate systems, then it transforms as a mixed second-rank tensor.

2.9 Given a second-rank contravariant tensor $M^{\mu\nu}$, show that its matrix inverse is a second-rank covariant tensor. Assume that the matrix inverse relation holds in any coordinate system by construction, that is, if you have $\bar M^{\mu\nu}$, then you construct the matrix inverse $\bar M_{\alpha\beta}$ in that new coordinate system.

2.10 Using cylindrical coordinates from (2.17), write out the line element ds^2 from (2.50).

2.11 Show that a second-rank tensor $G_{\mu\nu}$ can be written explicitly as the sum of a symmetric tensor $S_{\mu\nu} = S_{\nu\mu}$ and an antisymmetric one $A_{\mu\nu} = -A_{\nu\mu}$: $G_{\mu\nu} = S_{\mu\nu} + A_{\mu\nu}$. Hint: make linear combinations of $G_{\mu\nu}$ and $G_{\nu\mu}$ to isolate the symmetric and antisymmetric pieces.

2.12 Show that for a symmetric tensor $T^{\mu\nu} = T^{\nu\mu}$ and a generic $G_{\mu\nu} = S_{\mu\nu} + A_{\mu\nu}$ from the previous problem with $S_{\mu\nu} = S_{\nu\mu}$ and $A_{\mu\nu} = -A_{\nu\mu}$, $T^{\mu\nu}G_{\mu\nu} = T^{\mu\nu}S_{\mu\nu}$.

2.13 In three dimensions, a scalar function ψ has "spherical symmetry" if $\psi(\mathbf{r}) = \psi(r)$. For vectors, like the electric field, spherical symmetry means that the vector points in the $\mathbf{r}$ direction with magnitude that depends only on r: $\mathbf{E} = E(r)\hat{\mathbf{r}}$. For a tensor field (focusing on the spatial piece) h^{ij}, spherical symmetry means that h^{ij} can depend only on the radial $\mathbf{r}$ (this is not the whole story, as we shall see later on) and must have entries that depend only on r: $h^{ij} = h(r)r^i r^j$ (where r^i is the ith component of $\mathbf{r}$, and $h(r)$ is a function of r). Write out this assumed form in Cartesian coordinates and then transform it to spherical.

2.14 What is $g^\mu{}_\nu =$?

2.15 We can make a scalar out of a combination of the metric and its inverse. From the contravariant form's definition in (2.55), what is $g^{\mu\nu}g_{\mu\nu} =$?

2.16 Using the transformation to cylindrical coordinates in (2.17), write the cylindrical basis vectors in terms of the Cartesian ones. Find the covariant form, the contravariant form, and the normalized unit form as in Example 2.3.

2.17 Which of the following expressions are valid tensorial equations written in index notation? For the invalid ones, generate a correct expression (that is not a unique process) or indicate that it is impossible to do so with the ingredients present:

$$A^\mu B^\gamma{}_\gamma g_{\mu\nu} = T_\nu, \qquad \partial_\mu A^{\mu\nu} = F_\alpha,$$
$$A^\mu{}_\nu B^{\nu\rho} = F_\rho G^{\rho\mu}, \qquad A^\alpha{}_{\alpha\beta} = F^{\sigma\sigma}{}_{\sigma\beta}. \tag{2.201}$$

2.18 From (2.75) find the relativistic distance traveled by a particle at rest, $x(t) = x_0$ in a time t (use the metric associated with Cartesian coordinates and take $\sigma = t$). What is the Euclidean distance traveled in one cycle of the oscillatory $x(t) = a\cos(\omega t)$? Find the relativistic, Minkowski length, associated with this motion – express your result in terms of the elliptic function

$$E(k) = \int_0^{\pi/2} \sqrt{1 - k^2 \sin^2\theta}\, d\theta. \tag{2.202}$$

2.19 Show that the Euler–Lagrange equations in (2.88) applied to a Lagrangian $L = T - U$ (kinetic minus potential energy) give the triple: $m\ddot{x}(t) = -\frac{\partial U}{\partial x}$, $m\ddot{y}(t) = -\frac{\partial U}{\partial y}$, $m\ddot{z}(t) = -\frac{\partial U}{\partial z}$. What are the equations of motion in spherical coordinates, with potential energy $U(r, \theta, \phi)$? What if the potential energy is spherically symmetric, depending only on r, what are the equations of motion in that case? Show that there are two conserved quantities that can be identified from the equations of motion and give their physical interpretation.

2.20 Calculate the geodesic equations of motion for a sphere using the metric in (2.103) and the geodesic equations of motion from (2.100). Check that the proper time requirement in (2.105) is consistent with the equations of motion by taking the τ derivative of (2.105) (or easier, the square of both sides) and using the equations of motion to show that equality holds.

2.21 For the geodesic Lagrangian in an arbitrary parameterization (remember that the geodesic action is reparameterization invariant, so just take a generic $\dot{x}^\mu(\sigma)$ here):

$$L = -mc\sqrt{-\dot{x}^\alpha g_{\alpha\beta}\dot{x}^\beta} \tag{2.203}$$

in a $D = 3 + 1$ dimensional spacetime with metric $g_{\alpha\beta}$ (not necessarily Minkowski), construct the Hamiltonian by finding the canonical momentum, $p_\mu = \frac{\partial L}{\partial \dot{x}^\mu}$ (note that canonical momentum is a covariant first-rank object), then compute the Legendre transform to get the Hamiltonian

$$H = \dot{x}^\mu p_\mu - L. \tag{2.204}$$

What do you make of this? Try it again using temporal parameterization, and for concreteness, the Minkowski metric. Does this one make sense?

2.22 Show that the Lagrangian L_τ from (2.108) has Euler–Lagrange equations of motion that recover (2.100).

2.23 Show that for light, with Lagrangian $L = \sqrt{-\dot{x}^\mu(s)g_{\mu\nu}\dot{x}^\nu(s)} = 0$, you get the same equations of motion as for $L = (1/2)\dot{x}^\mu(s)g_{\mu\nu}\dot{x}^\nu(s)$.

2.24 For a two-dimensional space with Pythagorean line element $d\ell^2 = dx^2 + dy^2$, express the metric in polar coordinates s and ϕ with $x = s\cos\phi$, $y = s\sin\phi$. Using an arc length parameter σ such that $\dot{x}^\alpha(\sigma) \times g_{\alpha\beta}\dot{x}^\beta(\sigma) = 1$, write out the equations of motion that come from the geodesic Lagrangian $L = \sqrt{\dot{x}^\alpha(\sigma)g_{\alpha\beta}\dot{x}^\beta(\sigma)}$ with

$$\dot{x}^\alpha(\sigma) \doteq \begin{pmatrix} \dot{s}(\sigma) \\ \dot{\phi}(\sigma) \end{pmatrix} \tag{2.205}$$

(don't put in the defining relation for arc length, $\dot{x}^\alpha(\sigma)g_{\alpha\beta}\dot{x}^\beta(\sigma) = 1$, too soon, but don't wait until it's too late, either!).

2.25 For the line with coordinates $x(s) = s$, $y(s) = 2s$, $z(s) = s/2$ with $s = 0 \to 1$, find the representation of the line in spherical coordinates, that is, what are $r(s)$, $\theta(s)$, and $\phi(s)$?

2.26 A cone has the z axis as its central axis and makes an angle of α with respect to it (think of the points defined by a fixed polar angle, in spherical coordinates, at $\theta = \alpha$). Using the usual azimuthal angle ϕ measured with respect to the x axis and a distance along the cone, ρ (from origin to current location), as the coordinates on the cone (as shown in Figure 2.5), find the metric on this two-dimensional surface (imagine making infinitesimal moves in the ρ and ϕ directions, separately, and note that they are orthogonal in order to combine those infinitesimals in a quadratic expression for length from which you can extract the metric as in Example C.7). From the metric write out the geodesic Lagrangian in proper time parameterization, $\dot{x}^\mu g_{\mu\nu}\dot{x}^\nu = c^2$ (but working in a two-dimensional space). What are the equations of motion for ρ and ϕ that come from this Lagrangian? Compare with the geodesic equations of motion you get for flat, two-dimensional space written in polar coordinates.

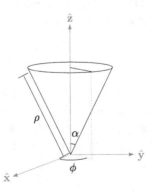

Conical setup for Problem 2.26.

2.27 Show that the Christoffel symbol defined in (2.140) is symmetric in its second two indices: $\Gamma_{\alpha\mu\nu} = \Gamma_{\alpha\nu\mu}$. Then show that this metric connection gives $g_{\alpha\beta;\mu} = 0$.

2.28 Show that if you have two connections $C^\mu{}_{\alpha\beta}$ and $D^\mu{}_{\alpha\beta}$, then there is a linear combination of the two that is a third-rank (mixed) tensor.

2.29 Find the nonzero Christoffel connection (of the first kind) coefficients in cylindrical and spherical coordinates.

2.30 Find the nonzero Christoffel connection (of the second kind) coefficients in cylindrical and spherical coordinates.

2.31 Given a time-independent A^μ, and working in spherical coordinates, evaluate the divergence from (2.145). You will first have to work out the nonzero Christoffel coefficients in spherical coordinates or use your result from (2.30). Write out $\nabla \cdot \mathbf{A}$ and compare with what you get from (2.145). Why are they different?

2.32 When we use semicolon notation for the covariant derivative, any index to the right of the semicolon is a derivative index, so that $\psi_{;\mu\nu} \equiv D_\mu D_\nu \psi$ and $F_{\mu;\nu\gamma} \equiv D_\gamma D_\nu F_\mu$. Write out $\psi_{;\mu\nu}$ and $F_{\mu;\nu\gamma}$ in terms of partial derivatives and connections.

2.33 Working in three-dimensional space and using spherical coordinates, write the ODE for parallel transport of a vector A^μ (with $A^0 = 0$) using the same path that was used in Example 2.10 (i.e., pick θ_0 and r_0 constant and use ϕ as the path parameter). The nonzero connection coefficients here can be found in (2.159). Solve for $A^\mu(\phi)$ using initial values at $\phi = 0$: $A^0 = 0$, $A^1(0) = A_0^r$, $A^2(0) = A_0^\theta$, $A^3(0) = A_0^\phi$, and evaluate the parallel-transported vector at $\phi = 2\pi$, back at the start. How does $A^\mu(2\pi)$ compare with the values of $A^\mu(0)$? Is this what you expected?

2.34 Show that $D_\mu A_\nu - D_\nu A_\mu = \partial_\mu A_\nu - \partial_\nu A_\mu$, the difference of the nontensorial partial derivatives here is itself a tensor.

2.35 Starting from (2.188) and working in Cartesian coordinates to avoid Christoffel complications, set $h_{\mu\nu} = A_{\mu,\nu} + A_{\nu,\mu}$ (to make $h_{\mu\nu}$ symmetric). Put this ansatz in to (2.188) and by requiring that no reference to A_α remain, set as many constants as you can. Once those constants are set, make sure that $h_{\mu\nu} = \psi_{,\mu\nu}$ also leads to insensitivity to ψ when put into (2.188).

2.36 Find the equations of motion for the Lagrangian in (2.196), leave your expressions in terms of the relativistic momentum,

$$\mathbf{p} \equiv \frac{m\mathbf{v}}{\sqrt{1 - v^2/c^2}}, \tag{2.206}$$

but write the forces in terms of the electric and magnetic fields.

3 The Riemann Tensor and Einstein's Equation

Almost everything we have done with tensors so far has been in the Minkowski spacetime setting. Our focus has been on the response of tensors (and connections) to transformations taking us from Cartesian coordinates to curvilinear ones, for example. Except for a few examples and a problem here or there, nothing has been done outside of the usual spacetime familiar to us from special relativity. So all of the work we have done on geodesics, has been purely notational. We know that the length-minimizing curves connecting any two points in spacetime are straight lines. All that changes is the representation of the line. The Minkowski metric that defines the distance between points has not changed, so the metric itself tells us only about the coordinate system in which we are expressing the underlying spacetime geometry. We have not yet seriously considered changing that geometry from Minkowski to something else.

In this chapter, we will see an expansion of the metric's role. It can define new and different ways to measure length, representing geometry beyond Minkowski, in addition to defining the coordinate system. These dual roles make it difficult to tell if we have a version of Minkowski spacetime written in unfamiliar coordinates, or a new geometry. Separating those two possibilities is important in general relativity. As we shall establish in this chapter, general relativity can be viewed as a theory with the metric as the central field, sourced by energy and moving energy. Particles respond to this metric field by moving along its geodesics. If a particular solution to the gravitational field equation is the metric of Minkowski spacetime, expressed in any coordinate system, then the geodesics are lines, and there is no gravitational effect. To obtain the elliptical orbits that we observe, as geodesics of some metric, that metric must *not* represent the Minkowski spacetime geometry.

The tool that distinguishes metrics that can be brought to the diagonal form of (2.53) via a coordinate transformation versus those that cannot is the "Riemann tensor." A metric that can be written as (2.53) by appropriate coordinate choice is associated with "flat" spacetime geometry, as opposed to "curved" (not flat). The Riemann tensor provides a test for flatness. For flat spacetime, the geodesic curves diverge from one another linearly, whereas for curved spacetimes, the geodesic curves can diverge in a variety of different ways, even converging in some cases. In addition to its role in testing for flatness, the Riemann tensor tells us how geodesics diverge in general, and that "geodesic deviation" forms the starting point for generating Einstein's field equations as successors to those of Newtonian gravity.

We will start by defining the Riemann tensor and generate some special coordinate systems in which to probe its structure. The four-index Riemann tensor has two important contractions of its indices. Contracting once gives the second-rank "Ricci" tensor, and contracting the Ricci tensor's indices gives the Ricci scalar. Each of these is important in general relativity, with the latter two making up the PDE content of Einstein's equation. In Appendix C, there is a discussion of curvature in one and two dimensions that can be used to build some intuition about the curvature discussed in this chapter, although there are also significant differences that we will highlight as we go. The geometry defined by, and interpretation of, the Riemann tensor defines the left-hand side of Einstein's equation, roughly, the "derivatives of the field" piece of the field equation. Sources are another matter, and we spend some time discussing the second-rank "stress tensor" source for gravity, building it up by analogy with the current density four-vector source from electricity and magnetism. Finally, we will generate the field equations and discuss their implications and interpretation before moving on to solutions and their physical content in Chapter 4.

3.1 The Riemann Tensor

In Section 2.8, while removing scalar dependence from the most general vector field equation, we assumed the equality of $D_\mu D_\nu \psi$ and $D_\nu D_\mu \psi$, a covariant derivative form of the cross-derivative equality we get from partials: $\partial_\mu \partial_\nu \psi = \partial_\nu \partial_\mu \psi$. To establish the assumed property, we'll work out the difference $(D_\mu D_\nu - D_\nu D_\mu)\psi$ in a general coordinate system. Let $A_\nu \equiv \psi_{;\nu}$. Then

$$D_\mu D_\nu \psi = A_{\nu;\mu} = A_{\nu,\mu} - \Gamma^\rho{}_{\mu\nu} A_\rho = \psi_{,\nu\mu} - \Gamma^\rho{}_{\mu\nu} \psi_{,\rho}. \qquad (3.1)$$

If we swap the index roles, then we get

$$D_\nu D_\mu \psi = \psi_{,\mu\nu} - \Gamma^\rho{}_{\nu\mu} \psi_{,\rho}. \qquad (3.2)$$

We know from the usual cross-derivative equality that $\psi_{,\mu\nu} = \psi_{,\nu\mu}$. Furthermore, from the form of $\Gamma^\alpha{}_{\mu\nu}$ in (2.139), the Christoffel symbol is symmetric in the interchange of its lower index pair, $\Gamma^\rho{}_{\nu\mu} = \Gamma^\rho{}_{\mu\nu}$. Then the expressions in (3.1) and (3.2) are equal, and we have shown that $D_\mu D_\nu \psi = D_\nu D_\mu \psi$ explicitly in any spacetime with a metric connection.

Let's go up a rank, does the covariant derivative commute with itself when acting on a tensor V^α? Proceeding as in the scalar case, let's construct the difference $(D_\mu D_\nu - D_\nu D_\mu)V^\alpha$. Take $A^\alpha{}_\mu \equiv D_\mu V^\alpha$; it's easier to keep track of the indices without the decorative punctuation. The covariant derivative of the mixed $A^\alpha{}_\mu$ is

$$A^\alpha{}_{\mu;\nu} = A^\alpha{}_{\mu,\nu} + \Gamma^\alpha{}_{\nu\rho} A^\rho{}_\mu - \Gamma^\rho{}_{\nu\mu} A^\alpha{}_\rho. \qquad (3.3)$$

Next, we put in the expression for $A^\alpha{}_\mu = V^\alpha{}_{,\mu} + \Gamma^\alpha{}_{\mu\sigma} V^\sigma$:

$$V^\alpha{}_{;\mu\nu} = V^\alpha{}_{,\mu\nu} + \Gamma^\alpha{}_{\mu\sigma,\nu} V^\sigma + \Gamma^\alpha{}_{\mu\sigma} V^\sigma{}_{,\nu} + \Gamma^\alpha{}_{\nu\rho} V^\rho{}_{,\mu}$$
$$- \Gamma^\rho{}_{\nu\mu} V^\alpha{}_{,\rho} + \left(\Gamma^\alpha{}_{\nu\rho} \Gamma^\rho{}_{\mu\sigma} - \Gamma^\rho{}_{\nu\mu} \Gamma^\alpha{}_{\sigma\rho} \right) V^\sigma. \tag{3.4}$$

Swapping $\mu \leftrightarrow \nu$ gives us the other derivative ordering,

$$V^\alpha{}_{;\nu\mu} = V^\alpha{}_{,\nu\mu} + \Gamma^\alpha{}_{\nu\sigma,\mu} V^\sigma + \Gamma^\alpha{}_{\nu\sigma} V^\sigma{}_{,\mu} + \Gamma^\alpha{}_{\mu\rho} V^\rho{}_{,\nu}$$
$$- \Gamma^\rho{}_{\mu\nu} V^\alpha{}_{,\rho} + \left(\Gamma^\alpha{}_{\mu\rho} \Gamma^\rho{}_{\nu\sigma} - \Gamma^\rho{}_{\mu\nu} \Gamma^\alpha{}_{\sigma\rho} \right) V^\sigma, \tag{3.5}$$

and then subtracting gives (using $\Gamma^\alpha{}_{\mu\nu} = \Gamma^\alpha{}_{\nu\mu}$ and $V^\alpha{}_{,\mu\nu} = V^\alpha{}_{,\nu\mu}$)

$$V^\alpha{}_{;\mu\nu} - V^\alpha{}_{;\nu\mu} = \left(\Gamma^\alpha{}_{\mu\sigma,\nu} - \Gamma^\alpha{}_{\nu\sigma,\mu} + \Gamma^\alpha{}_{\nu\rho} \Gamma^\rho{}_{\mu\sigma} - \Gamma^\alpha{}_{\mu\rho} \Gamma^\rho{}_{\nu\sigma} \right) V^\sigma. \tag{3.6}$$

The term in parentheses multiplying V^σ defines the "Riemann tensor":

$$\boxed{R^\alpha{}_{\sigma\nu\mu} \equiv \Gamma^\alpha{}_{\mu\sigma,\nu} - \Gamma^\alpha{}_{\nu\sigma,\mu} + \Gamma^\alpha{}_{\nu\rho} \Gamma^\rho{}_{\mu\sigma} - \Gamma^\alpha{}_{\mu\rho} \Gamma^\rho{}_{\nu\sigma}.} \tag{3.7}$$

The difference in derivative ordering is then directly proportional to the Riemann tensor, and the vector V^σ itself,

$$V^\alpha{}_{;\mu\nu} - V^\alpha{}_{;\nu\mu} = R^\alpha{}_{\sigma\nu\mu} V^\sigma. \tag{3.8}$$

Looking at the actual terms in the Riemann tensor, we have Christoffel symbols, which are not tensors, appearing as products (also presumably not tensorial) and with partial derivatives attached. You wouldn't look at the ingredients there and conclude that the combination transforms anything like a fourth-rank mixed tensor. Yet it is clear from (3.8), where the left side is manifestly tensorial, that the right side must be as well (and you can always check!).

The first index of the Riemann tensor can be lowered using the metric to give a fully covariant fourth-rank version written in terms of the Christoffel symbols of the first kind from (2.140),

$$\boxed{R_{\alpha\sigma\nu\mu} \equiv g_{\alpha\rho} \left(\Gamma^\rho{}_{\mu\sigma,\nu} - \Gamma^\rho{}_{\nu\sigma,\mu} \right) + \Gamma_{\alpha\nu\rho} \Gamma^\rho{}_{\mu\sigma} - \Gamma_{\alpha\mu\rho} \Gamma^\rho{}_{\nu\sigma}.} \tag{3.9}$$

In addition to the Riemann tensor itself, we will also be interested in its nontrivial index contractions. The second-rank symmetric "Ricci tensor" is defined by the one-three contraction of the Riemann tensor,

$$R_{\sigma\mu} \equiv g^{\alpha\nu} R_{\alpha\sigma\nu\mu} = R^\alpha{}_{\sigma\alpha\mu}. \tag{3.10}$$

The "Ricci scalar" is the trace of the Ricci tensor:

$$R \equiv g^{\sigma\mu} R_{\sigma\mu} = R^\sigma{}_\sigma. \tag{3.11}$$

Example 3.1 (The Riemann Tensor for a Sphere). We know the metric and connection coefficients (the second kind are listed in (2.165), for the first kind, from Problem 2.29) for the surface of a sphere of radius r (with coordinates $x^1 \equiv \theta, x^2 \equiv \phi$),

$$g_{\mu\nu} \doteq \begin{pmatrix} r^2 & 0 \\ 0 & r^2 \sin^2\theta \end{pmatrix},$$

$$\Gamma^1_{22} = -\sin\theta\cos\theta, \qquad \Gamma^2_{12} = \Gamma^2_{21} = \cot\theta, \tag{3.12}$$

$$\Gamma_{122} = -r^2\sin\theta\cos\theta, \qquad \Gamma_{212} = \Gamma_{221} = r^2\sin\theta\cos\theta.$$

To form the Riemann tensor, we need the nonzero derivatives of the connection coefficients. These only depend on θ, so the only nonzero entries are

$$\Gamma^1_{22,1} = -\cos(2\theta), \qquad \Gamma^2_{12,1} = \Gamma^2_{21,1} = -\frac{1}{\sin^2\theta}. \tag{3.13}$$

Finally, we need the terms quadratic in the connections, $\Gamma^\alpha_{\nu\rho}\Gamma^\rho_{\mu\sigma}$, for example. The nonzero components for these terms are

$$\Gamma^1_{2\rho}\Gamma^\rho_{12} = \Gamma^1_{2\rho}\Gamma^\rho_{21} = -\cos^2\theta, \qquad \Gamma^2_{1\rho}\Gamma^\rho_{12} = \Gamma^2_{1\rho}\Gamma^\rho_{21} = \frac{\cos^2\theta}{\sin^2\theta},$$

$$\Gamma^2_{2\rho}\Gamma^\rho_{22} = -\cos^2\theta. \tag{3.14}$$

Putting these all together, the nonzero Riemann tensor components are

$$R^1_{212} = -R^1_{221} = \sin^2\theta, \qquad R^2_{121} = -R^2_{112} = 1, \tag{3.15}$$

and you can compute the lower form in Problem 3.3.

From the Riemann tensor we can compute the Ricci tensor, $R_{\sigma\mu} = R^\alpha_{\sigma\alpha\mu}$, with nonzero components

$$R_{11} = 1, \qquad R_{22} = \sin^2\theta, \tag{3.16}$$

and finally, the Ricci scalar, $R \equiv g^{\sigma\mu}R_{\sigma\mu} = 2/r^2$. This scalar is most like the Gauss curvature from Appendix C in terms of scaling with r and units. Compare the value of the Ricci scalar with the $1/r^2$ Gauss curvature from Example C.5. One fundamental difference between the Gauss curvature and the Ricci scalar is that the Gauss curvature is defined in terms of a lower-dimensional surface's embedding in a higher-dimensional (flat) space, whereas the Ricci scalar does not require a higher-dimensional space for its definition and calculation.

3.1.1 Interpretation

The Riemann tensor is defined in terms of the difference of covariant derivatives acting on a first-rank tensor, and that does indicate its geometric significance, but a little work is needed to make its role in determining curvature clear. Going back to our one nonflat example, the surface of a sphere, think about the parallel transport story from Example 2.10: If we went around the

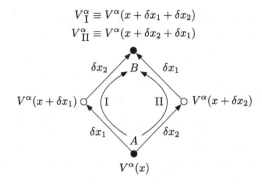

$$V_{\text{I}}^{\alpha} \equiv V^{\alpha}(x + \delta x_1 + \delta x_2)$$
$$V_{\text{II}}^{\alpha} \equiv V^{\alpha}(x + \delta x_2 + \delta x_1)$$

Fig. 3.1   We will parallel transport the vector V^{α} from point A with coordinates x to point B with coordinates $x + \delta x_1 + \delta x_2$ along the two paths shown. To compare the two vectors we get from the two paths, we will evaluate both $V_I^{\alpha} \equiv V^{\alpha}(x + \delta x_1 + \delta x_2)$ and $V_{II}^{\alpha} \equiv V^{\alpha}(x + \delta x_2 + \delta x_1)$ in terms of quantities defined at the starting point, x.

circle at polar angle θ_0 starting at $\phi = 0$, the transported vector comes back (at $\phi = 2\pi$) rotated through an angle $\psi = 2\pi \cos \theta_0$ with respect to the original orientation of the vector as shown in Figure 2.4. That does not happen in a flat spacetime, where parallel transport is just a vector moved along parallel to itself. Parallel transport along the same circular path, if it occurred in three dimensions and not trapped on the surface of the sphere, would have initial and final vectors (going from $\phi = 0$ to $\phi = 2\pi$) that are identical as you showed in Problem 2.33. Path dependence of parallel transport is typical of "curved" spacetimes. In this case, the paths are "staying at $\phi = 0$" and "going around the circle at θ_0 from $\phi = 0$ back to $\phi = 2\pi$."

Let's work out the more general path dependence of parallel transport. We'll take two nearby points A and B and take the two simple paths from A to B shown in Figure 3.1. To compare the result of the parallel transport, we need to pick a location and compute all quantities with respect to that single location.[1] Take point A to have coordinate values given by x (omitting the index to avoid clutter). Then we have two infinitesimal vectors δx_1 and δx_2 taking us along paths I and II. Working along path I, take V^{α} from x to $x + \delta x^1$ by parallel transport, meaning that

$$\frac{DV^{\alpha}}{D\tau} = 0 = V^{\alpha}{}_{;\nu}(x)\delta x_1^{\nu} \longrightarrow V^{\alpha}{}_{,\nu}(x)\delta x_1^{\nu} = -\Gamma^{\alpha}{}_{\nu\gamma}(x)V^{\gamma}(x)\delta x_1^{\nu}, \quad (3.17)$$

where I am highlighting that everything is evaluated at x. At $x + \delta x_1$, we can Taylor expand $V^{\alpha}(x + \delta x_1)$ and use the relationship between the derivatives of V^{α} and the connection from parallel transport above,

$$V^{\alpha}(x + \delta x_1) \approx V^{\alpha}(x) + V^{\alpha}{}_{,\nu}(x)\delta x_1^{\nu} = V^{\alpha}(x) - \Gamma^{\alpha}{}_{\nu\gamma}(x)V^{\gamma}(x)\delta x_1^{\nu}. \quad (3.18)$$

[1]   That sounds familiar, you can't compare vectors except at a single point.

From $x + \delta x_1$, continuing along path I, parallel transport gives

$$V^\alpha_{;\mu}\delta x_2^\mu = 0 \longrightarrow V^\alpha_{,\mu}(x + \delta x_1)\delta x_2^\mu = -\Gamma^\alpha_{\mu\rho}(x + \delta x_1)V^\rho(x + \delta x_1)\delta x_2^\mu,$$
(3.19)

and using this expression for the derivative of V^α at $x + \delta x_1$ in the Taylor expansion of $V^\alpha(x + \delta x_1 + \delta x_2)$, we obtain the approximation

$$\begin{aligned}
V^\alpha&(x + \delta x_1 + \delta x_2) \\
&\approx V^\alpha(x + \delta x_1) + V^\alpha_{,\mu}(x)\delta x_2^\mu \\
&= V^\alpha(x + \delta x_1) - \Gamma^\alpha_{\mu\rho}(x + \delta x_1)V^\rho(x + \delta x_1)\delta x_2^\mu.
\end{aligned}$$
(3.20)

Finally, we can rewrite references to $V^\alpha(x + \delta x_1)$ using (3.18) and Taylor expand the connection $\Gamma^\alpha_{\mu\rho}(x + \delta x_1)$ to express $V_I^\alpha \equiv V^\alpha(x + \delta x_1 + \delta x_2)$ entirely in terms of quantities evaluated at x, and we therefore omit the explicit x dependence below to keep our focus on the form of the sums:

$$V_I^\alpha \approx V^\alpha - \Gamma^\alpha_{\nu\gamma}V^\gamma\delta x_1^\nu - \left(\Gamma^\alpha_{\mu\rho} + \Gamma^\alpha_{\mu\rho,\nu}\delta x_1^\nu\right)\left(V^\rho - \Gamma^\rho_{\gamma\sigma}V^\sigma\delta x_1^\gamma\right)\delta x_2^\mu.$$
(3.21)

Dropping the term cubic in the small path lengths δx_1 and δx_2 and collecting terms,

$$V_I^\alpha \approx V^\alpha - \Gamma^\alpha_{\nu\gamma}V^\gamma\delta x_1^\nu - \Gamma^\alpha_{\mu\rho}V^\rho\delta x_2^\mu + \left(\Gamma^\alpha_{\mu\rho}\Gamma^\rho_{\nu\sigma} - \Gamma^\alpha_{\mu\sigma,\nu}\right)V^\sigma\delta x_1^\nu\delta x_2^\mu.$$
(3.22)

All right, so much for path I. Fortunately, we can get the corresponding expression for $V_{II}^\alpha \equiv V^\alpha(x + \delta x_2 + \delta x_1)$ just by reversing the roles of δx_1 and δx_2:

$$V_{II}^\alpha \approx V^\alpha - \Gamma^\alpha_{\nu\gamma}V^\gamma\delta x_2^\nu - \Gamma^\alpha_{\mu\rho}V^\rho\delta x_1^\mu + \left(\Gamma^\alpha_{\mu\rho}\Gamma^\rho_{\nu\sigma} - \Gamma^\alpha_{\mu\sigma,\nu}\right)V^\sigma\delta x_2^\nu\delta x_1^\mu.$$
(3.23)

Thinking ahead to the subtraction, the first three terms in each of these expressions will simply cancel out. The remaining terms are quadratic, depending on the product $\delta x_1 \delta x_2$. These remaining terms, the last ones in (3.22) and (3.23) are related by relabeling $\mu \leftrightarrow \nu$ since the first has $\delta x_1^\nu \delta x_2^\mu$ and the second $\delta x_2^\nu \delta x_1^\mu$. Subtracting (3.22) from (3.23) (and performing the dummy relabeling $\mu \leftrightarrow \nu$ on that expression),

$$V_{II}^\alpha - V_I^\alpha = \left(\Gamma^\alpha_{\nu\rho}\Gamma^\rho_{\mu\sigma} - \Gamma^\alpha_{\nu\sigma,\mu} - \Gamma^\alpha_{\mu\rho}\Gamma^\rho_{\nu\sigma} + \Gamma^\alpha_{\mu\sigma,\nu}\right)V^\sigma\delta x_1^\nu\delta x_2^\mu. \quad (3.24)$$

Compare the term in parenthesis to the definition of the Riemann tensor in (3.7). The difference in a vector under parallel transport along two nearby paths is

$$\boxed{V_{II}^\alpha - V_I^\alpha = R^\alpha_{\sigma\nu\mu}V^\sigma\delta x_1^\nu\delta x_2^\mu.}$$
(3.25)

The entries in the Riemann tensor are related to the choices we have in probing the path dependence of parallel transport, the particular paths selected by δx_1^ν and δx_2^μ.

If we are using a metric that is flat, then there is a coordinate system in which it takes the form (2.53), and in that case, both the Christoffel connection and its derivatives are zero, so that the Riemann tensor is zero in Cartesian coordinates, $R^\alpha_{\ \sigma\nu\mu} = 0$. Since this is a tensor statement, it must be zero in all other coordinate systems as well. A flat spacetime has parallel transport of vectors that is independent of the path we choose. The other direction is true as well: If parallel transport is path-independent, then the spacetime is flat. We'll sketch the proof, by construction, below. Most of this introduction, including the argument below, can be found in classic differential geometry texts like [28].

3.1.2 Flat Spacetimes and the Riemann Tensor

Referring to (3.25), if the Riemann tensor vanishes everywhere, then parallel transport is path-independent: $V^\alpha_{II} - V^\alpha_I = 0$, and

$$\frac{DV^\alpha}{D\tau} = V^\alpha_{\ ;\beta}\dot{x}^\beta = 0 \tag{3.26}$$

must hold for any tangent vector $\dot{x}^\beta$, so that $V^\alpha_{\ ;\beta} = 0$, or, using the lower form, $V_{\alpha;\beta} = 0$. Writing out this relation,

$$V_{\alpha;\beta} \equiv V_{\alpha,\beta} - \Gamma^\sigma_{\ \alpha\beta}V_\sigma = 0 \longrightarrow V_{\alpha,\beta} = \Gamma^\sigma_{\ \alpha\beta}V_\sigma. \tag{3.27}$$

Since the Christoffel connection is symmetric in the interchange of its lower indices, we also have $V_{\beta,\alpha} = \Gamma^\sigma_{\ \alpha\beta}V_\sigma$, so that $V_{\alpha,\beta} - V_{\beta,\alpha} = 0$. This condition implies that the vector V_α can be written as the gradient of a scalar,[2] $V_\alpha = \phi_{,\alpha}$, where the expression $V_{\alpha,\beta} = V_{\beta,\alpha}$ becomes

$$\phi_{,\alpha\beta} = \phi_{,\beta\alpha}, \tag{3.28}$$

the usual statement of cross-derivative equality.

Given a path P and a vector V_α with $V_{\alpha;\beta} = 0$, we can construct the integral

$$I = \int_P V_\alpha dx^\alpha. \tag{3.29}$$

Taking $V_\alpha = \phi_{,\alpha}$ and denoting the tangent vector to the path P as $\dot{x}^\alpha(\sigma)$ (for path parameter σ), the path integral becomes

$$I = \int_{\sigma_0}^{\sigma_f} \phi_{,\alpha}\dot{x}^\alpha d\sigma = \int_{\sigma_0}^{\sigma_f} \frac{d\phi}{d\sigma}d\sigma = \phi\big(x(\sigma_f)\big) - \phi\big(x(\sigma_0)\big), \tag{3.30}$$

from which it is clear that the integral itself is path independent, depending only on the evaluation of the scalar ϕ at the endpoints. Imagine that V^α is a force of some sort. Then the integral I is interpretable, physically, as the work done along some path. The statement $V^\alpha = \phi_{,}^{\ \alpha}$ is like saying that the force is derivable from a potential, so the force is conservative, at which point we know that the work integral is path independent. The logic here is familiar.

---

[2] The argument here is a generalization of the familiar statement, in three-dimensional vector calculus, that $\nabla \times \mathbf{V} = 0 \to \mathbf{V} = \nabla\phi$.

Now we'll construct a set of coordinates using the path independence of the integral I. Take a path P that starts at the origin and ends at an arbitrary location x^α. In terms of the curve defining the path, the endpoints are $x^\alpha(\sigma_0) = 0$ and $x^\alpha(\sigma_f) = x^\alpha$. In this context, the content of (3.30) is $I = \phi(x^\alpha) - \phi(0)$, and we can take the derivative of the integral I with respect to the end-point x^β (using β instead of α to index the point, to avoid confusion),

$$\frac{\partial I}{\partial x^\beta} = \frac{\partial \phi(x^\alpha)}{\partial x^\beta} = V^\beta(x^\alpha), \tag{3.31}$$

where $V^\beta(x^\alpha)$ is obtained by starting with a vector $V^\beta(0)$ at the origin and parallel transporting (along any path) to x^α. Pick a set of four vectors at the origin, V^a_σ for $a = 0, 1, 2, 3$, chosen so that $V^a_\sigma V^b_\rho g^{\sigma\rho} = \eta^{ab}$, the diagonal Minkowski metric with entries as in (2.53). We can always accomplish this, since the manifold can be taken to have Minkowski metric at a point (as we will see in Section 3.2). Now parallel transport the collection V^a_σ from the origin to x^α, an arbitrary point in the manifold. We know from Section 2.7 that parallel transport does not change lengths or angles, so that at every point along the way from the origin to x^α, we retain $V^a_\sigma V^b_\rho g^{\sigma\rho} = \eta^{ab}$.

We'll generate a coordinate transformation using the set V^a_μ. Define the new coordinate value at x^α to be

$$\bar{x}^a \equiv \int_P V^a_\rho dx^\rho = \phi^a(x^\alpha) - \phi^a(0) \tag{3.32}$$

for ϕ^a the scalar whose gradient is V^a_α. This equation is four copies of (3.30), so we know from (3.31) that

$$\frac{\partial \bar{x}^a}{\partial x^\mu} = \frac{\partial \phi^a}{\partial x^\mu} = V^a_\mu(x^\alpha). \tag{3.33}$$

The contravariant form of the metric in the new coordinates becomes

$$\bar{g}^{ab} = \frac{\partial \bar{x}^a}{\partial x^\mu} \frac{\partial \bar{x}^b}{\partial x^\nu} g^{\mu\nu} = \phi^a_{,\mu} \phi^b_{,\nu} g^{\mu\nu} = V^a_\mu V^b_\nu g^{\mu\nu} = \eta^{ab}, \tag{3.34}$$

and the metric has taken the flat Minkowski form at all points x^α; the spacetime is flat.

We have now shown both directions of the theorem: "If a spacetime has zero Riemann tensor, it is flat." First by showing that flat spacetimes have zero Riemann tensor, and in this section, by establishing that a zero Riemann tensor gives path-independent parallel transport, which can be used to explicitly construct a coordinate system in which the metric is of the form (2.53). The theorem gives the Riemann tensor its main interpretation as an indicator of flatness. Next, we'll look at other properties of the Riemann tensor that will be useful, and these can be developed using a specially adapted coordinate system.

3.2 Riemannian Coordinates

The space in which we are working, with a metric and connection related to the metric as in (2.139), has been specially chosen for its relevance to gravity, where the metric is the only field.[3] There are a variety of nice properties in such a metric space. One such property is that at any point P, we can choose a specific set of coordinates such that the metric takes on its diagonal form from (2.53) and the connection coefficients at P vanish.[4] In spaces with a metric connection, then, the equivalence principle, which says that locally, gravity is undetectable, is automatically true (this is the reason these spaces are central to general relativity). At a point, and in its infinitesimal vicinity, we have a Minkowski spacetime with no force or observable gravitational effect.

Let's construct a coordinate system that gives $\bar{g}_{\mu\nu}|_P$ its diagonal form and that has $\bar{\Gamma}^{\alpha}{}_{\mu\nu}|_P = 0$. Start with coordinates x^α and take P to be the origin, where $x^\alpha|_P = 0$. Then define the new coordinates as

$$\bar{x}^\mu = x^\nu + \frac{1}{2}\Gamma^\mu{}_{\alpha\beta}x^\alpha x^\beta, \tag{3.35}$$

where $\Gamma^\mu{}_{\alpha\beta}$ is the connection of the original coordinates evaluated at P. The derivatives of this coordinate transformation are

$$\frac{\partial \bar{x}^\mu}{\partial x^\gamma} = \delta^\mu_\gamma + \Gamma^\mu{}_{\gamma\beta}x^\beta, \qquad \frac{\partial^2 \bar{x}^\mu}{\partial x^\gamma \partial x^\sigma} = \Gamma^\mu{}_{\gamma\sigma}. \tag{3.36}$$

Now the connection transforms according to (2.125), and we have

$$\bar{\Gamma}^\rho{}_{\alpha\beta} = \frac{\partial \bar{x}^\rho}{\partial x^\mu}\frac{\partial x^\gamma}{\partial \bar{x}^\alpha}\frac{\partial x^\sigma}{\partial \bar{x}^\beta}\Gamma^\mu{}_{\gamma\sigma} - \frac{\partial^2 \bar{x}^\rho}{\partial x^\gamma \partial x^\sigma}\frac{\partial x^\gamma}{\partial \bar{x}^\alpha}\frac{\partial x^\sigma}{\partial \bar{x}^\beta}. \tag{3.37}$$

Evaluating this connection at P and using the derivatives from (3.36), we learn that the new connection is zero at P, by construction. We also know that $\bar{g}_{\mu\nu;\alpha} = 0$ for any coordinate transformation, so that

$$0 = \bar{g}_{\mu\nu,\alpha} - \bar{\Gamma}^\rho{}_{\alpha\mu}\bar{g}_{\rho\nu} - \bar{\Gamma}^\rho{}_{\alpha\nu}\bar{g}_{\mu\rho}, \tag{3.38}$$

when evaluated at P, gives $\bar{g}_{\mu\nu,\alpha}|_P = 0$, that is, the metric's derivatives vanish at P. Since the metric has constant value at P, we can perform an additional coordinate transformation to bring it to the diagonal form of (2.53) (see Problem 3.4). Note that we also have $\bar{g}^{\mu\nu}{}_{;\alpha}|_P = 0$, which gives $\bar{g}^{\mu\nu}{}_{,\alpha}|_P = 0$, too.

---

[3] We could imagine a theory of gravity in which the metric and, say, connection are the fields, each with their own sources. In general relativity, we only have source for the metric, and all other quantities, connections, curvature, etc. follow from it alone.

[4] More than that, you cannot do. In particular, you cannot always choose the coordinates so that the connection and its derivatives vanish at the point P, so the Riemann tensor is not, in general, zero. This dependence on the second derivatives of the metric is similar to a quadratic minimum. For a function $f(x) = k(x - p)^2$, the value of the function and its derivative at $x = p$ is zero, but the second derivative is not.

The existence of these "Riemannian coordinates" is quite natural for our physical requirements. After all, to good approximation, we live in a universe that has at its "points" (like terrestrial laboratories) a Minkowski metric written in Cartesian coordinates with metric derivatives that are zero.

The Riemann tensor at point P takes on the simplified form, working from the lowered version in (3.9),

$$\bar{R}_{\alpha\sigma\nu\mu}|_P = \bar{g}_{\alpha\rho}\left(\bar{\Gamma}^\rho_{\ \mu\sigma,\nu} - \bar{\Gamma}^\rho_{\ \nu\sigma,\mu}\right)\big|_P = \bar{g}_{\alpha\rho}\left[\left(\bar{g}^{\rho\gamma}\bar{\Gamma}_{\gamma\mu\sigma}\right)_{,\nu} - \left(\bar{g}^{\rho\gamma}\bar{\Gamma}_{\gamma\nu\sigma}\right)_{,\mu}\right]\big|_P. \tag{3.39}$$

We can write the Riemann tensor in terms of the second derivatives of the metric by expanding the connection derivatives,

$$\begin{aligned}
\bar{R}_{\alpha\sigma\nu\mu}|_P &= \bar{g}_{\alpha\rho}\left(\bar{g}^{\rho\gamma}_{\ \ ,\nu}\bar{\Gamma}_{\gamma\mu\sigma} - \bar{g}^{\rho\gamma}_{\ \ ,\mu}\bar{\Gamma}_{\gamma\nu\sigma}\right)\big|_P \\
&\quad + \frac{1}{2}(\bar{g}_{\alpha\mu,\sigma\nu} - \bar{g}_{\mu\sigma,\alpha\nu} - \bar{g}_{\alpha\nu,\sigma\mu} + \bar{g}_{\nu\sigma,\alpha\mu})|_P,
\end{aligned} \tag{3.40}$$

and the first term, involving the first derivatives of the metric, vanishes at P leaving (in Riemannian coordinates)

$$\boxed{\bar{R}_{\alpha\sigma\nu\mu}|_P = \frac{1}{2}(\bar{g}_{\alpha\mu,\sigma\nu} - \bar{g}_{\mu\sigma,\alpha\nu} - \bar{g}_{\alpha\nu,\sigma\mu} + \bar{g}_{\nu\sigma,\alpha\mu})|_P.} \tag{3.41}$$

Now it is easy to see the symmetries of the Riemann tensor from the symmetries of the metric and its second derivative: $\bar{g}_{\alpha\beta} = \bar{g}_{\beta\alpha}$ and $\bar{g}_{\mu\nu,\alpha\beta} = \bar{g}_{\mu\nu,\beta\alpha}$. First, note that the Riemann tensor, in these coordinates, is antisymmetric in the interchange of its first two indices:

$$\bar{R}_{\sigma\alpha\nu\mu}|_P = \frac{1}{2}(\bar{g}_{\sigma\mu,\alpha\nu} - \bar{g}_{\alpha\mu,\sigma\nu} - \bar{g}_{\sigma\nu,\alpha\mu} + \bar{g}_{\alpha\nu,\sigma\mu})|_P = -\bar{R}_{\alpha\sigma\nu\mu}|_P. \tag{3.42}$$

Since we have a tensor statement $\bar{R}_{\sigma\alpha\nu\mu}|_P = -\bar{R}_{\alpha\sigma\nu\mu}|_P$, it is true in any coordinate system, so $R_{\sigma\alpha\nu\mu}|_P = -R_{\alpha\sigma\nu\mu}|_P$ for an arbitrary coordinate system. The point P can be anywhere, so we have established that $R_{\sigma\alpha\nu\mu} = -R_{\alpha\sigma\nu\mu}$ in general, for any coordinate system, at any point.

Using similar logic, you will show in Problem 3.5 that $R_{\alpha\sigma\mu\nu} = -R_{\alpha\sigma\nu\mu}$ (antisymmetry under the interchange of the last pair of indices), $R_{\nu\mu\alpha\sigma} = R_{\alpha\sigma\nu\mu}$ (symmetry in the interchange of the first and second pairs), and the cyclic identity

$$R_{\alpha\sigma\nu\mu} + R_{\alpha\mu\sigma\nu} + R_{\alpha\nu\mu\sigma} = 0. \tag{3.43}$$

This last identity in (3.43) is redundant with the other symmetries unless each of the four values, α, ν, μ, and σ is distinct. To see this, take $\nu = \alpha$, for example; then (with no sum in the equation below)

$$R_{\alpha\sigma\alpha\mu} + R_{\alpha\mu\sigma\alpha} + R_{\alpha\alpha\mu\sigma} = R_{\alpha\sigma\alpha\mu} - R_{\alpha\sigma\alpha\mu} = 0 \text{ no sum,} \tag{3.44}$$

where we used antisymmetry to get $R_{\alpha\alpha\mu\sigma} = 0$, pair interchange symmetry to write $R_{\alpha\mu\sigma\alpha} = R_{\sigma\alpha\alpha\mu}$, and antisymmetry in the first pair to swap $\sigma \leftrightarrow \alpha$ and pick up a minus sign. This example, with $\nu = \alpha$, already has the cyclic combination evaluate to zero without any additional constraint implied. The number of unique constraints present in (3.43) is the number of ways to distribute the dimension of the spacetime over the four indices, or

$$\binom{D}{4} = \frac{1}{24}D(D-1)(D-2)(D-3) \tag{3.45}$$

using binomial coefficients.

We can use these symmetries to count the independent elements of the Riemann tensor. Let's start by thinking about how many independent choices we have for the first two indices in $R_{\alpha\sigma\nu\mu}$. Since $R_{\sigma\alpha\nu\mu} = -R_{\alpha\sigma\nu\mu}$, we can pick α and σ to be any pair of the four coordinates that are not the same and sorted in increasing order, so that $\alpha = 0$, $\sigma = 1$ is "the same as" $\alpha = 1$, $\sigma = 0$, those elements are not independent, and we count them once. That yields $N = 1/2D(D-1)$ combinations or the same number of independent entries as a $D \times D$ antisymmetric matrix. We have the same number of independent entries for the second pair of indices by the same logic.

Now we can think of the pair interchange $R_{\alpha\sigma\nu\mu} = R_{\nu\mu\alpha\sigma}$ in blocks as $R_{AB} = R_{BA}$, where A and B each have N independent entries. The elements of R_{AB} can be thought of as living in an $N \times N$ symmetric matrix, and those have $1/2N(N+1)$ independent components. The total number of unique elements of the Riemann tensor is then $1/2N(N+1)$ minus the number in (3.45) from the cyclic constraint,

$$\boxed{T(D) \equiv \frac{1}{2}N(N+1) - \binom{D}{4} = \frac{1}{12}D^2(D^2-1).} \tag{3.46}$$

In four-dimensional spacetime, $T(4) = 20$, in $2 + 1$ dimensions, $T(3) = 6$, and in a $(1 + 1)$-dimensional spacetime, $T(2) = 1$, only one independent component. This pattern of dimensional degrees of freedom will constrain the dimension in which a metric theory of gravity can exist.

The full set of symmetries of the Riemann tensor developed in this section is

$$\boxed{\begin{aligned} R_{\mu\nu\alpha\beta} &= -R_{\nu\mu\alpha\beta}, & R_{\mu\nu\alpha\beta} &= -R_{\mu\nu\beta\alpha}, & R_{\mu\nu\alpha\beta} &= R_{\alpha\beta\mu\nu}, \\ 0 &= R_{\mu\nu\alpha\beta} + R_{\mu\beta\nu\alpha} + R_{\mu\alpha\beta\nu}. \end{aligned}} \tag{3.47}$$

3.3 Geodesic Deviation

The path dependence of parallel transport in a curved spacetime also has an effect on geodesics. For Minkowski spacetime, working in Cartesian coordinates, it is clear that a pair of geodesics emanating from the same point, take it to be the origin, have linearly increasing distance between them. We can describe a family of geodesics in the xy plane using a parameter σ that gives the

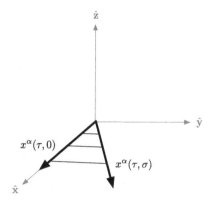

Two geodesics in Minkowski spacetime. The distance between them at equal times τ grows linearly in time according to (3.49).

angle that the geodesic makes with respect to the x axis,

$$x^\alpha(\tau, \sigma) \doteq \frac{1}{\sqrt{1 - v_0^2/c^2}} \begin{pmatrix} c\tau \\ v_0\tau \cos\sigma \\ v_0\tau \sin\sigma \\ 0 \end{pmatrix}. \tag{3.48}$$

For one geodesic at $\sigma = 0$ and another at an arbitrary $\sigma \neq 0$, the distance between the pair, as a function of τ, is

$$\sqrt{\left(x^\alpha(\tau, \sigma) - x^\alpha(\tau, 0)\right) g_{\alpha\beta} \left(x^\beta(\tau, \sigma) - x^\beta(\tau, 0)\right)} = \frac{v_0\tau\sqrt{2(1 - \cos\sigma)}}{\sqrt{1 - v_0^2/c^2}}, \tag{3.49}$$

which grows linearly in the proper time of either geodesic. You can see a sketch of the situation in Figure 3.2.

Contrast this growth behavior with two geodesics on a sphere of radius R. Just pictorially, if you start at the north pole of the sphere and move along the great circles that define the geodesics down to the south pole, then the distance between the curves grows initially, but after passing the equator, these curves get closer together as shown in Figure 3.3. A more quantitative expression can be developed by starting with the geodesic curves connecting the north and south poles,

$$x^\alpha(\tau, \sigma) = \frac{1}{\sqrt{1 - (\omega R/c)^2}} \begin{pmatrix} c\tau \\ \omega\tau \\ \sigma \end{pmatrix}, \tag{3.50}$$

in proper time parameterization τ with curve selection parameter $\sigma = \phi$ (the azimuthal angle). For this curve, ω is a constant with units of inverse time (so that $\omega\tau$ represents an angle).

For two different geodesics with two different values of σ, computing the distance between them requires us to integrate along a path connecting points on each curve. Rather than deal with that complication, we can compute the infinitesimal displacement between two nearby geodesics, one at $\sigma = 0$ and

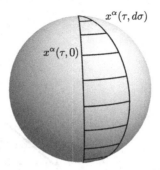

Fig. 3.3 Two geodesics run from the north to the south pole of a sphere. The separation distance between them grows initially but then shrinks.

one at $d\sigma \ll 1$. Then the infinitesimal displacement vector is

$$dx^\alpha(\tau) \equiv x^\alpha(\tau, d\sigma) - x^\alpha(\tau, 0) \doteq \frac{1}{\sqrt{1 - (\omega R/c)^2}} \begin{pmatrix} 0 \\ 0 \\ d\sigma \end{pmatrix}, \qquad (3.51)$$

and this vector is a constant. Its length is not, since the metric on the surface of the sphere depends on $\sin^2 \theta$ with $\theta = \omega\tau$,

$$dx = \sqrt{dx^\alpha(\tau) g_{\alpha\beta} dx^\beta(\tau)} = \frac{R d\sigma \sin(\omega\tau)}{\sqrt{1 - (\omega R/c)^2}}. \qquad (3.52)$$

As advertised in Figure 3.3, the distance between the geodesics starts at zero, grows, and then decreases.

With these two cases in mind, let's work out the deviation of nearby geodesics in a general setting. Suppose we have a family of geodesics $x^\alpha(\tau, \sigma)$ as above, with geodesic parameter τ, the proper time, and σ selecting the particular geodesic of interest from the family. For any fixed value of σ, the resulting curve is a geodesic with

$$\frac{D\dot{x}^\alpha}{D\tau} = \dot{x}^\alpha{}_{;\gamma} \dot{x}^\gamma = 0. \qquad (3.53)$$

The tangent to the geodesic $\dot{x}^\alpha$ points along the curve and has magnitude set by the proper time condition. But there is another derivative we can take here, $\frac{\partial x^\alpha}{\partial\sigma}$, pointing from the geodesic at σ to the one at $\sigma + d\sigma$, as shown in Figure 3.4. These two derivative "directions" are orthogonal (check that this is true for the two examples of this section in Problem 3.8).

Our goal is an expression for the evolution of the vector pointing from the geodesic at σ to the one at $\sigma + d\sigma$, call it s^α:

$$s^\alpha(\tau, \sigma) = \lim_{d\sigma \to 0} \left[x^\alpha(\tau, \sigma + d\sigma) - x^\alpha(\tau, \sigma) \right] = \frac{\partial x^\alpha(\tau, \sigma)}{\partial\sigma} d\sigma. \qquad (3.54)$$

Because s^α is a first-rank tensor, to study its τ-evolution, we will use the covariant $\frac{D}{D\tau}$ derivative,

$$\frac{Ds^\alpha}{D\tau} = s^\alpha{}_{;\rho} \frac{\partial x^\rho}{\partial\tau}. \qquad (3.55)$$

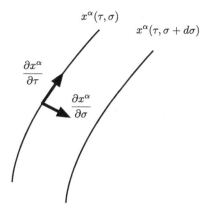

Fig. 3.4 Two nearby geodesics in a family of geodesics selected by σ. Each geodesic trajectory is in proper time parameterization, τ. The two derivatives form directions along and perpendicular to the geodesics.

Ultimately, for comparison with Newtonian gravity, we will need the relative acceleration $\frac{D^2 s^\alpha}{D\tau^2}$, and it turns out that this is related to the Riemann tensor, our topic for this chapter.

We'll start with a relation between the covariant τ and σ derivatives. Define the covariant σ derivative by analogy with the τ one. For contravariant tensor A^α,

$$\frac{DA^\alpha}{D\sigma} \equiv A^\alpha{}_{;\rho} \frac{\partial x^\rho}{\partial \sigma}. \tag{3.56}$$

Taking the τ derivative of $\frac{\partial x^\alpha}{\partial \sigma}$ and the σ derivative of $\frac{\partial x^\alpha}{\partial \tau}$,

$$\frac{D}{D\tau}\left(\frac{\partial x^\alpha}{\partial \sigma}\right) = \left(\frac{\partial x^\alpha}{\partial \sigma}\right)_{;\rho} \frac{\partial x^\rho}{\partial \tau} = \overbrace{\left(\frac{\partial x^\alpha}{\partial \sigma}\right)_{,\rho} \frac{\partial x^\rho}{\partial \tau}}^{=(\partial^2 x^\alpha)/(\partial\sigma\,\partial\tau)} + \Gamma^\alpha{}_{\gamma\rho} \frac{\partial x^\gamma}{\partial \sigma} \frac{\partial x^\rho}{\partial \tau},$$
$$\frac{D}{D\sigma}\left(\frac{\partial x^\alpha}{\partial \tau}\right) = \left(\frac{\partial x^\alpha}{\partial \tau}\right)_{;\gamma} \frac{\partial x^\gamma}{\partial \sigma} = \underbrace{\left(\frac{\partial x^\alpha}{\partial \tau}\right)_{,\gamma} \frac{\partial x^\gamma}{\partial \sigma}}_{=(\partial^2 x^\alpha)/(\partial\tau\,\partial\sigma)} + \Gamma^\alpha{}_{\rho\gamma} \frac{\partial x^\rho}{\partial \tau} \frac{\partial x^\gamma}{\partial \sigma}, \tag{3.57}$$

it is clear that these are equal: since $\Gamma^\alpha{}_{\rho\gamma} = \Gamma^\alpha{}_{\gamma\rho}$,

$$\frac{D}{D\tau}\left(\frac{\partial x^\alpha}{\partial \sigma}\right) = \frac{D}{D\sigma}\left(\frac{\partial x^\alpha}{\partial \tau}\right). \tag{3.58}$$

Back to our τ derivative of s^α, we can use the equality in (3.58) to rewrite it as the σ derivative of $\dot{x}^\alpha \equiv \frac{\partial x^\alpha}{\partial \tau}$,

$$\frac{Ds^\alpha}{D\tau} = d\sigma \frac{D}{D\tau}\left(\frac{\partial x^\alpha}{\partial \sigma}\right) = d\sigma \frac{D}{D\sigma}\left(\frac{\partial x^\alpha}{\partial \tau}\right), \tag{3.59}$$

and then taking the second derivative,

$$\frac{D^2 s^\alpha}{D\tau^2} = \left[\frac{D}{D\tau}\frac{D\dot{x}^\alpha}{D\sigma} - \frac{D}{D\sigma}\underbrace{\frac{D\dot{x}^\alpha}{D\tau}}_{=0}\right] d\sigma, \tag{3.60}$$

where we have added zero in the form $\frac{D\dot{x}^\alpha}{D\tau}d\sigma = 0$ by the geodesic assumption (3.53). What was the point of that zero? We now have the difference of two covariant derivatives, and we know how to handle that difference from (3.8). To see the relation emerge, let $V^\alpha \equiv \dot{x}^\alpha$. Then the two terms in (3.60) are

$$
\frac{D}{D\tau}\frac{DV^\alpha}{D\sigma} = \frac{D}{D\tau}\left(V^\alpha_{\;;\mu}\frac{\partial x^\mu}{\partial \sigma}\right) = V^\alpha_{\;;\mu\nu}\frac{\partial x^\mu}{\partial \sigma}\frac{\partial x^\nu}{\partial \tau} + V^\alpha_{\;;\mu}\frac{D}{D\tau}\left(\frac{\partial x^\mu}{\partial \sigma}\right),
$$
$$
\frac{D}{D\sigma}\frac{DV^\alpha}{D\tau} = \frac{D}{D\sigma}\left(V^\alpha_{\;;\nu}\frac{\partial x^\nu}{\partial \tau}\right) = V^\alpha_{\;;\nu\mu}\frac{\partial x^\nu}{\partial \tau}\frac{\partial x^\mu}{\partial \sigma} + V^\alpha_{\;;\nu}\frac{D}{D\sigma}\left(\frac{\partial x^\nu}{\partial \tau}\right).
$$
$$(3.61)$$

The second term in the final equality of each expression is the same using (3.58), so when we subtract, those will go away. Putting the derivatives into (3.60),

$$
\frac{D^2 s^\alpha}{D\tau^2} = \left(V^\alpha_{\;;\mu\nu} - V^\alpha_{\;;\nu\mu}\right)\underbrace{\frac{\partial x^\mu}{\partial \sigma}d\sigma}_{s^\mu}\frac{\partial x^\nu}{\partial \tau}, \tag{3.62}
$$

or, finally employing (3.8) and using $V^\alpha = \dot{x}^\alpha$,

$$
\boxed{\frac{D^2 s^\alpha}{D\tau^2} = R^\alpha_{\;\sigma\nu\mu}\dot{x}^\sigma \dot{x}^\nu s^\mu.} \tag{3.63}
$$

The acceleration of the deviation is the main result we need to develop Einstein's equation. After a short discussion of relativistic sources, we'll work out the acceleration of nearby gravitationally forced trajectories and reinterpret those trajectories as geodesics, giving us both the equation of motion and field equations of general relativity.

3.4 Gravitational Sources

The gravitational source for Einstein's field equation is a generalization of ρ, the mass density of Newtonian gravity. Back in Section 1.5, I suggested that we should really consider the energy density u as the source since mass and energy are equivalent. Then, in Section 1.6, we introduced an additional source $\mathbf{J} = \rho\mathbf{v} = u\mathbf{v}/c$, the "energy current density." Finally, in Section 1.7, we saw that the energy density u must be the zero component of a second-rank tensor object. In this section, we will build that second-rank tensor source, and it will come as no surprise that we'll use the four-vector current density, J^μ with $J^0 = \rho c$, and spatial components $\mathbf{J} = \rho\mathbf{v}$ (discussed in Section 1.7) from E&M as our model. Let's review that current density, first for a particle, then for a set of continuous sources.

3.4.1 Charge Sources

A point charge q sits at rest in a frame $\bar{L}$ that moves through a lab (another frame) L along a shared x axis with speed v; the lab setup is shown in Figure A.1. The charge density in the rest frame is $\bar{\rho} = q\delta(\bar{x})\delta(\bar{y})\delta(\bar{z})$ with $\bar{J}^x = 0$. The density and current in L is $\rho = q\delta(x-vt)\delta(y)\delta(z)$ with $J^x = \rho v$.

If these components indeed form a four-vector, J^μ with $J^0 = \rho c$ and $J^x = \rho v$, then the relation between the barred and unbarred coordinates should be $J^\mu = \tilde{\Lambda}^\mu_\alpha \bar{J}^\alpha$, where $\tilde{\Lambda}^\mu_\alpha$ is the inverse Lorentz transformation matrix (with $(+v/c)\gamma$ off-diagonal entries). The transformation gives us

$$\rho(x) = \gamma \bar{\rho}\big(\bar{x}(x)\big) = \gamma q\delta\big(\gamma(x-vt)\big)\delta(y)\delta(z) = q\delta(x-vt)\delta(y)\delta(z),$$
$$J^x(x) = \gamma v\bar{\rho}\big(\bar{x}(x)\big) = qv\delta(x-vt)\delta(y)\delta(z).$$
$$\text{(3.64)}$$

We get the "right" answer using the properties of the Dirac delta distribution: $\bar{\delta}^3(\bar{x}) = \frac{1}{\gamma}\delta^3(x)$, which comes from $\delta(kx) = \frac{1}{|k|}\delta(x)$. This relation between the charge and current densities in L and $\bar{L}$ establishes the four-vector character (under Lorentz boost) of J^μ; at least with respect to charges at rest, you can work out the result for a charge that is moving in $\bar{L}$ in Problem 3.10 to finish the job.

In four-dimensional spacetime, the elements of the current density for a point charge traveling along the trajectory $\mathbf{w}(t)$ are $J^0 = qc\delta^3(\mathbf{r} - \mathbf{w}(t))$ with $\mathbf{J} = J^0\mathbf{v}/c$ in L. For more general functions, combining the charge density ρ and velocity field $\mathbf{v}$ in

$$J^\mu \doteq \begin{pmatrix} \rho c \\ \rho \mathbf{v} \end{pmatrix} \qquad\qquad \text{(3.65)}$$

also gives a four-vector. Strictly speaking, we can appeal to superposition for the point source argument above, but we can also show that J^μ can be constructed from manifestly tensorial building blocks in the continuous case, which provides a nice model for the stress tensor source of gravity.

Our algorithm for constructing J^μ is to find ρ and $\mathbf{v}$ in a lab (either L or $\bar{L}$) and then generate ρc and $\rho \mathbf{v}$. Looking at the structure of (3.65), consider the "four-velocity" $\eta^\mu = \frac{dx^\mu}{d\tau}$ for the (infinitesimal) box containing the charge. In L, the four-velocity can be written as

$$\eta^\mu = \frac{dx^\mu}{d\tau} = \frac{dx^\mu}{dt}\underbrace{\frac{dt}{d\tau}}_{\equiv\gamma} \qquad\qquad \text{(3.66)}$$

with $\frac{dt}{d\tau} = \gamma$ from (A.30). At any point, there is a velocity vector $\mathbf{v}$ such that the charge at that point (really in an infinitesimal box surrounding the point) moves with velocity $\mathbf{v}$. Then in going from the charge rest frame to the frame in which the charge moves with $\mathbf{v}$ requires a boost in the $\hat{\mathbf{v}}$ direction, a one-dimensional Lorentz boost. That boost will contract one dimension of the volume surrounding the point, so we pick up one factor of γ: $\rho = \gamma\rho_0$. The relation between the spatial velocity components and the spatial proper velocity components is $\mathbf{v} = \gamma\boldsymbol{\eta}$, so the relation between the rest source density and the density in L is

$$\boxed{J^\mu = \rho_0\eta^\mu = \rho\frac{dx^\mu}{dt}.}$$

(3.67)

Since η^μ is a four-vector and ρ_0 a scalar, J^μ is a four-vector. The final equality in (3.67), involving ρ, which is not a scalar, and $\frac{dx^\mu}{dt}$, which is not a four-vector, is often useful in practice, since we typically know ρ and $\mathbf{v}$ in a particular frame L.[5]

3.4.2 Energy Sources

Returning to the sourcing relevant for gravity, we have an energy density u analogous to charge density ρ. In addition, we have a velocity field $\frac{dx^\mu}{dt}$ (not a tensor), and just as ρc is the zero component of J^μ, we expect u to be the zero–zero component of a second-rank tensor, call it $T^{\mu\nu}$. The natural update to the current density four-vector from E&M is then[6]

$$T^{\mu\nu} = \frac{u}{c^2}\frac{dx^\mu}{dt}\frac{dx^\nu}{dt}$$

(3.68)

and is called the "stress tensor." We have the same problem as before: Neither u nor $\frac{dx^\mu}{dt}$ is appropriately tensorial, u is not a scalar, and $\frac{dx^\mu}{dt}$ is not a vector, but the product in (3.68) is a tensor. Let's establish that, briefly, for a point particle at rest in $\bar{L}$ to see what is the same and what is different as compared with the charge case.

A massive point particle m is located at the origin of $\bar{L}$, with energy density $\bar{u} = mc^2\delta(\bar{x})\delta(\bar{y})\delta(\bar{z})$. From its definition in (3.68), $\bar{T}^{\mu\nu} = \bar{u}\frac{d\bar{x}^\mu}{dt}\frac{d\bar{x}^\nu}{dt}/c^2$, so that $\bar{T}^{00} = mc^2\delta(\bar{x})\delta(\bar{y})\delta(\bar{z})$ with all other components zero. In L, the energy density is $u = \gamma mc^2\delta(x - vt)\delta(y)\delta(z)$, and the nonzero components of the stress tensor from (3.68) are $T^{00} = mc^2\gamma\delta(x - vt)\delta(y)\delta(z)$, $T^{0x} = T^{00}v/c$, and $T^{xx} = T^{00}v^2/c^2$.

If $T^{\mu\nu}$ is a second-rank tensor, for a Lorentz boost in the x direction, we have $T^{\mu\nu} = \tilde{\Lambda}^\mu{}_\alpha\tilde{\Lambda}^\nu{}_\beta\bar{T}^{\alpha\beta}$, which yields the nonzero components

$$T^{00} = \gamma^2 mc^2\delta\big(\gamma(x - vt)\big)\delta(y)\delta(z) = \frac{mc^2}{\sqrt{1 - v^2/c^2}}\delta(x - vt)\delta(y)\delta(z),$$

$$T^{0x} = T^{00}v/c,$$

$$T^{xx} = T^{00}v^2/c^2,$$

(3.69)

---

[5]   Although neither ρ nor $\mathbf{v}$ behaves correctly as tensors of rank zero and one, respectively, their product does form the spatial piece of a four-vector as (3.67) shows.

[6]   We'll take our stress tensor, $T^{\mu\nu}$, to have entries with dimension of energy density, so we divide the $\frac{dx^\mu}{dt}$ by c.

matching what we got by applying (3.68) in both frames. We conclude that $T^{\mu\nu}$ is indeed a second-rank tensor. The generalization to functions u and $\frac{dx^\mu}{dt}$ that do not involve delta functions follows by superposition.

Just as in the charge source case, we can express a general $T^{\mu\nu}$ in terms of manifestly tensorial objects to establish its correct tensorial character. Call the energy density of a parcel of energy in its rest frame u_0; then an equivalent formulation of $T^{\mu\nu}$ is

$$T^{\mu\nu} = \frac{u_0}{c^2}\eta^\mu\eta^\nu = \frac{u}{c^2}\frac{dx^\mu}{dt}\frac{dx^\nu}{dt}. \qquad (3.70)$$

This equality follows from the tensorial nature of η^μ and the transformation character of u and $\frac{dx^\mu}{dt}$ in a manner similar to the argument that produced (3.67), and you should try your hand at it in Problem 3.11. When we want to highlight the tensorial nature of the stress tensor, we'll use the $u_0\eta^\mu\eta^\nu/c^2$ form. When we want to compute the stress tensor in a particular laboratory, for use in generating fields sourced by the stress tensor in that laboratory, we'll typically use the form $u\frac{dx^\mu}{dt}\frac{dx^\nu}{dt}/c^2$, as we do in E&M.[7] The stress tensor components are

$$T^{\mu\nu} \doteq \begin{pmatrix} u & uv^x/c & uv^y/c & uv^z/c \\ uv^x/c & u(v^x)^2/c^2 & uv^xv^y/c^2 & uv^xv^z/c^2 \\ uv^y/c & uv^yv^x/c^2 & u(v^y)^2/c^2 & uv^yv^z/c^2 \\ uv^z/c & uv^zv^x/c^2 & uv^zv^y/c^2 & u(v^z)^2/c^2 \end{pmatrix}. \qquad (3.71)$$

The conservation of the stress tensor takes the form $\partial_\mu T^{\mu\nu} = 0$, a set of four equations (one for each value of ν). Their physical content is familiar: working in a single lab frame, the equation $\nu = 0$ reads

$$\frac{1}{c}\frac{\partial T^{00}}{\partial t} + \frac{\partial T^{j0}}{\partial x^j} = 0 \rightarrow \frac{\partial(u/c)}{\partial t} + \nabla \cdot (u\mathbf{v}/c) = 0, \qquad (3.72)$$

which looks like the mass conservation we had in (1.122) if we set $u = \rho c^2$, and in this more general setting represents conservation of energy. The spatial $\nu = i = 1, 2, 3$ components are

$$\frac{1}{c}\frac{\partial T^{0i}}{\partial t} + \frac{\partial T^{ji}}{\partial x^j} = 0 \rightarrow \frac{\partial(uv^i/c^2)}{\partial t} + \frac{\partial}{\partial x^j}\left(uv^jv^i/c^2\right) = 0, \qquad (3.73)$$

which represents conservation of uv^i, which we identify as a momentum density based on the case of a point mass.

Our tensorial source is in place. We have a stress tensor $T^{\mu\nu}$ that encapsulates all forms of energy, and whose local conservation statement enforces energy and momentum conservation. We are now ready to relate the motion of nearby trajectories in Newtonian gravity to the geometric geodesic deviation, the process by which Einstein wrote down his field equation for general

---

[7]   In Maxwell's equations, the sources ρ and $\mathbf{J}$ are expressed in the lab frame that is the target for the calculation of $\mathbf{E}$ and $\mathbf{B}$, that is, you evaluate ρ and $\mathbf{J}$ where you want to know $\mathbf{E}$ and $\mathbf{B}$.

relativity. It's worth reading the original papers in which Einstein lays out the ideas that lead to the field equations of general relativity – they can be found in the collection [13].

3.5 Newtonian Deviation

Near the surface of the earth, gravity produces a uniform acceleration felt by all objects equally. If you and a friend were falling, then there would be no relative motion (locally). You could throw a ball back and forth, everything falling together with the same acceleration. It is this democracy that led Einstein to imagine that gravity is not really a force, since most of the familiar forces have different effects on different objects (neutral particles do not experience a force in an electric field, charged particles do, and the magnitude of charge effects the magnitude of the resulting acceleration). Instead, he thought of gravity as a feature of the arena in which motion occurs. So his target was a geometric theory of gravity.

Where should we look to apply the geometric notion? Locally, everything is falling together, so there isn't much we can do to detect the gravitational field. Indeed, you can eliminate the gravitational field at any point without significantly altering other physical predictions as we saw in Section 3.2 (eliminating the gravitational field amounts to using a freely falling reference frame with its Minkowski metric). Think of the larger structure of the gravitational field of the Earth. Everything falls with the same acceleration still, but the field lines point towards the center of the Earth, so that if you and a friend were to fall for a very long time, then you would notice yourselves getting closer together as you fall towards a common center (the tidal forces from Section 1.3.1 depicted in Figure 1.6 demonstrate this rapprochement). To detect gravity, then, you should look over large ranges and consider the relative motion of nearby falling bodies.

Not to let the cat out of the bag, but as a sketch of what will happen: We can describe the relative motion of masses that fall along nearby trajectories due to a nonuniform gravitational field, and we will take that description and demand that it be interpretable as the deviation of nearby geodesics governed by a metric. The metric will become the field of interest, particles will travel along geodesics, that's how gravity effects all particles equally (no mention of mass in the geodesic equation of motion), and the deviation of geodesics in this curved spacetime will come from (3.63). It is the direct comparison of Newtonian trajectory deviation with geodesic deviation that leads to Einstein's field equation.

We want an equation of motion for the separation of nearby falling bodies. Referring to Figure 3.5, mass m_1 moves along $\mathbf{r}_1(t)$, and mass m_2 moves along $\mathbf{r}_2(t)$. Let $\mathbf{s}(t) \equiv \mathbf{r}_2(t) - \mathbf{r}_1(t)$ be the (small) distance vector pointing from the

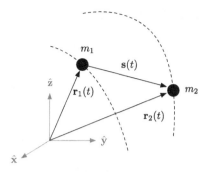

Fig. 3.5 Two masses travel along trajectories $\mathbf{r}_1(t)$ and $\mathbf{r}_2(t)$ moving under the influence of a gravitational field. At time t, the separation between the masses is given by $\mathbf{s}(t) \equiv \mathbf{r}_2(t) - \mathbf{r}_1(t)$.

first particle to the second. For a gravitational potential $\varphi(\mathbf{r})$, the equation of motion for each particle is

$$\ddot{\mathbf{r}}_1(t) = -\nabla\varphi\big(\mathbf{r}_1(t)\big), \qquad \ddot{\mathbf{r}}_2(t) = -\nabla\varphi\big(\mathbf{r}_2(t)\big), \tag{3.74}$$

and subtracting $\ddot{\mathbf{r}}_1(t)$ from $\ddot{\mathbf{r}}_2(t)$ gives us the second derivative of $\mathbf{s}(t)$:

$$\ddot{\mathbf{s}}(t) = -\nabla\varphi\big(\mathbf{r}_2(t)\big) + \nabla\varphi\big(\mathbf{r}_1(t)\big) = -\nabla\varphi\big(\mathbf{r}_1(t) + \mathbf{s}(t)\big) + \nabla\varphi\big(\mathbf{r}_1(t)\big). \tag{3.75}$$

Assuming that $\mathbf{s}(t)$ is small, we can Taylor expand the force about $\mathbf{r}_1(t)$ (you could pick either $\mathbf{r}_2(t)$ or $\mathbf{r}_1(t)$, of course) and let $\mathbf{F} = -\nabla\varphi$. Then

$$\mathbf{F}(\mathbf{r}_1 + \mathbf{s}) \approx \mathbf{F}(\mathbf{r}_1) + (\mathbf{s} \cdot \nabla)\mathbf{F}(\mathbf{r}_1) = -\nabla\varphi(\mathbf{r}_1) - \mathbf{s} \cdot \nabla\big(\nabla\varphi(\mathbf{r}_1)\big). \tag{3.76}$$

Putting this into the equation of motion governing $\mathbf{s}(t)$ in (3.75),

$$\ddot{\mathbf{s}}(t) \approx -\mathbf{s}(t) \cdot \nabla\big(\nabla\varphi\big(\mathbf{r}_1(t)\big)\big). \tag{3.77}$$

Using indices to clear up the order of operations and put the equation into a form comparable to (3.63),[8]

$$\frac{d^2 s^\alpha}{dt^2} = -s^\mu \partial_\mu \partial^\alpha \varphi. \tag{3.78}$$

The gravitational potential itself satisfies $\nabla^2\varphi = 4\pi Gu/c^2$ (refer to (1.83)) given an energy density u (which could involve mass), and we can write this using indices,

$$\partial^\mu \partial_\mu \varphi = \frac{4\pi G}{c^2} u. \tag{3.79}$$

We will now equate the Newtonian gravitational deviation with geodesic deviation by squinting: $\frac{d^2 s^\alpha}{dt^2} \sim \frac{D^2 s^\alpha}{D\tau^2}$. The differences aren't that extreme, the geodesic case is four-dimensional and involves proper time, whereas the gravitational case is necessarily three dimensional and uses t as a parameter, but

[8] Strictly speaking, we only have $\alpha = 1, 2$, and 3 in (3.78), since Newton's second law deals with the spatial trajectory, but we'll need to compare, ultimately, with four-vectors, so a little predictive expansion, to include $\alpha = 0$, is warranted.

t and τ coincide in the nonrelativistic limit, so the identification is natural. Equating the right-hand sides of (3.78) and (3.63) sets up

$$R^{\alpha}{}_{\sigma\nu\mu}\dot{x}^{\sigma}\dot{x}^{\nu}s^{\mu} = -s^{\mu}\partial_{\mu}\partial^{\alpha}\varphi, \qquad (3.80)$$

and then removing the common s^{μ} from both sides, lowering the α index, and using the antisymmetric interchange of the Riemann tensor to get rid of the minus sign,

$$R_{\alpha\sigma\mu\nu}\dot{x}^{\sigma}\dot{x}^{\nu} = \partial_{\mu}\partial_{\alpha}\varphi. \qquad (3.81)$$

We can connect this relation to the source by multiplying by $g^{\mu\alpha}$ to get the Laplacian on the right and giving the Ricci tensor $R_{\sigma\nu} = R^{\alpha}{}_{\sigma\alpha\nu}$ on the left,

$$R_{\sigma\nu}\dot{x}^{\sigma}\dot{x}^{\nu} = \frac{4\pi G}{c^2}u. \qquad (3.82)$$

Finally, we need an expression for the energy density u in terms of the stress tensor. We established in Section 1.7 that the energy density u is the 00 component of a second-rank tensor $T^{\mu\nu}$ on the basis of its transformation properties (transforming with two factors of the boost parameter γ instead of the one typical for the zero component of a first-rank tensor). We saw in Example 2.6 that we can use the tangent to the particle trajectory, written in proper time parameterization $\dot{x}^{\mu}(\tau)$, to find the 0 component of a first-rank tensor V_{μ}: $\dot{x}^{\mu}(\tau)V_{\mu}/c$ is a scalar that evaluates to V_0 in the rest frame of the particle moving along the curve with tangent $\dot{x}^{\mu}(\tau)$. We want to express u similarly as the scalar that is the 00 component of the tensor $T_{\mu\nu}$ in the (instantaneous) rest frame of the particle, so take $u = \dot{x}^{\sigma}\dot{x}^{\nu}T_{\sigma\nu}/c^2$. Using this value in (3.82),

$$R_{\sigma\nu}\dot{x}^{\sigma}\dot{x}^{\nu} = \frac{4\pi G}{c^4}T_{\sigma\nu}\dot{x}^{\sigma}\dot{x}^{\nu}, \qquad (3.83)$$

suggesting

$$R_{\sigma\nu} = \frac{4\pi G}{c^4}T_{\sigma\nu}. \qquad (3.84)$$

Now we have a general, tensorial expression, relating $R_{\sigma\nu}$, a nonlinear function of the metric, and its first and second derivatives to the general source of mass and energy distribution $T_{\sigma\nu}$: $R_{\sigma\nu} = (4\pi G/c^4)T_{\sigma\nu}$. We also have a story to tell about the motion of particles, since we obtained the expression in (3.83) by assuming that particles move along geodesics (in using (3.63)) with nearby particles diverging or converging by geodesic deviation, and that geodesic motion was designed to replace the forced trajectories of Newtonian gravity. So we have both a field equation, like Maxwell's equations, a PDE relating the sources to the field, and we have the analogue of the Lorentz force together with Newton's second law, the geodesic equation of motion.

We have realized the promise from Section 1.7, a theory of gravity that consists of a second-rank tensor field $g_{\mu\nu}$, together with a nonlinear (no superposition) field equation, and we have also made contact with the prediction that, based on the coupling of any second-rank tensor field with particles from Section 2.8.3, the resulting theory will have a purely geometric, in particular geodesic, effect on particles. Yet, there remains a problem.

The issue is similar in spirit to Maxwell's correction to Ampere's law from E&M. Prior to the correction, the equation reads

$$\nabla \times \mathbf{B} = \mu_0 \mathbf{J}, \tag{3.85}$$

and the problem was that if you take the divergence of both sides, the left side vanishes because the divergence of a curl is automatically zero, leaving us with $\nabla \cdot \mathbf{J} = 0$, which does not support charge conservation

$$\partial_\mu J^\mu = 0 = -\frac{\partial \rho}{\partial t} + \nabla \cdot \mathbf{J}. \tag{3.86}$$

Our current proposed field equation in (3.84) has the opposite problem. The statement of energy conservation (and more) is $D^\sigma T_{\sigma\nu} = 0$ (using the covariant derivative here to highlight the tensor statement); we assume that conservation of energy holds, but computing the covariant derivative of the Ricci tensor gives $D^\sigma R_{\sigma\nu} = 1/2 R_{;\nu}$ from Problem 3.12. We can easily fix this mismatch by subtracting off the offending portion to get a tensor on the left that is divergenceless. Doing that and taking $4\pi \rightarrow 8\pi$, an irrelevant bookkeeping factor (it is easiest to track down this factor later on after we have seen the "weak field" limit of the equation below, which you will do in Problem 3.20, or see [22], for example), we arrive at Einstein's equation

$$\boxed{R_{\mu\nu} - \frac{1}{2} g_{\mu\nu} R = \frac{8\pi G}{c^4} T_{\mu\nu}.} \tag{3.87}$$

The tensor combination on the left is known as the "Einstein tensor" $G_{\mu\nu} \equiv R_{\mu\nu} - \frac{1}{2} g_{\mu\nu} R$.

3.6 Einstein's Equation

General relativity is the combination of the field equation (3.87) that allows us to find the gravitational field $g_{\mu\nu}$ from its energetic source $T_{\mu\nu}$, and the geodesic equation of motion (in proper time parameterization, $\dot{x}^\mu \dot{x}^\nu g_{\mu\nu} = -c^2$),

$$g_{\alpha\nu} \ddot{x}^\nu + \frac{1}{2} \left(\frac{\partial g_{\alpha\nu}}{\partial x^\beta} + \frac{\partial g_{\alpha\beta}}{\partial x^\nu} - \frac{\partial g_{\beta\nu}}{\partial x^\alpha} \right) \dot{x}^\beta \dot{x}^\nu = 0, \tag{3.88}$$

telling us how free particles move in a spacetime with metric $g_{\mu\nu}$. The pair plays the role of Maxwell's equation and Newton's second law with the Lorentz force in E&M. The spacetime will not necessarily be flat. A flat spacetime has simple, dull, geodesics that are straight lines. To see interesting gravitational effects, we want the geodesics to be curves, and you cannot get those from a flat spacetime.

We have one solution to the field equations right off the bat. For an empty universe, with no energy sources anywhere, $T_{\mu\nu} = 0$, and the associated spacetime is just Minkowski, with metric that can be written in Cartesian coordinates as in (2.53). In that form, it is clear that the Riemann tensor, Ricci tensor, and Ricci scalar vanish, so Einstein's equation reads $0 = 0$, and the geodesics are straight lines. What if we solve Einstein's equation and end up with an unfamiliar metric – does that metric describe Minkowski spacetime? The Riemann tensor will tell us whether the metric is Minkowski written in an unfamiliar coordinate system.

The Minkowski metric is the unique solution for no matter or energy at all. We will be interested in solving Einstein's equation "in vacuum," meaning localized regions in which $T_{\mu\nu} = 0$, but in general, we will not mean that $T_{\mu\nu} = 0$ everywhere. In regions where $T_{\mu\nu} = 0$, Einstein's equation simplifies considerably. From

$$R_{\mu\nu} - \frac{1}{2}g_{\mu\nu}R = 0 \tag{3.89}$$

we can "take the trace" of the equation by multiplying by $g^{\mu\nu}$ (try it, but do Problem 3.14 first), and this results in $-R = 0$. Using that in (3.89), we get the equivalent vacuum field equation

$$R_{\mu\nu} = 0. \tag{3.90}$$

We say that solutions of this equation are "Ricci-flat" since they have vanishing Ricci tensor and, by extension, zero Ricci scalar. Does that mean that the spacetime is flat? It had better not, or we have no interesting gravity in vacuum. Flatness, remember, is determined by the Riemann tensor, and it is possible to have a nonzero Riemann tensor whose index contractions (Ricci tensor and scalar) are zero.

3.6.1 Weak Field Approximation

Before we study the full, nonlinear, Einstein equation, let's see what the "weak field" limit looks like. That linearized limit will allow us to see clearly how Newtonian gravity emerges, and also provides a point of contrast with the GEM structure we developed as a first, toy, theory in Chapter 1 (completed in Section 1.6).

Define a Cartesian coordinate system with Minkowski metric $\eta_{\mu\nu}$ that is (2.53). For a symmetric second-rank tensor field $h_{\mu\nu}$, form the metric $g_{\mu\nu} = \eta_{\mu\nu} + h_{\mu\nu}$. We'll assume that $h_{\mu\nu}$ is small, so that we keep only first-order terms in "the field" $h_{\mu\nu}$. The Christoffel symbol of the first kind is

$$\Gamma_{\alpha\mu\nu} = \frac{1}{2}(h_{\alpha\mu,\nu} + h_{\alpha\nu,\mu} - h_{\mu\nu,\alpha}) \tag{3.91}$$

with the symbol of the second kind given by

$$\Gamma^{\rho}{}_{\mu\nu} = \frac{1}{2}g^{\rho\alpha}(h_{\alpha\mu,\nu} + h_{\alpha\nu,\mu} - h_{\mu\nu,\alpha}) = \frac{1}{2}\eta^{\rho\alpha}(h_{\alpha\mu,\nu} + h_{\alpha\nu,\mu} - h_{\mu\nu,\alpha}), \tag{3.92}$$

where the second equality comes from dropping terms of order h^2 (see Problem 3.16 for motivation for the upper form of the metric). The Riemann tensor has only connection derivative terms, since the $\Gamma\Gamma$ pieces will be, again, of order h^2. So the Riemann tensor takes the same form

$$R_{\sigma\alpha\nu\mu} = \frac{1}{2}(h_{\sigma\mu,\alpha\nu} - h_{\alpha\mu,\sigma\nu} - h_{\sigma\nu,\alpha\mu} + h_{\alpha\nu,\sigma\mu}) \tag{3.93}$$

with vanishing derivatives of $\eta_{\alpha\beta}$.

To construct the Ricci tensor, we multiply the Riemann tensor by $g^{\sigma\nu} \approx \eta^{\sigma\nu} + O(h)$ and drop the terms that go like h^2,

$$R_{\alpha\mu} = \frac{1}{2}\left(h^\nu{}_{\mu,\alpha\nu} - \Box h_{\alpha\mu} - h_{,\alpha\mu} + h^\nu{}_{\alpha,\mu\nu}\right). \tag{3.94}$$

In these expressions, indices are raised using the Minkowski $\eta^{\mu\nu}$, which itself has zero derivative. Although the expressions are not written in general tensorial form (one cannot raise and lower indices on nontensorial objects), they are well-defined on the flat background in which they reside. To be concrete, we are setting

$$h \equiv h_{\mu\nu}\eta^{\mu\nu}, \qquad \Box \equiv \eta^{\mu\nu}\partial_\mu\partial_\nu, \qquad h^\mu{}_\nu \equiv \eta^{\mu\alpha}h_{\alpha\nu} \tag{3.95}$$

with commas representing derivatives as always.

The Ricci scalar comes from multiplying the Ricci tensor by $g^{\alpha\mu} \approx \eta^{\alpha\mu}$ and using $\eta^{\alpha\mu}$ to raise indices on $h_{\alpha\beta}$,

$$R = h^{\alpha\mu}{}_{,\alpha\mu} - \Box h. \tag{3.96}$$

Einstein's equation takes its "weak field" form, using ∂_μ instead of commas to make the comparison with (2.189) clear,

$$\frac{1}{2}\left[\partial_\alpha\partial_\nu h^\nu{}_\mu + \partial_\mu\partial_\nu h^\nu{}_\alpha - \Box h_{\alpha\mu} - \partial_\alpha\partial_\mu h - \eta_{\mu\alpha}\left(\partial_\gamma\partial_\nu h^{\gamma\nu} - \Box h\right)\right] = \frac{8\pi G}{c^4}T_{\alpha\mu}. \tag{3.97}$$

In making that comparison, note that any D_α acting on the field $h_{\mu\nu}$ and its contractions here becomes ∂_α, since, schematically,

$$D_\alpha h_{\mu\nu} = \partial_\alpha h_{\mu\nu} - \Gamma^\rho{}_{\alpha\mu}h_{\rho\nu} - \Gamma^\rho{}_{\alpha\nu}h_{\mu\rho} = \partial_\alpha h_{\mu\nu} + O\left(h^2\right). \tag{3.98}$$

Equation (3.97) is complicated, but we can simplify it by adopting a particular gauge for $h_{\mu\nu}$. As we found in Section 2.8.2, we can take $h_{\mu\nu} \to h_{\mu\nu} + (v_{\mu,\nu} + v_{\nu,\mu}) + \psi_{,\mu\nu}$ without changing the field equation and use that freedom to set the divergence of $h_{\mu\nu}$ to a convenient function. This is a recapitulation of the gauge fixing we did in Section 2.8, but it is worth doing it again for emphasis. For $h_{\mu\nu}$ with nonzero divergence $\partial^\mu h_{\mu\nu} = P_\nu$, take $\tilde{h}_{\mu\nu} = h_{\mu\nu} + (v_{\mu,\nu} + v_{\nu,\mu}) + \psi_{,\mu\nu}$. The divergence of this new field is

$$\partial^\mu\tilde{h}_{\mu\nu} = P_\nu + \partial^\mu v_{\mu,\nu} + \Box v_\nu + \Box\psi_{,\nu}. \tag{3.99}$$

Suppose our target divergence for $\tilde{h}_{\mu\nu}$ is F_ν: we want $\partial^\mu\tilde{h}_{\mu\nu} = F_\nu$ (which could be zero). Then pick ψ such that $\Box\psi = \partial^\mu v_\mu$ and $\Box v_\nu = -P_\nu + F_\nu$ by solving the Poisson problem four times, and then $\partial^\mu\tilde{h}_{\mu\nu} = F_\nu$. There are a variety of gauges of interest. The "Lorenz gauge" analogue from E&M looks like $\partial^\mu \tilde{h}_{\mu\nu} = 0$. We will use a different choice to make the field equation work out nicely; set

$$\partial^\mu \tilde{h}_{\mu\nu} = \frac{1}{2}\partial_\nu\left(\tilde{h}^{\alpha\beta}\eta_{\alpha\beta}\right). \tag{3.100}$$

Assume that we have achieved this gauge from the start so that $\partial^\mu h_{\mu\nu} = \partial_\nu h/2$ with $h \equiv h^{\alpha\beta}\eta_{\alpha\beta}$. This choice is called the "DeDonder," or "harmonic," gauge. For the "trace reversed" field, $H_{\mu\nu} \equiv h_{\mu\nu} - \eta_{\mu\nu}h/2$, it reads $\partial^\mu H_{\mu\nu} = 0$, and for this reason, the gauge is sometimes even called the "Lorenz" gauge (although we'll leave the field in the form that most naturally connects it to an effective metric, $h_{\mu\nu}$). Returning to the field equation (3.97), we are left with

$$\Box\left(h_{\alpha\mu} - \frac{1}{2}\eta_{\mu\alpha}h\right) = -\frac{16\pi G}{c^4}T_{\alpha\mu}. \tag{3.101}$$

There are three forms of source to consider from (3.71), shown schematically in block form (and lowering indices using $\eta_{\mu\nu}$):

$$T^{\mu\nu} \doteq \begin{pmatrix} u & \frac{uv^i}{c} \\ \frac{uv^i}{c} & \frac{uv^iv^j}{c^2} \end{pmatrix} \longrightarrow T_{\mu\nu} \doteq \begin{pmatrix} u & -\frac{uv^i}{c} \\ -\frac{uv^i}{c} & \frac{uv^iv^j}{c^2} \end{pmatrix}. \tag{3.102}$$

For a "weak" source, with source speeds that are small compared to the speed of light, the relevant pieces are u and $u\mathbf{v}/c$, just the energy and moving energy (momentum) pieces. The 3×3 block in the bottom right is of order $(v/c)^2$, which will be small for $v \ll c$. So we'll take u and $u\mathbf{v}/c$ to be the nonzero sources and approximate the T^{ij} piece as zero. We will further assume that the sources are static with $\frac{\partial}{\partial t}T^{\mu\nu} \approx 0$, giving fields that are also static, $\frac{\partial h_{\mu\nu}}{\partial t} = 0$, so we'll take $\Box \to \nabla^2$ in (3.101).

Returning to the left-hand side of the field equation, since $T_{ij} = 0$, the spatial $i, j = 1 \to 3$ pieces are

$$\nabla^2\left(h_{ij} - \frac{1}{2}\eta_{ij}h\right) = 0. \tag{3.103}$$

This equation has no source anywhere, and if we require the field $h_{\mu\nu}$ vanish at spatial infinity, then the solution to Laplace's equation is zero everywhere, so that

$$h_{ij} - \frac{1}{2}\eta_{ij}h = 0. \tag{3.104}$$

Multiplying this equation by η^{ij}, writing $h = h^0{}_0 + h^i{}_i$, and using $h_{00} = -h^0{}_0$ gives

$$h^i{}_i - \frac{3}{2}\left(h^0{}_0 + h^i{}_i\right) = 0 \longrightarrow h^i{}_i = 3h_{00} \tag{3.105}$$

with $h = 2h_{00}$.

Moving on to the sourced equations, the $\alpha = \mu = 0$ field equation reads

$$\nabla^2 h_{00} = -\frac{8\pi G}{c^4}u. \tag{3.106}$$

The $\alpha = 0$, $\mu = i$ (spatial) equation is

$$\nabla^2 h_{0i} = -\frac{16\pi G}{c^4} T_{0i} = \frac{16\pi G}{c^5} u v_i \qquad (3.107)$$

(numerically, $v^i = v_i$ since these are spatial indices raised and lowered with $\eta_{\mu\nu}$ and its identical inverse). Along with (3.106) and (3.107), we need to record the gauge condition

$$\partial_\mu h^{\mu\nu} = \frac{1}{2}\partial^\nu h = \partial^\nu h^{00} \qquad (3.108)$$

and the spatial solution

$$h_{ij} = \eta_{ij} h_{00}. \qquad (3.109)$$

The equations in (3.106) and (3.107) are more recognizable by letting $h^i \equiv h^{0i} = -h_{0i}$ (we're using the upper form since those are the ones that come from the components of vectors, so that the elements of $\mathbf{h}$ are really h^i) and expressing the energy density as a mass density using $u = \rho c^2$; then

$$\nabla^2\left(-\frac{h_{00}c^2}{2}\right) = 4\pi G\rho, \qquad \nabla^2\left(\frac{-\mathbf{h}c}{4}\right) = \frac{4\pi G}{c^2}\rho\mathbf{v}. \qquad (3.110)$$

These are the static Newtonian and gravitomagnetic equations we had way back at (1.134) with $\mathbf{g} = -\nabla\varphi$ and $\mathbf{b} = \nabla \times \mathbf{a}$ there. The effective gravitational potential coming from the metric perturbation is $\varphi = -h_{00}c^2/2$, with gravitomagnetic vector potential $\mathbf{a} = -\mathbf{h}c/4$ (and $\nabla \cdot \mathbf{a} = 0$).

The geodesic equation of motion, from (3.88), is

$$\eta_{\alpha\nu}\ddot{x}^\nu + \frac{1}{2}\left(\frac{\partial h_{\alpha\nu}}{\partial x^\beta} + \frac{\partial h_{\alpha\beta}}{\partial x^\nu} - \frac{\partial h_{\beta\nu}}{\partial x^\alpha}\right)\dot{x}^\beta \dot{x}^\nu = 0. \qquad (3.111)$$

For $h_{0i} = 0$ but nonzero h_{00} (and, of course, nonzero h_{ij} according to (3.109)), the spatial components of the equation of motion, indexed by $j = 1, 2, 3$, are

$$\ddot{x}_j(\tau) + \frac{1}{2}\left(-\frac{\partial h_{00}}{\partial x^j}\dot{x}^0\dot{x}^0 + 2\frac{\partial h_{00}}{\partial x^k}\dot{x}^k\dot{x}_j - \dot{x}^k\dot{x}_k\frac{\partial h_{00}}{\partial x^j}\right) = 0 \qquad (3.112)$$

or, using $\varphi = -h_{00}c^2/2$ and $\dot{x}^0 = c$ (in the nonrelativistic limit, where $\tau \approx t$),

$$\ddot{x}_j(t) + \frac{\partial\varphi}{\partial x^j} - 2\frac{\dot{x}^k\dot{x}^j}{c^2}\frac{\partial\varphi}{\partial x^k} + \frac{\dot{x}^k\dot{x}_k}{c^2}\frac{\partial\varphi}{\partial x^j} = 0. \qquad (3.113)$$

In vector form, the spatial acceleration is

$$\ddot{\mathbf{x}}(t) = -\nabla\varphi - \frac{v^2}{c^2}\nabla\varphi + \frac{2}{c^2}\mathbf{v}(\mathbf{v} \cdot \nabla\varphi). \qquad (3.114)$$

We dropped source terms of size v^2/c^2 from the stress tensor, so that if we similarly ignore terms of order v^2/c^2 on the motion side, then we recover $\ddot{\mathbf{x}}(t) = -\nabla\varphi$, which we recognize as the equation of motion for Newtonian gravity with potential φ. You can see what happens in the other case, with $h_{00} = 0$ and nonzero h_{0i} in Problem 3.18.

The weak field equations and nonrelativistic limit of the geodesic equation of motion provide a compelling way to set the proportionality constant appear-

ing on the right-hand side of Einstein's equation; by direct comparison with
the results of Newtonian gravity, try it out in Problem 3.20.

Example 3.2 (Spherically Symmetric Source). For a sphere of radius R with
total mass M, the gravitational potential for $r > R$ is $\varphi = -GM/r$. The
metric perturbation has Cartesian components $h_{00} = -2\varphi/c^2 = 2GM/(rc^2)$
and, from (3.109), $h_{11} = h_{22} = h_{33} = h_{00}$. All other components are zero,
and the metric, with perturbation in place, has the line element

$$ds^2 = -\left(1 - \frac{2GM}{rc^2}\right)c^2dt^2 + \left(1 + \frac{2GM}{rc^2}\right)\left(dx^2 + dy^2 + dz^2\right). \quad (3.115)$$

We can transform from Cartesian coordinates to spherical, and doing that gives
the line element

$$ds^2 = -\left(1 - \frac{2GM}{rc^2}\right)c^2dt^2 + \left(1 + \frac{2GM}{rc^2}\right)\left(dr^2 + r^2d\theta^2 + r^2\sin^2\theta d\phi^2\right). \quad (3.116)$$

In this form, we can pick out the components of the field expressed in spherical
coordinates,

$$h_{\mu\nu} \doteq \begin{pmatrix} \frac{2GM}{c^2r} & 0 & 0 & 0 \\ 0 & \frac{2GM}{c^2r} & 0 & 0 \\ 0 & 0 & \frac{2GMr}{c^2} & 0 \\ 0 & 0 & 0 & \frac{2GMr}{c^2}\sin^2\theta \end{pmatrix}. \quad (3.117)$$

In a more general spherical setting, we have $\varphi(r)$ as the Newtonian potential,
which can be found by solving $\nabla^2\varphi = 4\pi Gu/c^2$. Then the perturbation is just
(writing it in upper form)

$$h^{\mu\nu} \doteq \begin{pmatrix} -2\varphi/c^2 & 0 & 0 & 0 \\ 0 & -2\varphi/c^2 & 0 & 0 \\ 0 & 0 & -\frac{2\varphi/c^2}{r^2} & 0 \\ 0 & 0 & 0 & -\frac{2\varphi/c^2}{r^2\sin^2\theta} \end{pmatrix}. \quad (3.118)$$

Try using a different spherical source in Problem 3.24.

Remember that if the metric $g_{\mu\nu} = \eta_{\mu\nu} + h_{\mu\nu}$ has vanishing Riemann tensor,
then it is "flat" and therefore has geodesics that are straight lines. If we want
nontrivial motion to occur under the influence of this metric, then we must
have nonzero Riemann tensor. Computing the Riemann tensor for $\eta_{\mu\nu} + h_{\mu\nu}$
in spherical coordinates, we can easily verify that it is nonzero: the spacetime
is not flat.

Example 3.3 (A Slowly Rotating Sphere). If we take our sphere from the
previous example and set it spinning about the $\hat{\mathbf{z}}$ axis with constant angular
speed ω, then we pick up a term in $h_{\mu\nu}$ from the gravitomagnetic source in
(3.110). To solve

$$\nabla^2\mathbf{a} = \frac{4\pi G}{c^2}\rho\mathbf{v}, \quad (3.119)$$

think of the magnetic vector potential for a spinning sphere of charge. The equation the vector potential must solve is

$$\nabla^2 \mathbf{A} = -\mu_0 \mathbf{J}. \tag{3.120}$$

Outside the sphere, the magnetic field is a dipole with vector potential

$$\mathbf{A} = \frac{\mu_0 Q \omega R^2 \sin\theta}{20\pi r^2} \hat{\boldsymbol{\phi}}. \tag{3.121}$$

We can take $Q \to M$ and $\mu_0 \to -4\pi G/c^2$ to find the dipole solution to (3.119),

$$\mathbf{a} = -\frac{G}{c^2} \frac{M\omega R^2 \sin\theta}{5r^2} \hat{\boldsymbol{\phi}}. \tag{3.122}$$

We need the components with respect to the natural basis in spherical coordinates, $\hat{\boldsymbol{\phi}} = \frac{\mathbf{e_3}}{r\sin\theta}$, so

$$a^3 = -\frac{G}{c^2} \frac{M\omega R^2}{5r^3}. \tag{3.123}$$

Using $a^3 = -h^{03}c/4$, we can get the inverse metric component directly:

$$h^{03} = \frac{4G}{c^3} \frac{M\omega R^2}{5r^3}. \tag{3.124}$$

Finally, we'd like the lowered form of this in order to see what the metric, with this field on top, will look like. Using $h_{\mu\nu} = \eta_{\alpha\mu}\eta_{\beta\nu}h^{\alpha\beta}$, in spherical coordinates,

$$h_{03} = -\frac{4G}{c^3} \frac{M\omega R^2 \sin^2\theta}{5r}. \tag{3.125}$$

The moment of inertia of a spinning massive sphere is $I = 2/5MR^2$, and if we define the length $a \equiv I\omega/(Mc)$, then the field component is

$$h_{03} = -\frac{2GMa \sin^2\theta}{c^2 r}. \tag{3.126}$$

What we have in $\mathbf{h}$ is similar to the magnetic dipole field outside of a spinning sphere of charge. The effect of the field on a particle, from (3.111) (made explicit in Problem 3.18), is similar to the effect of a magnetic field on a charged particle. We can predict the effect of the gravitomagnetic field on a spinning test particle by carrying over the effect of a magnetic field on a magnetic dipole. Place a spinning test charge within a magnetic field, and the dipole will precess (as you showed in Problem 1.35). By the structural analogy set up in this example we expect that if you place a spinning test mass in the field of a spinning massive source, then the test mass spin will precess. The Gravity Probe B experiment [15] (see [16] for a complete set of papers) measured this spin precession in gyroscopes orbiting the Earth.[9]

---

[9]   Spin precession also happens in the context of static spherical sources, and is called "geodetic precession" in that setting. The Gravity Probe B experiment measured a combination of precession effects.

These last two examples provide weak field limits with which we can compare the solutions developed in Chapter 4, so they will prove especially helpful in building intuition about the spacetimes set up by spherically symmetric, spinning sources. Our final example has a nonzero energy density away from the source itself, and so is relevant for comparison with solutions to Einstein's equation that are not in vacuum.

Example 3.4 (A Charged Spherical Source). We found an approximation to the gravitational field associated with the electric field of a charge in Section 1.5. There we assumed that energy acted as a source in Newtonian gravity. We are now in position to find the "actual" weak field limit of the gravitational field set up by the energy density of the electric field of a point charge. Since the weak field equations are themselves linear, we'll focus on the field coming from the energy content, and then we can add back in the $h_{00} = 2GM/(rc^2)$ from the mass of the sphere itself.

Starting with $u = \epsilon_0 E^2/2$ for a point source, we have to solve

$$\nabla^2 h_{00} = -8\pi G \frac{u}{c^4} = -\frac{GQ^2}{4\pi \epsilon_0 c^4 r^4}, \qquad (3.127)$$

where Q is the charge of the central body. We know that h_{00} should be spherically symmetric, and then the Laplacian becomes $(1/r)(rh_{00})''$, where primes refer to r derivatives. Solving by integration, the 00 component is

$$h_{00} = -\frac{GQ^2}{8\pi \epsilon_0 c^4 r^2} + \frac{A}{r} + B, \qquad (3.128)$$

where we can take $A = 2GM/c^2$; that's just going to give us back the mass piece. We'll set $B = 0$ to get the field to vanish at spatial infinity. If we introduce the two length scales $R_M = 2GM/c^2$ and $R_Q^2 = GQ^2/(4\pi\epsilon_0 c^4)$, as we did in Section 1.5, then the metric perturbation is

$$h_{00} = \frac{R_M}{r} - \frac{R_Q^2}{2r^2}. \qquad (3.129)$$

What gravitational potential φ would we associate with this term? From $\varphi = -h_{00}c^2/2$ we would expect to have obtained

$$\varphi = c^2\left(-\frac{R_M}{2r} + \frac{R_Q^2}{4r^2}\right), \qquad (3.130)$$

just as we had in (1.87).

Exercises

3.1   This problem works through a one-dimensional curvature example; see Appendix C for definitions and discussion.

Part a. Working in two-dimensional Cartesian coordinates, take the curve

$$\mathbf{r}(u) = R\cos(\omega u)\hat{\mathbf{x}} + R\sin(\omega u)\hat{\mathbf{y}} \qquad (3.131)$$

and find the "arc length" parameterization, that is, find $\mathbf{r}(s)$ such that $\mathbf{r}'(s) \cdot \mathbf{r}'(s) = 1$, which still describes the same curve in two dimensions. The derivative of the tangent vector, $\mathbf{r}''(s)$, should be orthogonal to the tangent vector, and its magnitude $\kappa \equiv \sqrt{\mathbf{r}''(s) \cdot \mathbf{r}''(s)}$ is called the "curvature" of the curve. Does that make sense here? The inverse of κ is called the "radius of curvature."

Part b. Next consider the similar problem, in $D = 1 + 1$ (one spatial, one temporal coordinate) with metric

$$g_{\mu\nu} \doteq \begin{pmatrix} -1 & 0 \\ 0 & 1 \end{pmatrix}. \qquad (3.132)$$

Take the curve to be, setting c "=" 1 for this (and only this!) problem,

$$\mathbf{x}(t) = t\hat{\mathbf{t}} + \sqrt{b^2 + t^2}\hat{\mathbf{x}}. \qquad (3.133)$$

Sketch the curve and find it's "proper time parameterization," with

$$\frac{dx^\mu}{d\tau} g_{\mu\nu} \frac{dx^\nu}{d\tau} = -1.$$

In this setting, compute the derivative of the tangent vector, $\frac{d^2 x^\mu}{d\tau^2}$, and find its magnitude. Compare with part a.

3.2   Starting from the surface description of the vectors $\mathbf{X}_1$ and $\mathbf{X}_2$ in Figure C.4 of Appendix C, write down the surface area element $d\mathbf{a}$ in terms of these. Compute its magnitude da. Suppose you wanted to make a surface that minimized area, cook up an action that will work, and express the integrand entirely in terms of dot products of $\mathbf{X}_1$ and $\mathbf{X}_2$ (with each other and themselves). Show that the integrand can be written as the determinant of a matrix and identify that matrix (see Appendix C for inspiration).

Once you have the action, there are two neat "extensions": (1) Develop the extremization condition for the action to give a set of area-minimizing equations of motion (the area version of geodesic equations of motion); (2) Explore what changes when the two-dimensional surface has a timelike and a spacelike coordinate. This latter extension is the starting point for string theory.

3.3   From the components of the Riemann tensor $R^\alpha{}_{\sigma\nu\mu}$ for a sphere (see (3.15)), find the nonzero components of the lower form $R_{\alpha\sigma\nu\mu}$.

3.4 Show that for a constant metric at a point P with zero Christoffel connection there (achieved using the transformation in (3.35)), you can perform an additional, diagonalizing transformation that does not change the value of the connection at P.

3.5 Establish the following three properties of the Riemann tensor using Riemannian coordinates:

$$R_{\alpha\sigma\mu\nu} = -R_{\alpha\sigma\nu\mu}, \qquad R_{\nu\mu\alpha\sigma} = R_{\alpha\sigma\nu\mu},$$
$$R_{\alpha\sigma\nu\mu} + R_{\alpha\mu\sigma\nu} + R_{\alpha\nu\mu\sigma} = 0. \tag{3.134}$$

3.6 Prove the Bianchi identity using Riemannian coordinates,

$$R_{\alpha\sigma\nu\mu;\gamma} + R_{\alpha\sigma\gamma\nu;\mu} + R_{\alpha\sigma\mu\gamma;\nu} = 0. \tag{3.135}$$

3.7 Find the expressions for a family of geodesics $\theta(\tau, \sigma)$ and $\phi(\tau, \sigma)$ on a sphere where you start at the "front" pole ($\theta = \pi/2, \phi = 0$) and go to the "back" ($\theta = \pi/2, \phi = \pi$) with σ the initial angle of the tangent to the curve with respect to the $\hat{\mathbf{z}}$ axis. These geodesics avoid the north pole singularity and are a little nicer to work with. Show that at the end points, the geodesics are independent of σ, like the ones in (3.48).

3.8 For a family of geodesics, $x^\mu(\tau, \sigma)$, where τ is proper time, and σ selects a particular curve, the derivatives of $x^\mu(\tau, \sigma)$ with respect to τ and σ can be taken to be orthogonal (in the generalized sense of angle (2.162)). Show that this is the case for the flat space geodesics from (3.48) and the geodesics on the sphere from (3.50).

3.9 Work out the response of the object $v^\mu \equiv \frac{dx^\mu}{dt}$ with $v^0 = c$ and spatial components $\mathbf{v}$ that represent a particle's velocity under a Lorentz boost in the x direction. Is v^μ a four-vector?

3.10 A point charge q is moving with velocity $\bar{v}$ in the $\hat{\bar{\mathbf{x}}}$ direction in a lab $\bar{L}$ that is moving along the shared $\hat{\mathbf{x}}$ axis in L with speed v. Find the components of the E&M source, $\bar{J}^\mu$ and J^μ, and show that $J^\mu = \tilde{\Lambda}^\mu_{\ \alpha} \bar{J}^\alpha$, where $\tilde{\Lambda}^\mu_{\ \alpha}$ is the inverse Lorentz transformation (see Section 3.4.1). This case is done in Section A.6 in one dimension if you want to check your work. Do the same thing for a charge q that is moving in the $\hat{\bar{\mathbf{y}}}$ direction in $\bar{L}$.

3.11 Find the relation between the energy density u in a lab frame L and the rest frame energy density u_0. Using that and the relation $\eta^\mu = \frac{dx^\mu}{dt}\gamma_v$, show that

$$T^{\mu\nu} = \frac{u_0}{c^2}\eta^\mu\eta^\nu = \frac{u}{c^2}\frac{dx^\mu}{dt}\frac{dx^\nu}{dt} \tag{3.136}$$

directly, as was done for charge at the end of Section 3.4.1.

3.12 Use the Bianchi identity to show that $R^{\sigma\mu}_{\ \ ;\mu} = \frac{1}{2}R^{\ \sigma}_{;}$. Show that the covariant divergence of $R_{\mu\nu} - (1/2)g_{\mu\nu}R$ is zero.

3.13 From Einstein's equation (3.87) and the geodesic equation of motion (3.88), is the field of general relativity, $g_{\mu\nu}$, more like the electromagnetic $\mathbf{E}$ and $\mathbf{B}$ or more like the potentials V and $\mathbf{A}$? Is there a "gauge choice" here for the metric, some choice we can make that doesn't change physical predictions but makes calculation easier?

3.14 What is the scalar $g_{\alpha\beta}g^{\alpha\beta} = ?$

3.15 From Einstein's equation,

$$R_{\mu\nu} - \frac{1}{2}g_{\mu\nu}R = \frac{8\pi G}{c^4}T_{\mu\nu}, \tag{3.137}$$

in D dimensions, multiply by $g^{\mu\nu}$ and write the equation in terms of $R_{\mu\nu}$, $T_{\mu\nu}$, and the trace of the stress tensor, $T \equiv g^{\mu\nu}T_{\mu\nu}$, that is, swap the trace of the Ricci tensor for the trace of the stress tensor.

3.16 For $g_{\mu\nu} = \eta_{\mu\nu} + h_{\mu\nu}$ with the Minkowski metric $\eta_{\mu\nu}$ in Cartesian coordinates and a small "perturbation" $h_{\mu\nu}$ on top of that, show that $g^{\mu\nu} = \eta^{\mu\nu} + O(h)$, that is, the inverse form is, to first order in h, just the inverse of $\eta_{\mu\nu}$.

3.17 Suppose we have a metric $g_{\mu\nu}$ in the x coordinates and we make a new set of coordinates $\bar{x}^\mu = x^\mu + \epsilon f^\mu(x)$, where $\epsilon \ll 1$, and $f^\mu(x)$ is an arbitrary function of x. Assuming that this induces a change in the metric of the form $\bar{g}_{\mu\nu} = g_{\mu\nu} + \epsilon\sigma_{\mu\nu}$, use the scalar line element $ds^2 \equiv dx^\mu g_{\mu\nu}dx^\nu = d\bar{x}^\mu \bar{g}_{\mu\nu}d\bar{x}^\nu = d\bar{s}^2$ to find an expression for $\sigma_{\mu\nu}$ in terms of f^μ and its derivatives. This problem is meant to establish that "the gauge choice of general relativity," which we encountered in Section 3.6.1 for the metric perturbation $h_{\mu\nu} \rightarrow g_{\mu\nu} + v_{\mu,\nu} + v_{\nu,\mu}$, is precisely the choice of coordinates in which to express the metric.

3.18 Evaluate the geodesic equation of motion in the weak field limit (3.111) for a pure gravitomagnetic field h_{0i} with $h_{00} = h_{ij} = 0$. Use temporal parameterization and focus on the spatial components of the equation of motion.

3.19 What is the contribution to the geodesic equation of motion coming from the spatial perturbation h_{ij}?

3.20 Suppose you have the form of Einstein's equation but do not know the constant appearing on the right-hand side:

$$R_{\mu\nu} - \frac{1}{2}g_{\mu\nu}R = \kappa T_{\mu\nu}. \tag{3.138}$$

Take a point source of mass M sitting at rest at the origin; by evaluating the weak field limit of the field equation and the equation of motion for a (nonrelativistic) test particle that comes from the geodesic equation of motion, and demanding that these match the Newtonian prediction, show that $\kappa = 8\pi G/c^4$.

3.21 For a metric perturbation $h_{\mu\nu}$, the geodesic Lagrangian is $L = m/2(g_{\mu\nu} + h_{\mu\nu})\dot{x}^\mu\dot{x}^\nu$. Take $g_{\mu\nu}$ to be the Minkowski metric in spherical coordinates. What must you set $h_{\mu\nu}$ to in order to get a Lagrangian that looks like $L = T - (-GMm/r)$, that is, what $h_{\mu\nu}$ leads to a mechanical Lagrangian with potential energy given by a spherical central mass M?

3.22 Leave the time dependence in u and $\mathbf{v}$ for the stress tensor in the weak field limit. What are the field equations now? Do they match our prediction from Section 1.6?

3.23 For the spherically symmetric solution to the weak field form of Einstein's equation in (3.118), show that $\partial_\mu h^{\mu\nu} = \partial^\nu h^{00}$.

3.24 What is the Newtonian gravitational potential inside a uniformly distributed ball of mass with total mass M and radius R (set the zero of the potential at $r = 0$)? From the potential, write out the perturbation form of the metric, $g_{\mu\nu} = \eta_{\mu\nu} + h_{\mu\nu}$, in spherical coordinates.

The following problems benefit from a computer algebra program with a package for computing connections, the Riemann tensor, and Ricci tensor and scalar given a metric. The author's `EinsteinVariation.` `package.m` *for use in* `Mathematica` *is available from the book website.*

3.25 Using the package, compute the Christoffel connection for Minkowski spacetime written in cylindrical coordinates. Compute the Riemann tensor – is it what you expect?

3.26 Referring to Example C.7 in Appendix C, using the coordinates θ and ϕ shown there, develop the two-dimensional metric on the surface of the torus. From that metric, find the Ricci scalar (use the package) – which is it more like, the mean curvature or the Gauss curvature?

3.27 The metric for the surface of a cone was worked out in Problem 2.26 – using the coordinates found there and the resulting metric, is the surface of a cone flat or curved?

3.28 A space is called "conformally flat" if there exist coordinates x such that the metric takes the form

$$g_{\mu\nu} \doteq \Phi(x)\mathbb{I} \tag{3.139}$$

for a function of the two coordinates $\Phi(x)$, where $\mathbb{I}$ is the D-dimensional identity matrix (no temporal coordinate here, but it's easy to add one in by putting a minus sign in for one entry). Show that for two-dimensional conformally flat spaces with

$$g_{\mu\nu} \doteq \Phi(u, v) \begin{pmatrix} 1 & 0 \\ 0 & 1 \end{pmatrix}, \tag{3.140}$$

if the Ricci scalar vanishes, so does the Riemann tensor.

3.29 Find the Riemann tensor for a metric with general spherically symmetric perturbation field $h^{\mu\nu}$ from (3.118).

3.30 Einstein's equation in vacuum (away from sources) reads $R_{\mu\nu} = 0$. Starting with a metric of the form

$$g_{\mu\nu} \doteq \begin{pmatrix} -a(r) & 0 & 0 & 0 \\ 0 & b(r) & 0 & 0 \\ 0 & 0 & r^2 & 0 \\ 0 & 0 & 0 & r^2 \sin^2\theta \end{pmatrix} \tag{3.141}$$

for coordinates $x^0 = ct$, $x^1 = r$, $x^2 = \theta$, and $x^3 = \phi$, it is clear from the lower two-by-two block that for fixed r and t, the angles θ and ϕ describe a spherical surface. Compute $R_{\mu\nu}$ for this metric using the `Mathematica` package. How many independent entries do you have? Using algebraic linear combinations of the equations $R_{\mu\nu} = 0$, isolate differential equations governing $a(r)$ and $b(r)$ that you can solve – do so by hand (although you may use `Mathematica` to generate the simplifying linear combinations). Once you have the solution (with any relevant constants of integration), impose the "boundary condition" that $g_{\mu\nu}$ becomes the Minkowski metric (expressed in spherical coordinates) as $r \to \infty$ to set one of the constants of integration. That should leave you with one constant of integration left over – set it by comparing g_{00} to the metric perturbation with $h_{00} = -2\varphi/c^2$ for gravitational potential φ associated with a point source mass M at the origin. Check that the resulting fully specified metric really does have $R_{\mu\nu} = 0$ and $R = 0$ but that the spacetime it describes is not flat.

We have studied the weak-field form and solutions of Einstein's equation. Now we come to exact solutions and their implications for particle motion. Since the field equations of general relativity are nonlinear, superposition does not hold. Fortunately, there aren't many physically relevant source configurations to consider, so we don't really need to be able to build up solutions for arbitrary distributions of mass/energy out of point solutions.

The most relevant astrophysical "point" source is a spinning massive sphere. We'll start with a static sphere and find the most general spherically symmetric vacuum (away from the source) solution to Einstein's equation. This "Schwarzschild solution" already has a lot of interesting physics to probe. So we'll pause in our enumeration of vacuum solutions to think about its physical predictions. In particular, its geodesic trajectories for both particles and light are relevant, especially since they can be directly compared with the Newtonian results from Section 1.3 and Section 1.4.

After working through some implications of the Schwarzschild solution, we will simply quote the spinning massive spacetime, the Kerr metric (you are invited to check that it satisfies Einstein's equation in vacuum!). We will think about its geodesics and some of the exotic properties that adding a slight "twist" to spacetime incurs. We could also add charge to our spinning spheres, but because that generates an energy density throughout space and is therefore not an example of a vacuum solution, we'll save it for later. A charged sphere is also less common over large time scales, since the presence of charge attracts the opposite charge, "quickly" neutralizing the sphere.

Finally, we'll develop the "Weyl" metrics. These are axially symmetric, and so of less obvious utility, but their structure is interesting, and there are certain speculative configurations of energy that would generate spacetimes like these. In all cases, we work away from the sources, in vacuum, where Einstein's equation simplifies to $R_{\mu\nu} = 0$ as in (3.90). If the field equation is always the same, how do we accommodate different types of sources, spheres versus spinning spheres versus lines? The same problem exists in electrostatics, where $\nabla \cdot \mathbf{E} = 0$ in vacuum. The way we select the source is by imposing symmetry on the form of the electric field or, if using $\nabla^2 V = 0$, imposing symmetry assumptions on V. These assumptions are ultimately (although often implicitly) used to enforce boundary conditions as in Problem 4.1. We'll start with the implementation of spherical symmetry for a metric ansatz.

4.1 Spherical Symmetry

For an electric potential $V(\mathbf{r})$, spherical symmetry means that V depends only on the distance to the origin, $V(\mathbf{r}) = V(r)$. For a vector like $\mathbf{E}(\mathbf{r})$, spherical symmetry means that the magnitude is a function only of r, and the direction is radial so that rotating the coordinate system about any axis does not change the form of the field, $\mathbf{E}(\mathbf{r}) = E(r)\hat{\mathbf{r}}$ (we could use the nonunit vector $\mathbf{r}$ to express the same dependence). These functional structures for V and $\mathbf{E}$ are motivated by form invariance under an arbitrary rotation of the coordinate axes, which is really the definition of spherical symmetry.

How about the spatial components of a second-rank tensor, how should we enforce spherical symmetry in this case? Following the pattern, we'll take $h^{ij} = b(r)r^i r^j$, so that the magnitude of the tensor, $b(r)r^2$, depends only on the distance to the origin, and the direction is, for any choice of i or j, radial. If we start off in Cartesian coordinates with

$$h^{ij} \doteq b\left(\sqrt{x^2 + y^2 + z^2}\right) \begin{pmatrix} x^2 & xy & xz \\ xy & y^2 & yz \\ xz & yz & z^2 \end{pmatrix} \tag{4.1}$$

and transform to spherical coordinates,

$$\bar{h}^{ij} = \frac{\partial \bar{x}^i}{\partial x^k} \frac{\partial \bar{x}^j}{\partial x^\ell} h^{k\ell} \doteq b(r) \begin{pmatrix} r^2 & 0 & 0 \\ 0 & 0 & 0 \\ 0 & 0 & 0 \end{pmatrix}, \tag{4.2}$$

then as expected, the magnitude appears as the "rr" component.

There is another spherically symmetric term we can add in here. In the second-rank tensor setting, we have the coordinate-independent Kronecker delta tensor δ^i_j. A term like $f(r)\delta^i_j$ is spherically symmetric in that it is insensitive to rotations of the coordinate system: the Kronecker delta takes on values 1 and 0 in all coordinate systems, and an arbitrary function $f(r)$ is likewise unchanged by a rotation since it depends only on an invariant of that transformation. In order to add $f(r)\delta^i_j$ to h^{ij}, we have to raise an index on the delta using the metric, so let's agree to start with

$$h^{ij} = b(r)r^i r^j + f(r)\delta^i_k g^{kj} = b(r)r^i r^j + f(r)g^{ij}. \tag{4.3}$$

In spherical coordinates, the most general spherically symmetric spatial tensor is

$$h^{ij} \doteq \begin{pmatrix} b(r)r^2 + f(r) & 0 & 0 \\ 0 & f(r)/r^2 & 0 \\ 0 & 0 & f(r)/(r^2\sin^2\theta) \end{pmatrix}. \tag{4.4}$$

We can simplify slightly by performing a coordinate transformation with $\bar{r} = r/\sqrt{f(r)}$, $\bar{\theta} = \theta$, $\bar{\phi} = \phi$; then there is a new function $a(\bar{r})$, a combination of

$f(r)$, its derivatives, and $b(r)$ appearing in the "rr" spot:

$$h^{ij} \doteq \begin{pmatrix} a(\bar{r}) & 0 & 0 \\ 0 & 1/(\bar{r}^2) & 0 \\ 0 & 0 & 1/(\bar{r}^2 \sin^2 \theta) \end{pmatrix}. \tag{4.5}$$

The point is that we can redefine the radial coordinate to subsume $f(r)$. In this form, it is clear that h^{ij} really depends only on one independent function of a coordinate.

Moving to four-dimensional spacetime, we'll start off with a static tensor, meaning that there is no t dependence, limiting us to new entries only in the 00 component, which we could make an arbitrary function of r. Our starting point, then, will be a covariant second-rank symmetric tensor with two independent functions, the $a(\bar{r})$ above and a new $h_{00}(\bar{r})$ function. Omitting the bars, since the coordinates won't take on their spherical meaning anyway, we have an initial spherically symmetric metric form

$$g_{\mu\nu} \doteq \begin{pmatrix} -p(r) & 0 & 0 & 0 \\ 0 & q(r) & 0 & 0 \\ 0 & 0 & r^2 & 0 \\ 0 & 0 & 0 & r^2 \sin^2 \theta \end{pmatrix}. \tag{4.6}$$

The lower two-by-two subblock is the metric of a 2-sphere (see Example 2.7 and Example C.5), reminding us of the spherical symmetry here. Our goal now is to use Einstein's equation in vacuum to find $p(r)$ and $q(r)$. I do not recommend calculating the Ricci tensor by hand (although it can, of course, be done). Using a simple package to compute the elements,[1] the nonzero components are

$$R_{00} = \frac{1}{4rp(r)q(r)^2}\big[rq(r)p'(r)^2$$
$$+ p(r)\big(rp'(r)q'(r) - 2q(r)\big(2p'(r) + rp''(r)\big)\big)\big],$$

$$R_{11} = \frac{1}{4rp(r)^2q(r)}\big[p(r)\big(4p(r) + rp'(r)\big) + rq(r)\big(p'(r)^2 - 2p(r)p''(r)\big)\big],$$

$$R_{22} = \frac{1}{2}\bigg[2 - \frac{2 + rp'(r)/p(r)}{q(r)} + \frac{rq'(r)}{q(r)^2}\bigg],$$

$$R_{33} = \frac{\sin^2 \theta}{2p(r)q(r)^2}\big[p(r)\big(2q(r)^2 - 2q(r) + rq'(r)\big) - rq(r)p'(r)\big], \tag{4.7}$$

where primes mean r derivatives.[2] These equations look worse than they are, but it takes some clever linear combinations to see the simplifications. Just working from the denominators in the R_{00} and R_{11} terms suggests

$$R_{00} - R_{11}\frac{p(r)}{q(r)} = -\frac{q(r)p'(r) + p(r)q'(r)}{rq(r)^2} = 0, \tag{4.8}$$

---

[1] The "EinsteinVariation.m" Mathematica package can be found at the author's website.
[2] The form of the initial ansatz can simplify (or complexify) the Ricci tensor components, as you will see in Problem 4.2.

from which we get $q(r) = -1/p(r)$. Using this relation, we are left with two equations for $p(r)$,

$$1 + \big(rp(r)\big)' = 0, \qquad 2p'(r) + rp''(r) = 0, \tag{4.9}$$

both of which are solved by

$$p(r) = -1 + \frac{\alpha}{r}. \tag{4.10}$$

The lone constant of integration, α, can be set using the weak field limit from Example 3.2. There we found $h_{00} = 2GM/(rc^2)$, and for our $g_{00} = -1+\alpha/r$, it is clearly α/r that is the weak field piece; then

$$\frac{2GM}{rc^2} = \frac{\alpha}{r} \longrightarrow \alpha = \frac{2GM}{c^2}. \tag{4.11}$$

Putting this in, we arrive at the "Schwarzschild metric," the unique, spherically symmetric solution to Einstein's equation in vacuum [44],

$$g_{\mu\nu} \doteq \begin{pmatrix} -(1 - \frac{2GM}{rc^2}) & 0 & 0 & 0 \\ 0 & \frac{1}{1-2GM/(rc^2)} & 0 & 0 \\ 0 & 0 & r^2 & 0 \\ 0 & 0 & 0 & r^2\sin^2\theta \end{pmatrix}. \tag{4.12}$$

It describes the spacetime outside of a static spherical central body of mass M.

The Schwarzschild solution to Einstein's equation is relatively simple, with limits that are easy to identify and physically interpret. There is a natural length scale built in to the solution $R_M \equiv 2GM/c^2$, the so-called "Schwarzschild radius," which we encountered already in Section 1.3. For locations far from this radius, $r \gg R_M$, the solution becomes

$$g_{\mu\nu} \approx \begin{pmatrix} -1 & 0 & 0 & 0 \\ 0 & 1 & 0 & 0 \\ 0 & 0 & r^2 & 0 \\ 0 & 0 & 0 & r^2\sin^2\theta \end{pmatrix} + \frac{R_M}{r}\begin{pmatrix} 1 & 0 & 0 & 0 \\ 0 & 1 & 0 & 0 \\ 0 & 0 & 0 & 0 \\ 0 & 0 & 0 & 0 \end{pmatrix}. \tag{4.13}$$

The first term here is just the Minkowski metric written in spherical coordinates, with the second term a perturbation on top of that flat background.

The Schwarzschild metric is *not* flat, as you can check in Problem 4.4, although by construction it is "Ricci-flat," with $R_{\mu\nu} = 0$ and $R = 0$. The metric is undefined for both $r = 0$ and $r = R_M$. This may not be too concerning, since $r = 0$ is certainly *inside* the massive source, where this solution does not apply. The location R_M also typically resides within the source: For the Sun, $R_M = 3\,\text{km}$, whereas the Sun's radius is $\approx 700\,000\,\text{km}$, so that, again, the solution is not valid at R_M. The only way for R_M to be "seen" would be if all the mass of the central body were packed in to a sphere of radius less than R_M, and that configuration is called a "black hole."

Before moving on to the physical implications of the Schwarzschild metric, you should try finding spherically symmetric solutions to the vacuum Einstein equation in different dimensions in Problem 4.7, Problem 4.9, and Problem 4.11. Einstein's theory changes dramatically in different dimensions, and in interesting ways.

4.2 How Far Apart Are Points?

The radial coordinate r in the Schwarzschild geometry does not measure the distance to the origin of the coordinate system, as it does in flat geometry. Take $\theta = \pi/2$ and $\phi = 0$, and two different radial positions r_1 and $r_2 > r_1$, with the same temporal coordinate t. For a flat spacetime, written in spherical coordinates, the distance between these locations is $r_2 - r_1$, but that won't hold in a curved spacetime. The distance between r_1 and r_2 is given by the integral of

$$ds^2 = \left(1 - \frac{R_M}{r}\right)^{-1} dr^2. \tag{4.14}$$

Taking the square root and integrating from r_1 to r_2 give

$$s = r_2 \sqrt{1 - \frac{R_M}{r_2}} - r_1 \sqrt{1 - \frac{R_M}{r_1}}$$
$$+ \frac{1}{2} R_M \log \left[\frac{1 + \sqrt{1 - R_M/r_2}}{1 - \sqrt{1 - R_M/r_2}} \frac{1 - \sqrt{1 - R_M/r_1}}{1 + \sqrt{1 - R_M/r_1}} \right]. \tag{4.15}$$

This is the "proper distance" (using the instantaneous $dt = 0$) between r_1 and r_2, and you should check that in the $R_M \to 0$ limit, you recover the flat space result in Problem 4.12.

Although the Schwarzschild radial coordinate is not the same as the flat, spherical coordinate, when it comes to radial distances, it does play the same role in computing the circumference of circles. Fix the Schwarzschild coordinate r at some value and take $\theta = \pi/2$; what is the circumference of a circle in this geometry for $\phi = 0 \to 2\pi$? We'll set $dt = 0$ to again compute the proper distance. In addition, set $dr = 0$ and $d\theta = 0$. Then the line element for this curve is just

$$ds^2 = r^2 d\phi^2 \longrightarrow s = 2\pi r, \tag{4.16}$$

the usual circumference of a circle.

4.3 Radial Geodesics

Let's think about the simplest geodesics associated with (4.12) (for much more on Schwarzschild geodesics, see [7]). Rather than work from the geodesic equation of motion, we'll go back to the Lagrangian itself, in proper time parameterization from (2.108) with the metric in place:

$$L = \frac{m}{2}\left[-\left(1 - \frac{R_M}{r(\tau)}\right)c^2\dot{t}(\tau)^2 + \left(1 - \frac{R_M}{r(\tau)}\right)^{-1}\dot{r}(\tau)^2\right.$$
$$\left. + r(\tau)^2\left(\dot{\theta}(\tau)^2 + \sin^2\theta(\tau)\dot{\phi}(\tau)^2\right)\right], \tag{4.17}$$

where proper time parameterization means the same thing it always does, $\dot{x}^\mu g_{\mu\nu}\dot{x}^\nu = -c^2$, so that the Lagrangian is itself a constant of the motion, $L = -mc^2/2$. This is different from the usual situation in classical mechanics, in which the Hamiltonian but *not* the Lagrangian is constant.

Radial motion means that θ and ϕ are fixed. We might as well take $\theta = \pi/2$ and $\phi = 0$; then the Lagrangian reads

$$L = \frac{m}{2}\left[-\left(1 - \frac{R_M}{r(\tau)}\right)c^2\dot{t}(\tau)^2 + \left(1 - \frac{R_M}{r(\tau)}\right)^{-1}\dot{r}(\tau)^2\right] = -\frac{1}{2}mc^2. \tag{4.18}$$

First, we notice that the coordinate time is "ignorable," meaning that it doesn't appear by itself. Then the Euler–Lagrange equation for $t(\tau)$ is

$$-\frac{d}{d\tau}\frac{\partial L}{\partial\dot{t}(\tau)} + \underbrace{\frac{\partial L}{\partial t(\tau)}}_{=0} = 0. \tag{4.19}$$

The derivative of L with respect to $\dot{t}(\tau)$ is a constant of the motion,

$$\frac{\partial L}{\partial\dot{t}(\tau)} = -mc^2\left(1 - \frac{R_M}{r(\tau)}\right)\dot{t}(\tau). \tag{4.20}$$

In the absence of a source ($R_M = 0$), this is just the negative of the relativistic energy, and it has that same interpretation with a source present,[3] so we'll let the constant E be defined by

$$E = mc^2\left(1 - \frac{R_M}{r(\tau)}\right)\dot{t}(\tau) \longrightarrow \dot{t}(\tau) = \frac{E/(mc^2)}{1 - R_M/r(\tau)}, \tag{4.21}$$

and this can be used directly in (4.18) to eliminate $\dot{t}(\tau)$,

$$L = \frac{-E^2/(mc^2) + m\dot{r}(\tau)^2}{2(1 - R_M/r(\tau))} = -\frac{1}{2}mc^2. \tag{4.22}$$

Solving for $\dot{r}(\tau)^2$ gives

$$\dot{r}(\tau)^2 = c^2\left(-1 + \frac{R_M}{r(\tau)} + \left(\frac{E}{mc^2}\right)^2\right). \tag{4.23}$$

---

[3] What we really have is the momentum conjugate to $x^0 \equiv ct$ here, $p_0 = \frac{\partial L}{\partial(c\dot{t})} = -E/c$ in proper time parameterization. The minus sign is showing up because we typically think of $p^0 = E/c$; the contravariant form gets the physical interpretation. Then the canonical momentum, which is naturally covariant, has $p_0 = -E/c$ in Minkowski spacetime.

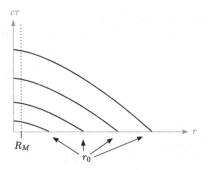

Fig. 4.1 The Minkowski diagram above shows plots of $c\tau$ versus r for particles starting at different r_0 values at time $c\tau = 0$ that then fall in towards the center of the central body, right through the Schwarzschild radius at R_M.

If you take the τ derivative of both sides of this equation, you can get the radial acceleration in this proper time setting,

$$\ddot{r}(\tau) = -\frac{R_M c^2}{2r(\tau)^2} = -\frac{GM}{r(\tau)^2}, \tag{4.24}$$

precisely what we would have in Newtonian gravity for $\ddot{r}(t)$. One thing we already know, then, is that the solution $r(\tau)$ is continuous until the particle gets to zero. That is interesting and suggests that in some fundamental way, there is no problem at the Schwarzschild radius. If you were falling towards a sphere of mass M with radius $R < R_M$, you would pass through R_M at some finite proper time, with nothing dramatic happening.

We can solve for $r(\tau)$ for arbitrary initial conditions, but it suffices to consider the solution for the special case in which a massive body is at rest at spatial infinity and falls inward towards the central sphere. Out at spatial infinity, the Schwarzschild metric reduces to the usual Minkowski one, with only the first term in (4.13) contributing. So out there, we can interpret the coordinates as spherical, and going back to (4.23), if we start the particle off in this way, then $E = mc^2$ which makes sense. Now taking the square root of (4.23) with the minus sign (for infall) and the constant energy set to the rest energy at infinity, we have to solve

$$\dot{r}(\tau) = -c\sqrt{\frac{R_M}{r(\tau)}}, \tag{4.25}$$

and if we take $r(0) = r_0$, the solution is

$$r(\tau) = \left(\frac{r_0^{3/2}}{\sqrt{R_M}} - \frac{3c\tau}{2}\right)^{2/3} R_M^{1/3}. \tag{4.26}$$

A Minkowski diagram of $c\tau$ versus r is shown in Figure 4.1, and there we can see that, as advertised, the solution is continuous for any starting location. There is no discontinuity at R_M or anywhere else, and the particle falls in towards location $r = 0$ in finite proper time.

With $r(\tau)$ in place from (4.26), we can solve (4.21) for $t(\tau)$ with $E = mc^2$:

$$ct(\tau) = c\tau + 2\sqrt{R_M}\left(\sqrt{r_0} - \sqrt{r(\tau)}\right)$$
$$+ R_M \log\left(\frac{(1 + \sqrt{r(\tau)/R_M})(1 - \sqrt{r_0/R_M})}{(1 - \sqrt{r(\tau)/R_M})(1 + \sqrt{r_0/R_M})}\right). \tag{4.27}$$

What is it that we really want out of the radial geodesics? If this were "normal" special relativistic physics, then we would be more interested in $r(t)$, the radial distance as a function of the coordinate time. From the functional form, we could draw the trajectory on Minkowski diagrams of ct versus r, and this would give us a clear picture of what the trajectory looks like as measured by a clock in a lab through which the particle is moving. The same is basically true of $r(t)$ for the Schwarzschild case. Out at spatial infinity, the metric is flat and written in spherical coordinates, so we would interpret r as the radial coordinate, with t measured with respect to our clock at rest in our "lab" infinitely far away from where the motion is occurring.

To find $r(t)$, we first need $\frac{dr}{dt}$, which we can get from the ratio $\dot{r}(\tau)/\dot{t}(\tau)$ (both of which we have from (4.21) and the square root of (4.23)),

$$\frac{dr}{dt} = \frac{dr}{d\tau}\frac{d\tau}{dt} = \frac{\dot{r}(\tau)}{\dot{t}(\tau)} = \pm\frac{mc^3}{E}\left(1 - \frac{R_M}{r}\right)\left[-\left(1 - \left(\frac{E}{mc^2}\right)^2\right) + \frac{R_M}{r}\right]^{1/2}. \tag{4.28}$$

There are a number of things we could do at this point. The most useful, in terms of thinking about the motion, is a Minkowski diagram of radially in- and out-going trajectories. For a Minkowski diagram, we want $ct(r)$, the coordinate time as a function of position. Flipping over $\frac{dr}{dt}$, we can write

$$\frac{dct(r)}{dr} = \pm\frac{E}{mc^2(1 - R_M/r)}\left[-\left(1 - \left(\frac{E}{mc^2}\right)^2\right) + \frac{R_M}{r}\right]^{-1/2}, \tag{4.29}$$

and again focusing on the solutions that are at rest at spatial infinity, with $E = mc^2$,

$$\frac{dct(r)}{dr} = \pm\frac{\sqrt{r/R_M}}{1 - R_M/r} \tag{4.30}$$

with solution, taking the minus sign to get infall,

$$ct(r) = -\frac{2}{3\sqrt{R_M}}\left(r^{3/2} - r_0^{3/2} + 3R_M(\sqrt{r} - \sqrt{r_0})\right)$$
$$+ R_M \log\left(\frac{(1 + \sqrt{r/R_M})(1 - \sqrt{r_0/R_M})}{(1 - \sqrt{r/R_M})(1 + \sqrt{r_0/R_M})}\right), \tag{4.31}$$

where we have set $ct(r_0) = 0$. A Minkowski diagram of the trajectories for a few different starting locations r_0 is shown in Figure 4.2. There you can see that the particle approaches but never crosses the Schwarzschild radius. From the point of view of an observer far from the central body, infalling particles get closer and closer to the Schwarzschild radius while going slower and slower, eventually stopping at $r = R_M$ for $ct \to \infty$.

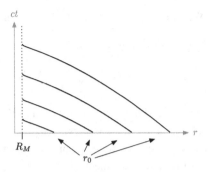

Fig. 4.2 A Minkowski diagram for ct versus r showing radial infall of particles starting at different initial positions r_0. Here particles never pass through the Schwarzschild radius of the central body.

Example 4.1 (There and Back Again). The proper time equation of motion for Schwarzschild geometry from (4.24) is identical to the Newtonian radial infall equation of motion, $\ddot{r}(t) = -GM/r(t)^2$. We used an alternate parameterization of ct and r in the latter (Newtonian) case, and it is beneficial here as well (this is done in [35]). Since it is τ that provides the connection, we'll take the following ψ-parameterized radial description, reminiscent of (1.55) from Example 1.3,

$$c\tau(\psi) = \frac{r_0}{2}\sqrt{\frac{r_0}{R_M}}(\psi + \sin\psi), \qquad r(\psi) = \frac{r_0}{2}(1 + \cos\psi), \qquad (4.32)$$

where $\psi = -\pi \to \pi$ describes a particle leaving $r = 0$ at $c\tau = -\pi/2r_0 \times \sqrt{r_0/R}$, reaching r_0 at $\tau = 0$, and then returning to $r = 0$ at $c\tau = \pi/2r_0 \times \sqrt{r_0/R}$.

To make the appropriate spacetime diagram, we need to find $ct(\psi)$ to go along with $r(\psi)$. We can get the relation from (4.21) using the constant energy determined by setting $r(0) = r_0$, $\dot{r}(0) = 0$ in (4.23),

$$\frac{dt}{d\tau} = \frac{dt}{d\psi}\frac{d\psi}{d\tau} = \frac{E/(mc^2)}{1 - R_M/r} \longrightarrow \frac{dct}{d\psi} = \frac{r_0^2\sqrt{-1 + r_0/R_M}(1 + \cos\psi)^2}{2(r_0 - 2R_M + r_0\cos\psi)} \tag{4.33}$$

and this can be solved for $ct(\psi)$:

$$ct(\psi) = \frac{1}{2}\sqrt{-1 + \frac{r_0}{R_M}}\left(\psi(r_0 + 2R_M) + r_0\sin\psi\right) \\ + R_M\log\left(\left|\frac{\sqrt{-1 + r_0/R_M}\cos(\psi/2) + \sin(\psi/2)}{\sqrt{-1 + r_0/R_M}\cos(\psi/2) - \sin(\psi/2)}\right|\right). \tag{4.34}$$

Strictly speaking, we solved for $ct(\psi)$ for $\psi = -\pi \to -\psi_0$, where $\psi_0 = \cos^{-1}(2R_M/r_0 - 1)$ has $r(\psi_0) = R_M$, and the right-hand side changes sign. Then we solved for the region $\psi = -\psi_0 \to 0$ for the points outside the event horizon and noted that both solutions can be captured using the log function with the absolute value of the argument, as shown.

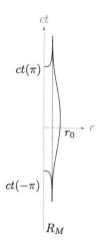

Fig. 4.3 A particle starts at $r = 0, c\tau = -\pi/2 r_0 \sqrt{r_0/R_M}$, at $ct(-\pi)$ from (4.35). In finite proper time, the particle makes it to r_0 at $\tau = 0$. Meanwhile, the coordinate time has become infinitely negative and then returned to match the coordinate time at r_0. From rest at this location, the particle is pulled back towards the center, again with finite proper time, but this time with the coordinate time becoming infinite and running backwards in the process. A backwards-running coordinate time is "perfectly fine" in the Schwarzschild geometry, even if it is not the stuff of our everyday experience.

Looking at the spacetime diagram of the motion in Figure 4.3, we can clearly see the problematic piece of the trajectory. The particle starts at

$$ct(-\pi) = -\frac{\pi}{2}\sqrt{-1 + \frac{r_0}{R_M}}(r_0 + 2R_M) \qquad (4.35)$$

and goes to $-\infty$ as $r(\psi) \to R_M$, all while τ (and, of course, ψ) remain finite. From negative temporal infinity, the particle comes into r_0 at $ct = 0$, where it is at rest. The process then proceeds in the other direction, with the particle falling back to R_M at $ct \to \infty$, then going from $t \to \infty$ back to the origin at $ct(\pi)$ (the negative of (4.35)). All of this occurs at increasing value of τ, but for values of ct with $r < R_M$, $\dot{t}(\tau) < 0$ from (4.33), so the coordinate time decreases as the proper time increases. When the particle is outside the event horizon, $\dot{t}(\tau) > 0$, and coordinate time increases with τ.

For massive particles, we use the proper time parameterization and $L = -1/2mc^2$ to solve for $ct(r)$. But we can use the same Lagrangian to study the radial geodesics associated with light just by taking $L = 0$ as in Section 2.4.1. With the lightlike geodesics, we can construct the light cone at different points in the ctr plane. Since $L = 0$,

$$L = \frac{1}{2}\left[-\left(1 - \frac{R_M}{r(\tau)}\right)c^2\dot{t}(\tau)^2 + \left(1 - \frac{R_M}{r(\tau)}\right)^{-1}\dot{r}(\tau)^2\right] = 0. \qquad (4.36)$$

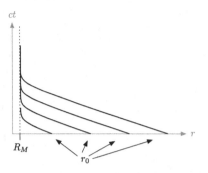

Fig. 4.4   Minkowski diagram, in ct versus r, for lightlike radial infall starting from a few different locations r_0 at $ct = 0$. The light never crosses the Schwarzschild radius in coordinate time parameterization, similar to the material infall shown in Figure 4.2.

Dividing both sides by $\dot{r}(\tau)^2$ and isolating $\frac{d(ct)}{dr}$,

$$\frac{d(ct)}{dr} = \pm\frac{1}{1 - R_M/r} \tag{4.37}$$

with solution, taking $ct(R_0) = 0$,

$$ct(r) = \pm\left(r - R_0 + R_M \log\left(\frac{r - R_M}{R_0 - R_M}\right)\right). \tag{4.38}$$

From the tangent defined by (4.37), it is clear that we recover the 45° lines that bound the "usual" light cones as $r \to \infty$. But closer in, the light cone structure changes dramatically. Plotting the radial infall curves in Figure 4.4, we see that light does not cross the special boundary at R_M, just as we had for massive particle radial geodesics parameterized by time in (4.31) and shown in Figure 4.2. But in that case, we know that the massive particles in fact cross right over the boundary at R_M and fall in towards the center, by analyzing the trajectories in proper time parameterization. Is the same true for light; can it fall in past the boundary? What we lack is a "proper time" parameterization with which to "see" (pun intended) the effect.

The immediate problem with the metric is the "rr" component g_{11}, which becomes infinite for $r \to R_M$. We'll look for a coordinate system that does not break down at this special radius. Because we are focused on the radial geodesics, we will think about coordinate transformations that take t and r to some new u and v set. The goal, again, is to remove the singularity at $r = R_M$, and to keep track of its location, we'll let $v = r$, so that it stays fixed. Referring to the covariant transformation law, with u and v the "barred" coordinates,

$$\bar{g}_{\mu\nu} = \frac{\partial x^\alpha}{\partial \bar{x}^\mu} \frac{\partial x^\beta}{\partial \bar{x}^\nu} g_{\alpha\beta}, \tag{4.39}$$

where the sums are really just over 0 and 1 in the two-dimensional subspace. The transformed metric is

$$\bar{g}_{\mu\nu} \doteq \begin{pmatrix} -(1 - \frac{R_M}{v})(\frac{\partial(ct)}{\partial u})^2 & -(1 - \frac{R_M}{v})\frac{\partial(ct)}{\partial v}\frac{\partial(ct)}{\partial u} \\ -(1 - \frac{R_M}{v})\frac{\partial(ct)}{\partial v}\frac{\partial(ct)}{\partial u} & \frac{1}{1-R_M/v} - (1 - \frac{R_M}{v})(\frac{\partial(ct)}{\partial v})^2 \end{pmatrix}. \tag{4.40}$$

Our goal is to write this transformed metric in the form $\bar{g}_{\mu\nu} = \eta_{\mu\nu} + h_{\mu\nu}$. The zero–zero component of the Schwarzschild metric is already in this form, so it is reasonable to set $ct = u + f(v)$ by additive separation of variables. Then $\bar{g}_{00} = g_{00}$ numerically, and we retain the h_{00} of the Schwarzschild solution. Now the transformed metric looks like

$$\bar{g}_{\mu\nu} \doteq \begin{pmatrix} -(1 - \frac{R_M}{v}) & -(1 - \frac{R_M}{v})f'(v) \\ -(1 - \frac{R_M}{v})f'(v) & \frac{1}{1-R_M/v} - (1 - \frac{R_M}{v})f'(v)^2 \end{pmatrix}. \tag{4.41}$$

Is it possible to also obtain a perturbative version of g_{11} that can be viewed as $1 + h_{11}$? In Schwarzschild coordinates, the large r limit would give $g_{11} \approx 1 + R_M/r$, and we'll make $h_{11} = R_M/r$ our target. Referring to (4.41), we want $f(v)$ to satisfy the ODE

$$f'(v) = \pm\frac{R_M}{v - R_M} \longrightarrow f(v) = B \pm R_M \log(R_M - v) \tag{4.42}$$

for constant of integration B. The coordinate transformation is then $u = ct - B \mp R_M \log(R_M - v)$, $v = r$, with metric

$$\bar{g}_{\mu\nu} \doteq \begin{pmatrix} -(1 - \frac{R_M}{v}) & \mp\frac{R_M}{v} \\ \mp\frac{R_M}{v} & 1 + \frac{R_M}{v} \end{pmatrix}. \tag{4.43}$$

This set of coordinates is called the "Eddington–Finkelstein" coordinate system (described in [12, 17]), with the plus choice the "ingoing" coordinates and minus the "outgoing" set.

Take the plus sign in $u = ct - B + R_M \log(R_M - v)$ but check out Problem 4.16 to see what happens if you keep the minus. Let's go back and compute the lightlike radial geodesics in this new coordinate system. Dropping the barred notation, we'll work in the two-dimensional subspace with coordinates u and v and a metric given by the entries in (4.43),

$$L = \frac{1}{2}\left(-\left(1 - \frac{R_M}{v(\tau)}\right)\dot{u}(\tau)^2 + \left(1 + \frac{R_M}{v(\tau)}\right)\dot{v}(\tau)^2 + 2\frac{R_M}{v(\tau)}\dot{u}(\tau)\dot{v}(\tau)\right) = 0. \tag{4.44}$$

This time, we'll use the ignorability of u to define a constant K of the motion via

$$\frac{\partial L}{\partial \dot{u}} \equiv -K = -\left(1 - \frac{R_M}{v(\tau)}\right)\dot{u}(\tau) + \frac{R_M}{v(\tau)}\dot{v}(\tau), \tag{4.45}$$

so that the Lagrangian itself is

$$L = \frac{v(\tau)(K^2 - \dot{v}(\tau)^2)}{2(R_M - v(\tau))} = 0. \tag{4.46}$$

Solving this equation for $\dot{v}(\tau)$, we have the pair (taking both roots for $\dot{v}(\tau)$):

$$\dot{v}(\tau) = -K, \qquad \dot{u}(\tau) = K,$$
$$\dot{v}(\tau) = K, \qquad \dot{u}(\tau) = \frac{K(R_M + v(\tau))}{v(\tau) - R_M}, \tag{4.47}$$

for ingoing and outgoing geodesics, respectively.

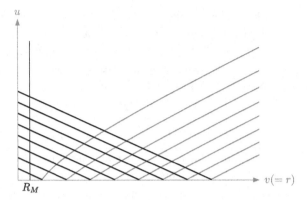

Fig. 4.5 The lightlike radial trajectories shown on a Minkowski diagram with a few different starting points $r_0 > R_M$. The ingoing trajectories are black, with the outgoing ones gray. Light goes through the points with $v = R_M$.

Switching to the Minkowski diagram format, we are interested in $\frac{du}{dv}$ for $\dot{v}(\tau) = \pm K$,

$$\dot{v}(\tau) = -K \longrightarrow \frac{du}{dv} = -1,$$
$$\dot{v}(\tau) = K \longrightarrow \frac{du}{dv} = \frac{R_M + v}{v - R_M}. \tag{4.48}$$

From these equations for the tangent to the radial geodesic curves in the Minkowski diagram (a plot, here, of motion with horizontal axis $v = r$, vertical axis u), the ingoing trajectories are just lines with slope -1, whereas the outgoing curves are bent, with solution

$$u(v) = (v - v_0) + 2R_M \log\left(\frac{v - R_M}{v_0 - R_M}\right). \tag{4.49}$$

The ingoing and outgoing trajectories for light are shown in Figure 4.5, where you can see the trajectory of light suffers no discontinuity as it goes in towards $v = 0$.

What happens to light emitted from $v = r < R_M$? We can use our pair of solutions for points with starting radius below R_M, and plots of those curves are shown in Figure 4.6. Light that starts out "inside" the radius R_M goes in towards the center, either along a line with unit slope (for light pointing "in") or along a logarithmic curve. The surface at R_M acts as a "one way membrane," allowing material particles (try working out the timelike geodesics in Eddington–Finkelstein coordinates in Problem 4.17) and light to enter but not exit. This behavior defines the "event horizon," and the radius R_M is known as the "Schwarzschild event horizon." In order for the event horizon to exist in vacuum, all the material producing the gravitational field must be inside the Schwarzschild radius. These sources are called "black holes" and were already a prediction of Newtonian gravity as discussed in Section 1.3. We now see them emerge in Einstein's theory of gravity.

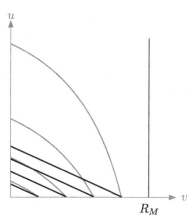

Fig. 4.6   The two types of radial lightlike geodesics for starting locations inside the Schwarzschild radius R_M.

4.4 Kruskal–Szekeres Coordinates

Schwarzschild coordinates are good when we are interested in describing physics as seen by an observer far away, at spatial infinity where the Schwarzschild metric is flat. They are not as good when we want to think about what happens near or below the event horizon at R_M, where the coordinate system breaks down. It is clear, from either the radial infall of massive particles or the radial motion of light in Eddington–Finkelstein coordinates that the "singularity" in Schwarzschild coordinates at $r = R_M$ is an artifact of the coordinate system itself and not a feature of the spacetime (if it were, then we would not be able to remove the singularity by making a change of coordinates). This type of "coordinate singularity" shows up at, for example, the north pole in spherical coordinates. One way of determining, at least roughly, if a singularity is a coordinate singularity or a "true" singularity is to compute scalars (coordinate-independent quantities) that involve the Riemann tensor. If the scalar is infinite at the singularity, then all coordinate systems share the singularity. In Problem 4.18, you will show that a particular scalar correctly sorts the Schwarzschild singularities at $r = R_M$ and $r = 0$ into "coordinate" and "true."

The Schwarzschild coordinates for points with $r < R_M$ are strange. For $r > R_M$, the t coordinate represents a time, and the r coordinate a position, just based on the signs associated with them in the line element,

$$ds^2 = -\left(1 - \frac{R_M}{r}\right)c^2 dt^2 + \left(1 - \frac{R_M}{r}\right)^{-1} dr^2, \qquad (4.50)$$

with $g_{00} < 0$ and $g_{11} > 0$. But just below the event horizon, we have $g_{00} > 0$ and $g_{11} < 0$, so that t is a spatial coordinate, and r a temporal one.[4] If you

---

[4] Using the signs in the metric to distinguish between spatial and temporal coordinate interpretations.

plot the light cones, the intersection of ingoing and outgoing lightlike radial geodesics, from lightlike radial geodesics that start at values below R_M, you find that they are tipped over sideways, and this happens abruptly at $r = R_M$. Contrast that with the light cones defined by the intersection of radial geodesics in Figure 4.6 – these also tip sideways, an expression of the inevitable eventual fall to $r = 0$ below the event horizon, but do so continuously.

The boundary at $r = R_M$, while real, and with physics that can be described in Schwarzschild coordinates, has broader implications for spacetime. To probe those, it is useful to define coordinates that are free of singularity, so that we use one set of coordinates everywhere. Focusing on the two-dimensional (ct, r) subspace of a spherically symmetric spacetime, there exists a coordinate system that is conformally flat,[5] meaning that there are coordinates u and v such that $ds^2 = \Phi(u, v)(-dv^2 + du^2)$ for some function $\Phi(u, v)$. Taking the original Schwarzschild subspace with its ct and r coordinates and using the contravariant transformation of $g^{\mu\nu}$, we can find the metric in the new coordinates u and v using

$$\bar{g}^{\mu\nu} = \frac{\partial \bar{x}^\mu}{\partial x^\alpha} \frac{\partial \bar{x}^\nu}{\partial x^\beta} g^{\alpha\beta} \tag{4.51}$$

(the contravariant version is useful in anticipation of the resulting form of the coordinates u and v, which are difficult to invert).

To avoid the coordinate singularity, we will generate a pair of coordinate transformations, one on each side of R_M, and require that they have the same conformally flat metric. That will lead to a discontinuity in the coordinate *transformation*, the price we pay to make the metric continuous. The "Kruskal–Szekeres" coordinate system [29, 48] proceeds by relating u and v to r everywhere as follows,

$$-v^2 + u^2 = \left(\frac{r}{R_M} - 1\right) e^{r/R_M}, \tag{4.52}$$

and exploiting the multiple solutions to this equation to define u and v differently on either side of R_M. With those choices in place, the transformation reads

$$
\left.
\begin{aligned}
v &= \sqrt{1 - \frac{r}{R_M}} e^{r/(2R_M)} \cosh\left(\frac{ct}{2R_M}\right) \\
u &= \sqrt{1 - \frac{r}{R_M}} e^{r/(2R_M)} \sinh\left(\frac{ct}{2R_M}\right)
\end{aligned}
\right\} \text{ for } r < R_M,
$$

$$
\left.
\begin{aligned}
v &= \sqrt{-1 + \frac{r}{R_M}} e^{r/(2R_M)} \sinh\left(\frac{ct}{2R_M}\right) \\
u &= \sqrt{-1 + \frac{r}{R_M}} e^{r/(2R_M)} \cosh\left(\frac{ct}{2R_M}\right)
\end{aligned}
\right\} \text{ for } r > R_M.
\tag{4.53}
$$

---

[5] This is true for any two-dimensional space; see [7].

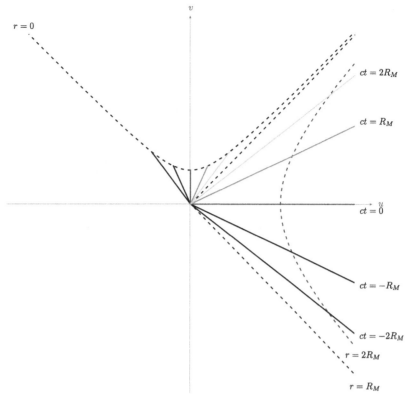

Fig. 4.7 Curves of constant r and ct (Schwarzschild coordinates) plotted in the uv plane of the Kruskal–Szekeres coordinate system. The dashed hyperbolae have constant value of r with solid lines representing constant ct values. The grayscale is tied to the constant value of ct and r here. Notice that the constant ct "curves" are piecewise lines, made up of both the $r < R_M$ and $r > R_M$ forms of u and v from (4.53).

In both of these cases, we get

$$\bar{g}^{\mu\nu} \doteq \frac{e^{r/R_M}r}{4R_M^3}\begin{pmatrix} -1 & 0 \\ 0 & 1 \end{pmatrix} \longrightarrow \bar{g}_{\mu\nu} \doteq \frac{4e^{-r/R_M}R_M^3}{r}\begin{pmatrix} -1 & 0 \\ 0 & 1 \end{pmatrix}, \qquad (4.54)$$

and conformal form has been achieved, albeit implicitly, with the function $\Phi(u, v)$ defined in terms of $r(u, v)$, itself determined by (4.52). In Figure 4.7, we have plotted curves of constant ct and r in the uv plane using (4.53). The conformal form of the metric in Kruskal–Szekeres coordinates means that the radial geodesic for light in the uv plane is $v = \pm u$, just as in the familiar Minkowski spacetime with $ct = \pm x$ shown in Figure A.4.

We can use the ψ-parameterized form for radial geodesics from Example 4.1 to see what infall looks like in Kruskal–Szekeres coordinates. The idea is to take $r(\psi)$ from (4.32) and $ct(\psi)$ from (4.34) and put them into (4.53) to get $v(\psi)$ and $u(\psi)$ using the appropriate set for points with $r < R_M$ and $r > R_M$, then plot the geodesics in the uv plane for $\psi = 0 \to \pi$ for a particle starting from rest at r_0 at $t = 0$ (noting that we have to switch from the outer form of

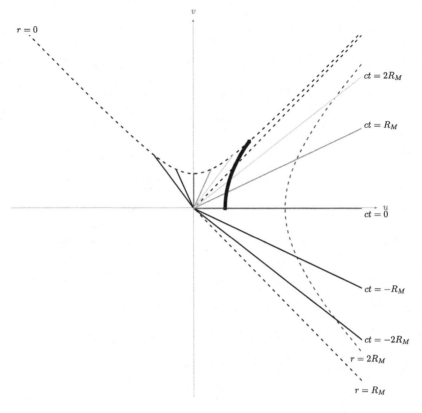

Fig. 4.8 A massive radial geodesic in Kruskal–Szekeres coordinates (thick line). The particle starts from rest at $1.25\,R_M$ on the horizontal axis and travels "upward" through increasing time and decreasing radial coordinate. The particle crosses the "point" $r = R_M$ and continues on to reach $r = 0$. Any light signal sent from along the trajectory would move along lines of unit slope, parallel to $r = R_M$, so that none would reach points with $r > R_M$ after the particle passes through the horizon.

the coordinates to the inner, once the particle reaches R_M at $\psi_0 = \cos^{-1}(-1 + 2R_M/r_0)$). The result for one radial geodesic is shown in Figure 4.8, where we can see the particle starting at $r_0 = 1.25\,R_M$ and moving in time until it crosses the event horizon (now a line) and continues on to $r = 0$. Any light signal sent by the traveler would move along the lightlike $45°$ lines, and it is clear that once the massive body has passed the event horizon, no light can be sent back across the R_M surface.

How about the full out and back story from Example 4.1? What does that look like in Kruskal–Szekeres coordinates? For $r_0 = 1.5\,R_M$, the plot is shown in Figure 4.9. It is interesting that the portion of the trajectory that goes from $r = 0 \rightarrow R_M$ is disconnected from the rest of the trajectory, in which the particle crosses the horizon at R_M at $ct < 0$ and then smoothly falls to $r = 0$. That's the nature of the coordinate system, though, with the "point" $r = R_M$ spread out over a line so that the negative time pieces of the trajectory that fall between $0 < r < R_M$ get displaced.

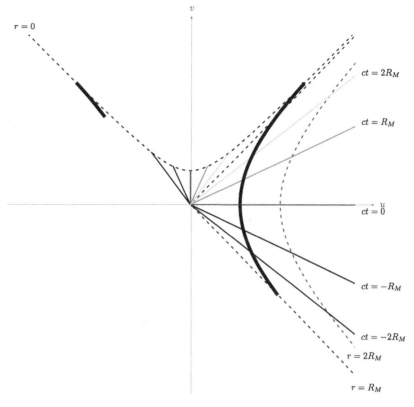

A massive radial geodesic in Kruskal–Szekeres coordinates (thick line). This time, the particle starts at $r = 0$ at $t < 0$, arriving at $r_0 = 1.5 R_M$ at $t = 0$, and then returns. The initial portion of the trajectory is on the left side of the coordinate system, where $r = 0$ at $ct < 0$ is. The trajectory moves over to the right half of the coordinate system when the particle passes through the event horizon (at $ct = -\infty$), then proceeds continuously until the particle ends up across the horizon (at $ct = \infty$), and eventually returns to $r = 0$.

4.5 Scattering and Bound Trajectories

Since we will be focusing on geodesics that at least start outside of the event horizon, we will use the Schwarzschild coordinates. These are adapted for observers far away (at infinity) and so are useful in matching predictions to what we observe when looking at faraway objects. We'll start with the Lagrangian

$$L = \frac{m}{2}\left[-\left(1 - \frac{R_M}{r(\tau)}\right)c^2\dot{t}(\tau)^2 + \left(1 - \frac{R_M}{r(\tau)}\right)^{-1}\dot{r}(\tau)^2 \right.$$
$$\left. + r(\tau)^2\left(\dot{\theta}(\tau)^2 + \sin^2\theta(\tau)\dot{\phi}(\tau)^2\right)\right] = \alpha, \tag{4.55}$$

where $\alpha = -mc^2/2$ for massive particles, zero for light. The structure of this Lagrangian is very similar to that of Newtonian gravity, and it shares the rotational symmetry that leads to conserved angular momentum. So we can again pick $\theta = \pi/2$ and $\dot{\theta} = 0$ as in Section 1.4, leaving

$$L = \frac{m}{2}\left[-\left(1 - \frac{R_M}{r(\tau)}\right)c^2\dot{t}(\tau)^2 + \left(1 - \frac{R_M}{r(\tau)}\right)^{-1}\dot{r}(\tau)^2 + r(\tau)^2\dot{\phi}(\tau)^2\right] = \alpha.$$

(4.56)

Neither t nor ϕ appears in L, so there are two constants of motion, naturally associated with the energy of the particle (as before) and its remaining z-component of angular momentum:

$$\frac{\partial L}{\partial \dot{t}(\tau)} = -E = -mc^2\left(1 - \frac{R_M}{r}\right)\dot{t}(\tau), \qquad \frac{\partial L}{\partial \dot{\phi}(\tau)} = L_z = mr(\tau)^2\dot{\phi}(\tau).$$

(4.57)

We can use this pair to eliminate reference to $\dot{t}(\tau)$ and $\dot{\phi}(\tau)$ in the Lagrangian and then solve for $\dot{r}(\tau)^2$:

$$\frac{1}{2}m\dot{r}(\tau)^2 = \alpha + \frac{E^2}{2mc^2} + \underbrace{\frac{L_z^2 R_M}{2mr^3} - \frac{L_z^2}{2mr^2} - \frac{\alpha R_M}{r}}_{\equiv -U_{\text{eff}}}.$$

(4.58)

The r-dependent term on the right defines an effective potential similar to the one in (1.61). We can rewrite (4.58) in the form of a one-dimensional Hamiltonian where the constant "energy" is $\bar{E} \equiv \alpha + E^2/(2mc^2)$:

$$\bar{E} = \frac{1}{2}m\dot{r}(\tau)^2 + U_{\text{eff}}.$$

(4.59)

4.5.1 Massive Particles

For massive particle trajectories, $\alpha = -mc^2/2$, and the effective potential is

$$U_{\text{eff}} = \frac{L_z^2}{2mr^2} - \frac{mc^2 R_M}{2r} - \frac{R_M L_z^2}{2mr^3}.$$

(4.60)

The first two terms are precisely the "centrifugal barrier" and Newtonian gravitational potential energy associated with a spherical central body of mass M. The third term is the only new element here. But its presence dominates the behavior of the effective potential for $r \to 0$. In Newtonian gravity, it is the positive $1/r^2$ term that sets the behavior near zero, as shown in Figure 1.7. In particular, the sharp upturn as $r \to 0$ shown there meant that nothing (with $L_z \neq 0$) makes it to the center of the central body. The effective potential for geodesics in Schwarzschild geometry, sketched in Figure 4.10, shows that there are many orbital trajectories that end below the event horizon. Any particle with "energy" above the peak (like $\bar{E}_f$ in the figure) will end up at $r = 0$ in finite time if it begins with $\dot{r} \leq 0$. We also retain the "scattering" trajectories (particle comes in, interacts, and goes back out) at $\bar{E}_s$, the bound elliptical orbits (these will now precess), shown in the figure at $\bar{E}_e$, and the degenerate circular orbits $\bar{E}_o$, all as before.

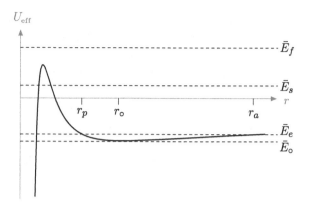

Fig. 4.10   A sketch of the effective potential from (4.60) showing three distinct regions: (1) near $r = 0$, the potential behaves like $-1/r^3$, (2) there is a barrier associated with the $1/r^2$ term, and (3) a tail for large r that is the usual $-1/r$. In addition to energy constants $\bar{E}$ from (4.59) associated with scattering ($\bar{E}_s$), elliptical ($\bar{E}_e$) and circular ($\bar{E}_o$) orbits, there is a new option $\bar{E}_f$, in which a particle ends up pulled into $r = 0$.

Let's start with the most familiar, circular orbits. There are two obvious places where you could get circular orbits. Referring to Figure 4.10, both the local minimum (at r_o) and the maximum support motion with $\dot{r}(\tau) = 0$. The maximum is unstable, though, and any particle orbiting at that inner radius wouldn't last long. The pair of extrema are at

$$\frac{dU_{\text{eff}}}{dr} = 0 \longrightarrow r = \frac{L_z^2}{m^2c^2R_M}\left(1 \pm \sqrt{1 - 3\left(\frac{mcR_M}{L_z}\right)^2}\right). \quad (4.61)$$

We lose the distinction between the maximum and minimum radii when the square root above is zero, and for radii that make the square root imaginary, the orbit is unstable, so there is an "innermost stable circular orbit" (or "ISCO") that occurs when $L_z = \sqrt{3}mcR_M$ giving $r = 3R_M$ in (4.61).

The rest of the geodesics benefit from our move to $\rho \equiv 1/r$, turning the expression on the right in (4.58) into a polynomial in ρ. In addition, we will parameterize the motion in ρ with the angle ϕ as in Section 1.4 to get a geometric picture of the trajectory. Using $\dot{\phi}(\tau) = L_z/(mr(\tau)^2)$, we can write (4.58) as

$$\left(\frac{d\rho(\phi)}{d\phi}\right)^2 = -\frac{m^2c^2}{L_z^2} + \frac{E^2}{c^2L_z^2} + R_M\rho(\phi)^3 - \rho(\phi)^2 + \frac{m^2c^2R_M}{L_z^2}\rho(\phi). \quad (4.62)$$

Taking the ϕ derivative of both sides, we can clear out the constants and solve for $\rho''(\phi)$ with an eye towards comparison with (1.67):

$$\frac{d^2\rho(\phi)}{d\phi^2} = -\rho(\phi) + \frac{m^2c^2R_M}{2L_z^2} + \frac{3}{2}R_M\rho(\phi)^2. \quad (4.63)$$

The first two terms on the right are precisely the ones from (1.67), only the third term is new.

There are several ways to make progress with the nonlinear equation of motion in (4.63). We'll look at numerical solutions for geodesics later in this chapter, but for now, let's focus on some approximate solutions that let us characterize the new piece of the motion brought by the ρ^2 term. We'll assume that the particle motion occurs far from the Schwarzschild radius and is approximately circular. Call that circular radius R, and we'll take $R_M/R \ll 1$. There is a natural length scale built into (4.63), $\ell \equiv L_z/(mc)$; this will end up being of order R, so we will also have $R_M/\ell \ll 1$. Finally, define the dimensionless $u(\phi) \equiv R\rho(\phi)$ allowing us to keep track of the small quantities. Then (4.63) becomes

$$\frac{d^2u(\phi)}{d\phi^2} = -u(\phi) + \frac{RR_M}{2\ell^2} + \frac{3}{2}\frac{R_M}{R}u(\phi)^2. \tag{4.64}$$

There is a circular orbit when the right-hand side of this equation is zero, occurring at

$$u = \frac{R}{3R_M}\left(1 \pm \sqrt{1 - \frac{3R_M^2}{\ell^2}}\right). \tag{4.65}$$

If we expand on the right in powers of $R_M/\ell \ll 1$, then

$$u \approx \frac{R}{3R_M}\left[1 \pm \left(1 - \frac{3}{2}\left(\frac{R_M}{\ell}\right)^2\right)\right], \tag{4.66}$$

and the goal is to get $u \approx 1$; then $\rho(\phi) \approx 1/R$ as desired. If we take the plus sign, then the leading term for u is $\sim R/R_M \gg 1$, so take the minus sign, leaving us with leading order $u \approx RR_M/(2\ell^2)$. For this to be of order unity, set $R \equiv 2\ell^2/R_M$, which is much larger than ℓ itself.

We will assume that $u(\phi) = 1 + \epsilon f(\phi)$ for $\epsilon \equiv R_M/R \ll 1$, so that the motion is dominated by the circular orbit at radius R with corrections that are consistent with $R \gg R_M$. Putting our form for $u(\phi)$ into (4.64), the equation of motion for $f(\phi)$ is

$$\epsilon\frac{d^2f(\phi)}{d\phi^2} = -1 - \epsilon f(\phi) + 1 + \frac{3}{2}\epsilon\left(1 + 2\epsilon f(\phi) + \epsilon^2 f(\phi)^2\right). \tag{4.67}$$

We can cancel a factor of ϵ, and we'll drop the $f(\phi)^2$ term (which goes like ϵ^2 even after canceling an ϵ) to get

$$\frac{d^2f(\phi)}{d\phi^2} = -f(\phi)(1 - 3\epsilon) \longrightarrow \tag{4.68}$$
$$f(\phi) = A\cos(\phi\sqrt{1 - 3\epsilon}) + B\sin(\phi\sqrt{1 - 3\epsilon}).$$

To compare with the Newtonian case, we only need to keep the cosine term. Going from $u = R\rho$ to $r = 1/\rho$ gives

$$r(\phi) = \frac{R}{1 + \epsilon A\cos(\phi\sqrt{1 - 3\epsilon})}, \tag{4.69}$$

a form that we recognize. The numerator defines the semilatus rectum of an ellipse, $p = R$, and the dimensionless constant in the denominator defines the eccentricity $e = \epsilon A$. We'll just measure the orbital parameters directly using

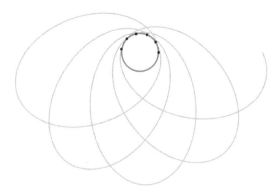

Fig. 4.11 Five orbits of a precessing elliptical trajectory with ellipticity exaggerated for clarity. The circle at the center is defined by the perihelion radius and is sketched for reference. The points on the circle show the shift in the closest approach point starting at $\phi = 0$, the right-most point on the perihelion circle, and ending at $5 \times (2\pi + 3\pi R_M/p)$ on the left.

(1.73) to relate the geometric p and e to the easily observable semimajor and semiminor axes of planets in our solar system, for example.

Finally, expanding the argument of cosine in $\epsilon \ll 1$ and writing everything in terms of p and e, we get

$$r(\phi) = \frac{p}{1 + e\cos(\phi(1 - (3/2)(R_M/p)))}. \tag{4.70}$$

Solution (4.70) is an "ellipse" that precesses as it goes around the central body. If you start at the point of closest approach, $\phi = 0$, then you return to that point not at $\phi_r = 2\pi$, but instead at

$$\left(1 - \frac{3}{2}\frac{R_M}{p}\right)\phi_r = 2\pi \longrightarrow \phi_r = \frac{2\pi}{1 - (3/2)(R_M/p)} \approx 2\pi\left(1 + \frac{3}{2}\frac{R_M}{p}\right). \tag{4.71}$$

Each orbit picks up an additional shift of $\Delta\phi = 3\pi R_M/p$, as shown in Figure 4.11. You can calculate the effect for Mercury, which as the closest planet to the sun exhibits the biggest angular shift, in Problem 4.19.

4.5.2 Light

To compute the geodesics for light, we have to go back to (4.56) and use $\alpha = 0$, dividing through by m while we're at it. We'll work in Schwarzschild coordinates here, since we are focused, for now, on orbits that don't fall into the central body. Finally, to highlight that there is no proper time for light, we'll switch the name of the curve parameter to s. The starting Lagrangian is

$$L = \frac{1}{2}\left[-\left(1 - \frac{R_M}{r(s)}\right)c^2\dot{t}(s)^2 + \left(1 - \frac{R_M}{r(s)}\right)^{-1}\dot{r}(s)^2 + \frac{1}{2}r(s)^2\dot{\phi}(s)^2\right] = 0. \tag{4.72}$$

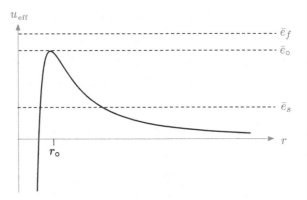

A sketch of the effective potential for lightlike geodesics with nonzero angular momentum from (4.75). There are three types of motion characterized by the "energies" $\bar{e}_f$, in which light falls in to the center of the source mass, $\bar{e}_o$, a circular orbit, and the "scattering" $\bar{e}_s$ (for geodesics starting to the right of the maximum).

We again have ignorable t and ϕ coordinates and can use those to define "energy" e and the z component of "angular momentum," ℓ_z. I put these in quotes since they do not have the correct units, even though they play the role of E and L_z for massive particles. Using

$$e = \left(1 - \frac{R_M}{r(s)}\right)c^2 \dot{t}(s), \qquad \ell_z = r(s)^2 \dot{\phi}(s) \tag{4.73}$$

as in (4.57), we can replace $\dot{t}(s)$ and $\dot{\phi}(s)$, set $L = 0$, and solve for $\dot{r}(s)^2$ as we did to get (4.58):

$$\frac{1}{2}\dot{r}(s)^2 = \frac{e^2}{2c^2} - \frac{\ell_z^2}{2r(s)^2} + \frac{\ell_z^2 R_M}{2r(s)^3}. \tag{4.74}$$

From this equation, we can extract an effective potential (per unit mass), as before, with $\bar{e} \equiv e^2/(2c^2)$ here, and one-dimensional "energy conservation"

$$\bar{e} = \frac{1}{2}\dot{r}(s)^2 + \underbrace{\frac{\ell_z^2}{2r(s)^2} - \frac{\ell_z^2 R_M}{2r(s)^3}}_{\equiv u_{\text{eff}}}. \tag{4.75}$$

A sketch of the potential, which is dominated by the negative $1/r^3$ term near $r \to 0$ and by the positive $1/r^2$ term for large r, is shown in Figure 4.12. There are three interesting types of motion we can get for light with nonzero angular momentum. There is a capture trajectory, in which light goes around the central body but eventually falls into $r = 0$, associated with $\bar{e}_f$ in the figure. If we start at $\bar{e}_f$ with $\dot{r} > 0$, then the geodesic heads away from the central body to infinity. At $\bar{e}_s$ we see a "scattering" geodesic with light coming in, bending around the source mass and going back out. Finally, there is an unstable circular orbit that occurs at $r_o = 3R_M/2$, where the effective potential has a maximum at $\bar{e}_o$.

We'll focus on the bending of light geodesics with "energy" $\bar{e}_s$ that start to the right of the maximum in Figure 4.12, since these were also predicted in Section 1.5.3. Again using an inverse radial coordinate parameterized by ϕ takes (4.74) to

$$\left(\frac{d\rho(\phi)}{d\phi}\right)^2 = \frac{e^2}{c^2 \ell_z^2} - \rho(\phi)^2 + R_M \rho(\phi)^3. \tag{4.76}$$

Taking the ϕ derivative to avoid the sign change in $\rho'(\phi)$,

$$\frac{d^2\rho(\phi)}{d\phi^2} = -\rho(\phi) + \frac{3}{2} R_M \rho(\phi)^2. \tag{4.77}$$

We start, as in Section 1.4.4, with the $R_M = 0$ case (no central mass), with solution $\rho(\phi) = \sin\phi/R$. The trajectory in this case is a line at $y = R$ for $x = -\infty \to \infty$. We'll again assume that the unperturbed length scale R is large compared to the Schwarzschild radius R_M, so take $R_M/R \equiv \epsilon \ll 1$. Define the dimensionless $u(\phi) = R\rho(\phi)$. Then (4.77) reads

$$\frac{d^2 u(\phi)}{d\phi^2} = -u(\phi) + \frac{3}{2}\epsilon u(\phi)^2. \tag{4.78}$$

We expect $u(\phi)$ to be dominated by the $R_M \to 0$ limit, so we'll set $u(\phi) = \sin\phi + \epsilon f(\phi)$ for unknown "correction" $f(\phi)$. Using this assumed form in (4.78),

$$-\sin\phi + \epsilon\frac{d^2 f(\phi)}{d\phi^2} = -\left(\sin\phi + \epsilon f(\phi)\right) + \frac{3}{2}\epsilon\left(\sin^2\phi + 2\epsilon f(\phi)\sin\phi + \epsilon^2 f(\phi)^2\right). \tag{4.79}$$

Just as we did for the massive elliptical trajectories, divide by ϵ, but this time, keep only the leading order terms,

$$\frac{d^2 f(\phi)}{d\phi^2} \approx -f(\phi) + \frac{3}{2}\sin^2\phi. \tag{4.80}$$

This equation looks like that of a "driven" harmonic oscillator. The solution is

$$f(\phi) = A\cos\phi + B\sin\phi + \frac{1}{2}\left(1 + \cos^2\phi\right), \tag{4.81}$$

and we'll set $A = B = 0$ since that behavior has already been taken care of by the leading $\sin\phi$ in u, so we just need the new term $(1 + \cos^2\phi)/2$,

$$f(\phi) = \frac{1}{2}\left(1 + \cos^2\phi\right), \tag{4.82}$$

and then

$$u(\phi) = \sin\phi + \frac{R_M}{2R}\left(1 + \cos^2\phi\right). \tag{4.83}$$

Referring back to Figure 1.11 with our current $r(\phi) = R/u(\phi)$, we have $r(\pi/2) = R/(1 + \epsilon/2) \approx R$ as the point of closest approach. The x axis crossing happens at $x_0 = r(0) = R/(\epsilon) = R^2/(R_M)$, leading to a "viewing" angle θ with

$$\theta \approx \tan\theta \approx \frac{R}{x_0} \longrightarrow \theta \approx \frac{R_M}{R}, \tag{4.84}$$

and then the deflection angle $\psi = 2\theta$ is

$$\psi = \frac{2R_M}{R}, \tag{4.85}$$

twice the result from (1.104), as advertised. As an example of the sizes involved here, the Sun has the Schwarzschild radius $R_M \approx 3$ km, and the *smallest* R could be is a grazing incidence at the solar surface, $R \approx 7 \times 10^5$ km, giving a deflection angle of $\psi \approx 8.6 \times 10^{-6}$ or $\psi \approx 1.8$ arcseconds. This light deflection effect was (arguably) measured during a solar eclipse in 1919 by Eddington [11].

4.5.3 Numerical Geodesics

There have been few exact results so far. We have been focused on exploring the new physics of the Schwarzschild geodesics, and for that and comparison purposes, perturbations of known trajectories are useful. But at some point, we would like to look at the strong field regime, where perturbation is unwieldy. There are also some exotic elements to the trajectories that we can probe numerically. So flip through Appendix B, and we'll begin by putting the geodesic equations of motion into the form relevant for Runge–Kutta numerical solution.

To handle a general metric, we'll keep the numerical solution focused on the geodesic equations of motion, so that we will solve

$$\frac{d\dot{x}^\sigma(\tau)}{d\tau} = -\Gamma^\sigma{}_{\beta\nu}\dot{x}^\beta(\tau)\dot{x}^\nu(\tau). \tag{4.86}$$

The first step is to write this set of four second-order equations as eight first-order ones, and we can do this using the embedding $f^0 \equiv x^0$, $f^1 \equiv x^1$, $f^2 \equiv x^2$, $f^3 \equiv x^3$, $f^4 \equiv h^0 \equiv \dot{x}^0$, $f^5 \equiv h^1 \equiv \dot{x}^1$, $f^6 \equiv h^2 \equiv \dot{x}^2$, and $f^7 \equiv h^3 \equiv \dot{x}^3$. Then the geodesic equation can be written as

$$\frac{d}{d\tau}\begin{pmatrix} f^\mu \\ h^\sigma \end{pmatrix} \doteq \begin{pmatrix} h^\mu \\ -\Gamma^\sigma{}_{\beta\nu}h^\beta(\tau)h^\nu(\tau) \end{pmatrix} \tag{4.87}$$

with position dependence found inside the connection coefficients.

For initial conditions we are in luck, since the Schwarzschild metric has geodesic motion that lies in a plane, and we have taken that to be the xy plane, so that initially (and at all other times), $\theta(0) = \pi/2$, $\dot{\theta}(0) = 0$. In addition, we have two constants of the motion, E (or e for light), which can be used to set $\dot{t}(0)$, and L_z (ℓ_z), which gives us $\dot{\phi}(0)$. We will choose L_z (ℓ_z) so as to include the variety found in Figure 4.10 and Figure 4.12, and E will be chosen to select a specific type of trajectory from those figures (using the relations $\bar{E} = -mc^2/2 + E/(2mc^2)$ and $\bar{e} = e/(2c^2)$). For the last of the initial conditions, start off at $\phi(0) = 0$ with $t(0) = 0$ and pick any r from a valid region for the chosen energy from the effective potential sketches. The only remaining initial condition is $\dot{r}(0)$, and we'll set this by requiring that we are

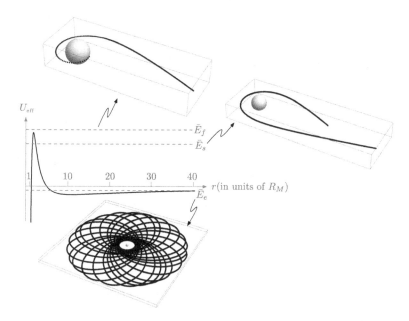

Fig. 4.13 Three example massive particle geodesics with the effective potential energy plot used to inform the initial conditions of each trajectory. In each case, the particle starts at a distance of $10R_M$ (the sphere represents the source mass and is given radius R_M for reference). For the top trajectory associated with energy $\bar{E}_f$, the particle goes around the central body and then falls into it, approaching $r = 0$ (obscured by the sphere). On the right, a particle comes in, goes around the sphere, then heads back out, escaping to infinity. The bottom is orbital motion occurring between roughly $4R_M$ and $40R_M$; you can clearly see the precession.

in proper time parameterization for particles, $\dot{x}^\mu(0)g_{\mu\nu}\dot{x}^\nu(0) = -c^2$ or, for light, $\dot{x}^\mu(0)g_{\mu\nu}\dot{x}^\nu(0) = 0$. In each case, we are presented with two options for $\dot{r}(0)$, positive (initially outgoing) and negative (initially ingoing), and we take the negative one.

Plots of points along the particle geodesics for each of the different types of trajectory shown in Figure 4.10 are displayed in Figure 4.13. To make these plots, we interpret the coordinates r, θ, and ϕ as spherical (appropriate if we are viewing from far away) and plot points using $x(\tau) = r(\tau)\sin(\theta(\tau))\cos(\phi(\tau))$, etc. at different values of τ.

For light, with geodesics shown in Figure 4.14, we can see a capture with $\bar{e}_f$ and an extreme case of bending at $\bar{e}_s$. As a warning, the circular trajectory at $\bar{e}_o$ is shown. In that case, the light circles around the central body many times before eventually "falling" off the maximum and heading outward. Here it is the numerical method that provides the perturbation that causes the particle to leave its unstable circular orbit. In a real physical situation, perturbations from interaction with other objects would have the same effect.

These trajectories are shown for equally spaced proper time, but we might also want to display them as functions of the coordinate time. That will change the overall appearance of the trajectory, since, for example, in coordinate time

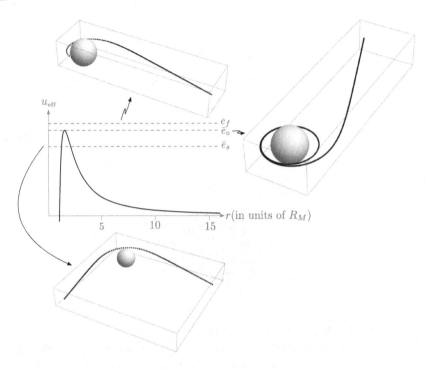

Fig. 4.14 Some representative geodesics for light and the effective potential energy governing them. In each plot, the spherical source of radius R_M is shown for reference. In the top trajectory with $\bar{e}_f$, the light starts at $10R_M$ and falls below R_M heading towards $r = 0$. The bottom trajectory at $\bar{e}_s$ is an extreme example of the bending of light that starts at $10R_M$ and gets deflected through roughly ninety degrees (this geodesic would not be described by the perturbative setup we used in Section 4.5.2). On the right, the unstable circular trajectory at $r_o = 3R_M/2$ is shown at $\bar{e}_o$. After many circular orbits, the light leaves, with its instability seeded by the numerical error.

parameterization we know that a massive particle never falls below the event horizon of a black hole, whereas in proper time parameterization it does, so whole sections of the trajectory (i.e., the portion with $r < R_M$) won't be probed in the coordinate time case. Using coordinate time to parameterize the motion also gives us a better picture of what we, as observers far from other astrophysical bodies of interest, would see since the Schwarzschild coordinate time is *our* proper time as stationary observers at spatial infinity (approximately).

Reviewing the motivation for switching to coordinate time takes longer than the procedure that accomplishes it. All we have to do is divide both sides of (4.87) by $(h^0/c) = \dot{t}(\tau)$ to get

$$\frac{d}{dt}\begin{pmatrix} f^\mu \\ h^\sigma \end{pmatrix} \doteq \begin{pmatrix} h^\mu c/h^0 \\ -\Gamma^\sigma{}_{\beta\nu}h^\beta(\tau)h^\nu(\tau)c/h^0 \end{pmatrix}. \tag{4.88}$$

Try making some of the trajectories from Figure 4.13 and Figure 4.14 in Problem 4.20.

4.6 Gravitational Time Dilation and Red Shift

In this section, we'll think more about the implications of the Schwarzschild solution, but ones that are not explorable using geodesics. Gravitational time dilation is the difference in the rate of clocks held at different locations in a gravitational field, an idea we saw early on in Section 1.8. The effect occurs for clocks at rest, unlike special relativistic time dilation. For a clock, or anything, to be at rest in a gravitational field, there must be some force holding it up. So when we construct a "clock at rest" at Schwarzschild radial coordinate r, for example, there must be rockets or some other mechanism keeping it there.

Suppose we have a pair of clocks at rest outside of a spherical massive body of mass M. The source sets up the Schwarzschild spacetime, and we'll take the clocks to be at $\theta = \pi/2$, $\phi = 0$, that is, along a line, with one at r_1 and the other at $r_2 > r_1$. Take a fixed coordinate time interval dt; then each clock measures a different local time. In a time dt, the clock at r_1 registers a proper time $d\tau_1$, related to dt by

$$d\tau_1 = \sqrt{1 - \frac{R_M}{r_1}}dt. \tag{4.89}$$

Similarly, the clock at r_2 measures $d\tau_2$ with

$$d\tau_2 = \sqrt{1 - \frac{R_M}{r_2}}dt. \tag{4.90}$$

Relating the two times through the common factor of dt,

$$\frac{d\tau_1}{\sqrt{1 - R_M/r_1}} = \frac{d\tau_2}{\sqrt{1 - R_M/r_2}} \longrightarrow d\tau_2 = \sqrt{\frac{1 - R_M/r_2}{1 - R_M/r_1}}d\tau_1, \tag{4.91}$$

and since everything is at rest, we can integrate to get the same relation between an elapsed time $\Delta\tau_2$ in terms of $\Delta\tau_1$,

$$\Delta\tau_2 = \sqrt{\frac{1 - R_M/r_2}{1 - R_M/r_1}}\Delta\tau_1. \tag{4.92}$$

You can expand this relation assuming that r_1 and r_2 are much larger than R_M in Problem 4.21. But just to get the direction right, take $r_1 = 2R_M$ and $r_2 = 20R_M$; then $\Delta\tau_2 = \sqrt{19/10}\Delta\tau_1$. If a year passes at r_1, then ≈ 1.4 years pass at r_2.

Using the weak field form of gravity with $g_{\mu\nu} = \eta_{\mu\nu} + h_{\mu\nu}$, we can generalize the time dilation result to spacetimes other than Schwarzschild. Suppose you have a static distribution of mass ρ that generates a Newtonian gravitational potential φ in the usual way via $\nabla^2\varphi = 4\pi G\rho$. This potential is related to h_{00} by $h_{00} = -2\varphi/c^2$ as we saw in Section 3.6.1. Then the relation between the time interval $d\tau_1$ elapsed on a clock at rest at vector location $\mathbf{r}_1$ and the coordinate time is

$$d\tau_1 = \sqrt{1 - h_{00}}dt = \sqrt{1 + \frac{2}{c^2}\varphi(\mathbf{r}_1)}dt, \tag{4.93}$$

and similarly for a clock at rest at $\mathbf{r}_2$, giving the general relation

$$\frac{d\tau_1}{\sqrt{1 + (2/c^2)\varphi(\mathbf{r}_1)}} = \frac{d\tau_2}{\sqrt{1 + (2/c^2)\varphi(\mathbf{r}_2)}}. \tag{4.94}$$

You can use this result to find the difference in time elapsed between the crust of the Earth and its center [50] in Problem 4.22.

The gravitational redshift of light is really just a manifestation of time dilation. We'll again take two stationary clocks at $\theta = \pi/2$, $\phi = 0$ and radial locations r_1 and r_2 as above. Now light with period T_1 as measured by the clock at r_1 is emitted at r_1, headed towards r_2. When it gets there, the clock at r_2 measures its period to be T_2. What is the relation between the frequency of light measured at r_1 where it is emitted and its received frequency at r_2? From (4.92), the two periods are related by

$$T_2 = \sqrt{\frac{1 - R_M/r_2}{1 - R_M/r_1}}T_1, \tag{4.95}$$

and since the frequency f is just one over the period, we have $f_1 = 1/T_1$ and $f_2 = 1/T_2$ with

$$f_2 = \sqrt{\frac{1 - R_M/r_1}{1 - R_M/r_2}}f_1. \tag{4.96}$$

This frequency change was measured in the Pound–Rebka experiment: gamma rays displayed frequency shifts as they moved to different heights within the gravitational field of the Earth [38].

Taking $r_2 \rightarrow \infty$, an observer far away from the source, and letting $r_1 \rightarrow r$ an arbitrary location, we have

$$f_\infty = \sqrt{1 - \frac{R_M}{r}}f_r, \tag{4.97}$$

the observed frequency "at infinity" is less than the emitted frequency at r, the frequency has been "redshifted."

4.7 The Kerr Metric

The Schwarzschild metric is an important solution to Einstein's equation in vacuum, but it describes a spherical source that does not move, and most astrophysical objects, while approximately spherical, are also spinning. The Kerr metric is the spacetime exterior to a spinning massive sphere. It is a vacuum solution, so that it has $R_{\mu\nu} = 0$ (check in Problem 4.26). Deriving it is not easy; the original paper by Kerr [27] uses sophisticated algebraic techniques but leaves the metric in a form that is difficult to interpret and use for physically relevant situations. The preferred coordinate system is "Boyer–Lindquist" from [5], in which the metric takes the form

$$
g_{\mu\nu} \doteq \begin{pmatrix} -(1 - \frac{R_M r}{\rho^2}) & 0 & 0 & -\frac{a R_M r \sin^2\theta}{\rho^2} \\ 0 & \frac{\rho^2}{\Delta} & 0 & 0 \\ 0 & 0 & \rho^2 & 0 \\ -\frac{a R_M r \sin^2\theta}{\rho^2} & 0 & 0 & ((r^2 + a^2) + \frac{a^2 R_M r \sin^2\theta}{\rho^2})\sin^2\theta \end{pmatrix},
$$

$$(4.98)$$

$$
R_M \equiv \frac{2GM}{c^2},
$$

$$
\rho^2 \equiv r^2 + a^2 \cos^2\theta,
$$

$$
\Delta \equiv a^2 - R_M r + r^2.
$$

You should work out the "large r" limit in Problem 4.27 to identify the effective $h_{\mu\nu}$ and compare with our predicted form from Example 3.3; that will provide the interpretation of the source as a spinning massive sphere with mass M and angular momentum-related length[6] $a = J/(Mc)$. You can explore the various limiting cases in Problem 4.28.

The Kerr metric has a number of interesting and new properties as compared to Schwarzschild. Let's start with the event horizon, where $g_{11} \to \infty$ in the Schwarzschild case. The Kerr metric has $g_{11} \to \infty$ for $\Delta \to 0$, which happens for radial coordinate

$$
r = \frac{1}{2}R_M \pm \sqrt{\left(\frac{R_M}{2}\right)^2 - a^2}.
$$

$$(4.99)$$

The outer radius is the relevant one here, since the inner radius is then obscured by it:

$$
r_h = \frac{1}{2}R_M + \sqrt{\left(\frac{R_M}{2}\right)^2 - a^2}.
$$

$$(4.100)$$

This surface is undefined for values of a that are larger than $R_M/2$, and that serves to define the maximum spin of a Kerr spacetime; you couldn't have a spherical body of mass M with $a > R_M/2$.[7]

In addition to the event horizon, there is something happening when $\rho^2 = 0$ (at $r = 0$, $\theta = \pi/2$), since then the inverses appearing in the metric are infinite,

---

[6]   You will also find the Kerr metric written in terms of $a = J/M$, an angular momentum per unit mass, as in [7]. Most renderings have the speed of light set to one, so there is no difference between the two forms.

[7]   Alternatively, for $a > R_M/2$, the complex r_h in (4.100) could simply mean that the horizon doesn't exist. There is still a singularity in the Kerr metric, and Penrose's "cosmic censorship hypothesis" states that such singularities should be covered by event horizons so that they are undetectable. From this point of view, there *must* be a real value of r_h.

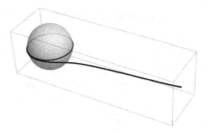

A spinning spherical central body sets up the Kerr geometry. A particle falls along a geodesic with zero angular momentum, yet has nonzero $\frac{d\phi}{dt}$ due to frame dragging. Here the geodesic is parameterized by t so that we can see the angular velocity from (4.104).

and we get zero contribution from the radial and θ directions. In that case, the line element for the Kerr metric is

$$ds^2 = -c^2 dt^2 + a^2 d\phi^2, \tag{4.101}$$

describing a ring of radius a. This singularity is called the "ring singularity." You should check to see if these singularities are coordinate or true singularities in Problem 4.31.

In the Kerr spacetime, as with any that have nontrivial g^{0i} components, it is possible to have a geodesic trajectory that has zero angular momentum but nonzero $\dot{\phi}(t)$. Consider geodesics in the plane with $\theta = \pi/2$ and $\dot{\theta} = 0$ (see [7] for a complete description of the geodesics of the Kerr spacetime). The geodesic Lagrangian is

$$L = \frac{1}{2}m\left(g_{00}c^2\dot{t}(\tau)^2 + 2g_{03}c\dot{t}(\tau)\dot{\phi}(\tau) + g_{11}\dot{r}(\tau)^2 + g_{33}\dot{\phi}(\tau)^2\right), \tag{4.102}$$

and the metric components are independent of t and ϕ, so the conjugate momentum of each is a constant. We'll use the usual names for these constants,

$$\frac{\partial L}{\partial \dot{t}} = -E = g_{00}mc^2\dot{t}(\tau) + mcg_{03}\dot{\phi}(\tau),$$
$$\frac{\partial L}{\partial \dot{\phi}} = L_z = mcg_{03}\dot{t}(\tau) + mg_{33}\dot{\phi}(\tau). \tag{4.103}$$

Taking $L_z = 0$ gives the relation $cg_{03}\dot{t}(\tau) = -g_{33}\dot{\phi}(\tau)$; we can form the angular speed by division:

$$\frac{d\phi(t)}{dt} = \frac{\dot{\phi}(\tau)}{\dot{t}(\tau)} = -\frac{cg_{03}}{g_{33}} = \frac{acR_M}{a^2r + r^3 + a^2R_M}. \tag{4.104}$$

A particle with no angular momentum will rotate around the central body in the same direction as its spin, a phenomenon known as "frame dragging." The effect is larger closer in to the source mass. In Figure 4.15, we see the trajectory of a particle that starts far from the source falling radially inward that gets turned around the central body – the effect is most pronounced near the event horizon at r_h from (4.100). This geodesic was created numerically using the geodesic equations of motion from Section 4.5.3 and appropriate ($L_z = 0$) initial conditions in coordinate time parameterization. You should try creating some of the other equatorial trajectories from Section 4.5.3 using the Kerr metric with various spin values. Because the Kerr geometry is not spherically symmetric, there are more than just planar orbits, and some of these can get pretty interesting.

4.8 The Weyl Metrics

The Weyl class of metrics is associated with axisymmetric vacuum spacetimes. Axisymmetry in flat space (we'll add in time at the end) is easiest to express in cylindrical coordinates, where the symmetry refers to rotation about the z axis. Scalars, then, are axisymmetric if they depend only on s (the radial coordinate) and z. Vectors can have magnitude that depends on s and z, and point in the $\hat{\mathbf{s}}$ and $\hat{\mathbf{z}}$ directions. The polar angle ϕ does not appear, nor can vectors point in that direction, since that would lead to angular dependence.

Let's start by constructing a general "axially symmetric" metric out of the components of $\hat{\mathbf{s}} = \cos\phi\hat{\mathbf{x}} + \sin\phi\hat{\mathbf{y}}$ and $\hat{\mathbf{z}}$ together with functions of s and z. For the three-dimensional spatial subspace, we take, in Cartesian coordinates,

$$g^{ij} = p\left(\sqrt{x^2 + y^2}, z\right)s^i s^j + q\left(\sqrt{x^2 + y^2}, z\right)z^i z^j$$
$$+ u\left(\sqrt{x^2 + y^2}, z\right)\left(s^i z^j + s^j z^i\right) + v\left(\sqrt{x^2 + y^2}, z\right)\delta^{ij}, \tag{4.105}$$

where $p(s, z)$, $q(s, z)$, $u(s, z)$, and $v(s, z)$ are arbitrary functions, and δ^{ij} is the flat spatial metric (inverse). We can then convert to cylindrical coordinates with ordering $\bar{x}^1 = s, \bar{x}^2 = z, \bar{x}^3 = \phi$:

$$\bar{g}^{ij} \doteq \begin{pmatrix} p(s, z) + v(s, z) & u(s, z) & 0 \\ u(s, z) & q(s, z) + v(s, z) & 0 \\ 0 & 0 & \frac{v(s,z)}{s^2} \end{pmatrix}. \tag{4.106}$$

The sz sector (top 2×2 subblock) of the metric forms a two-dimensional subspace, and then we know that there is a coordinate choice that will bring that sector into conformally flat form, with just an overall (unknown) function $w(s, z)$ sitting out front. So we could really start off with a spatial metric of the form (dropping the bar and redefining s and z to be the conformally flat set),

$$g^{ij} = \begin{pmatrix} w(s, z) & 0 & 0 \\ 0 & w(s, z) & 0 \\ 0 & 0 & \frac{v(s,z)}{s^2} \end{pmatrix}. \tag{4.107}$$

To this spatial metric, we will add in the temporal component, with yet another unknown function of s and z.[8] The initial line element has the form (removing the s^2 from $d\phi^2$ to simplify the calculation)

$$ds^2 = -e^{2a(s,z)}c^2dt^2 + e^{2y(s,z)}\left(ds^2 + dz^2\right) + e^{2f(s,z)}d\phi^2, \qquad (4.108)$$

where the use of exponentials is to simplify the Ricci tensor components. Those components can be computed, giving four equations governing the three unknown functions $a(s,z)$, $y(s,z)$, and $f(s,z)$. As it turns out, the R_{00} and $R_{\phi\phi}$ components are independent of $y(s,z)$, and taking a $\pm$ pair of linear combinations yields the simplified field equations (in vacuum)

$$e^{-2a+2b}R_{00} + e^{2b-2f}R_{\phi\phi}$$
$$= 0 = \left(\frac{\partial a}{\partial z}\right)^2 + \frac{\partial^2 a}{\partial z^2} + \left(\frac{\partial a}{\partial s}\right)^2 + \frac{\partial^2 a}{\partial s^2} - (a \to f),$$
$$e^{-2a+2b}R_{00} - e^{2b-2f}R_{\phi\phi}$$
$$= 0 = \left(\frac{\partial}{\partial s}(a+f)\right)^2 + \left(\frac{\partial}{\partial z}(a+f)\right)^2 + \frac{\partial^2}{\partial s^2}(a+f) + \frac{\partial^2}{\partial z^2}(a+f).$$
$$\qquad (4.109)$$

The second equation can be satisfied by setting $f = -a$, eliminating one of the metric components. Then the first equation becomes

$$e^{-2a+2b}R_{00} + e^{2b+2a}R_{\phi\phi} = 0 = 2\left(\frac{\partial^2 a}{\partial s^2} + \frac{\partial^2 a}{\partial z^2}\right), \qquad (4.110)$$

which doesn't look too bad. We can build $f = -a$ into the starting point, which is traditionally taken to be (putting the s^2 back on the $d\phi^2$ term)

$$ds^2 = -e^{2a(s,z)}c^2dt^2 + e^{-2a(s,z)}\left(e^{2b(s,z)}\left(ds^2 + dz^2\right) + s^2d\phi^2\right), \quad (4.111)$$

where now $a(s,z)$ and $b(s,z)$ are the independent unknown functions.

You will work out the details in Problem 4.32, but part of the solution to the vacuum Einstein equation is set by an equation (a linear combination of the Ricci components) governing a:

$$\nabla^2 a(s,z) \equiv \frac{\partial^2 a(s,z)}{\partial z^2} + \frac{1}{s}\frac{\partial a(s,z)}{\partial s} + \frac{\partial^2 a(s,z)}{\partial s^2} = 0, \qquad (4.112)$$

that is, the function $a(s,z)$ satisfies a flat space Laplace equation. Then $b(s,z)$ is determined from a. There is an infinite family of solutions for $a(s,z)$, and this is sensible since there are many sources that lead to axisymmetric spacetimes. The Schwarzschild geometry, for example, must be included, in addition to spacetimes with infinite line sources.

---

[8]   In this case, we are assuming that the source is static, so that we have zeroed the off-diagonal components involving the temporal coordinate, just as we did in the static, spherically symmetric starting point from Section 4.1.

Exercises

4.1 For the electric potential in vacuum, $\nabla^2 V = 0$, find $\mathbf{E}$ given the boundary condition

$$\lim_{\epsilon \to 0}\left(\left.\frac{\partial V}{\partial z}\right|_{z=\epsilon} - \left.\frac{\partial V}{\partial z}\right|_{z=-\epsilon}\right) = \alpha \qquad (4.113)$$

for a constant α and a small positive constant ϵ. What is the source of the electric field? What if, instead, we had

$$\lim_{\epsilon \to 0}\left(\left.\frac{\partial V}{\partial x}\right|_{x=\epsilon} - \left.\frac{\partial V}{\partial x}\right|_{x=-\epsilon}\right) = \alpha, \qquad (4.114)$$

what is $\mathbf{E}$ and its source here?

4.2 Using $p(r) = a(r)^2 b(r)$ and $q(r) = 1/b(r)$ in (4.6), find the nonzero Ricci tensor components and solve for $a(r)$ and $b(r)$.

4.3 Show that a spherically symmetric metric that satisfies Einstein's equation in vacuum must be time-independent (this is known as "Birkhoff's theorem"). One way to proceed is to start from (4.6) with $p(r, t)$ and $q(r, t)$ and then show that Einstein's equation implies that there is no time dependence in these functions.

4.4 Show that the Schwarzschild metric (4.12) does not represent a flat spacetime.

4.5 What is the energy associated with a particle that starts at $t = 0$ from rest at R_0 in the Schwarzschild geometry? Using this energy, evaluate the right-hand side of (4.29) and use that expression for the tangent to the curve $ct(r)$ to sketch the trajectory of a particle falling radially into a (spherical) central body of mass M for a few different starting locations. Make sure the slope at $r = R_0$ and $r = R_M$ is correct, but just a rough sketch at other points will do. Sketch a few outgoing trajectories as well.

4.6 Stable circular geodesics in Schwarzschild geometry are essentially just like the geodesics on a sphere that we saw in Example 2.20. If the circular orbit occurs at Schwarzschild coordinate r, how far does a massive particle go in one full orbit (use $ds = \sqrt{-dx^\mu g_{\mu\nu}dx^\nu}$ so that the total distance is real)? What is the relation between coordinate time and proper time? What is the spatial distance traveled in one orbit? If you viewed this as a circle, what would its radius be?

4.7 Show that there can be no gravity outside of spherically symmetric sources in $D = 1 + 1$ (one spatial, one temporal dimension), assuming that Einstein's equation in vacuum reads, here $R_{\mu\nu} = 0$. No gravity in this context means that all vacuum solutions associated with a spherically symmetric distribution of mass M are flat, and hence the geodesics are straight lines.

4.8 There was no gravity available in $D = 1 + 1$ dimensions because of the constraints of geometry. What does Newtonian gravity look like in one spatial dimension? Assume that the form of the field equation is unchanged: $\nabla^2 \varphi = 4\pi \tilde{G} \rho$ for $\tilde{G}$ a gravitational constant with the appropriate units. Is Newtonian gravity detectable in $D = 1 + 1$?

4.9 Starting from a spherically symmetric metric in $D = 2 + 1$ dimensions

$$g_{\mu\nu} = \begin{pmatrix} -a(r) & 0 & 0 \\ 0 & b(r) & 0 \\ 0 & 0 & r^2 \end{pmatrix} \tag{4.115}$$

for coordinates $x^0 = ct$, $x^1 = r$, $x^2 = \phi$, show that the vacuum solution is flat.

4.10 How about Newtonian gravity in $D = 2 + 1$ dimensions: Does $\nabla^2 \varphi = 4\pi \tilde{G} \rho$ (for $\tilde{G}$ the gravitational constant in $2 + 1$ dimensions) produce nontrivial, detectable gravity from spherically symmetric sources?

4.11 In four spatial dimensions with Cartesian coordinates $\{x, y, z, u\}$, the surface of a three-dimensional sphere has the constraint

$$x^2 + y^2 + z^2 + u^2 = R^2, \tag{4.116}$$

where R is the sphere's radius. If you were trapped on the surface of the three-sphere, what would the metric be? (hint: on a two sphere, fixing θ gives a one-sphere (circle) of radius $r \sin \theta$ – for three-dimensional spheres, introduce a new polar angle ψ such that for fixed ψ, you recover a two-sphere of radius $r \sin \psi$, take it from there). What is the Ricci scalar associated with the surface?

4.12 Make a sketch of the proper distance in (4.15) for a few different values of R_M and show that taking $R_M \to 0$ gives $s = r_2 - r_1$.

4.13 Einstein's equation with "cosmological constant" Λ is

$$R_{\mu\nu} - \frac{1}{2} g_{\mu\nu} R + g_{\mu\nu} \Lambda = \frac{8\pi G}{c^4} T_{\mu\nu}. \tag{4.117}$$

Show that vacuum solutions to this equation have $R_{\mu\nu} = 2 g_{\mu\nu} \Lambda$.

4.14 Find the spherically symmetric vacuum solution to (4.117) in $D = 3 + 1$ (feel free to use the `Mathematica` package). Your solution should have some constants of integration. Set those constants by demanding that you recover Schwarzschild when $\Lambda = 0$. Using the weak-field limit, what is the effective Newtonian potential associated with the cosmological constant? What Newtonian source generates this potential?

4.15 Find the spherically symmetric vacuum solution to Einstein's equation with cosmological constant Λ from (4.117), but in $D = 2 + 1$ dimensions. Is the spacetime flat?

4.16 Use the Eddington–Finkelstein coordinate definition with the minus choice in (4.42), so that $v = r$ and $u = ct + B - \log(R_M - v)$, to find the radial geodesics for light. Sketch the geodesics on a "Minkowski diagram" with u and v axes.

4.17 Find the massive particle geodesic equations of motion in Eddington–Finkelstein coordinates (use the plus sign in (4.42)). Sketch the geodesics on a "Minkowski diagram" of u versus v.

4.18 A coordinate singularity is one that comes from a coordinate choice rather than an intrinsic feature of the spacetime (think of the metric associated with the north pole of a sphere – that point appears to have metric problems because of the polar angle, yet there is nothing special happening there). There are two singularities in the Schwarzschild metric (places where it is potentially undefined); identify these points. One way to establish that a point is a coordinate or "true" singularity is to build scalars and see what happens at these "problematic" points for these scalars (a good idea, since scalars are coordinate independent). What is the simplest, nonzero, scalar that you can make from the metric (and its derivatives) that is *not* $g_{\mu\nu} g^{\mu\nu} = D$? Evaluate this scalar for the Schwarzschild metric (use the `Mathematica` package). Identify the two singularities as either "coordinate" or "true."

4.19 Look up the orbital parameters for Mercury, convert those to the p and e appearing in (4.70) (if necessary), and find the angular advance $\Delta\phi$ per century for Mercury. Look up the observed perihelion precession of Mercury and compare (if the theory of general relativity had appeared a few hundred years earlier, it would have been ignored, at least as far as explaining the precession of Mercury is concerned).

4.20 Using your Runge–Kutta routine and the geodesics equations of motion in temporal parameterization from (4.88), remake the highest energy plots from Figure 4.13 and Figure 4.14 and compare with the proper time plots.

4.21 Take $r_1 \gg R_M$ and $r_2 \gg R_M$ in (4.92) to find the time dilation result far from the Schwarzschild radius. If one year passes for a clock at rest on the surface of the earth, how much time has passed on a clock orbiting the Earth at a height (above the Earth's surface) of $20\,200$ km (don't worry about the orbital motion of the clock, pretend it is at rest, and ignore the special relativistic time dilation), where the global positioning system satellites are?

4.22 What is the difference in age between the crust and center of the Earth? Model the Earth as a sphere of total mass M spread uniformly throughout it and assume the crust is 4.5 billion years old [50].

4.23 Light of frequency f_1 is emitted at a location r_1 in a Schwarzschild spacetime (with R_M as the Schwarzschild radius). What is the minimum frequency that could be observed at a location r?

4.24 Using
$$\tau_2 = \sqrt{\frac{1 - R_M/r_2}{1 - R_M/r_1}}\,\tau_1, \tag{4.118}$$

find the time elapsed at infinity, given an elapsed time Δt on a stationary clock at Schwarzschild coordinate r.

4.25 In a Schwarzschild spacetime, what is the proper distance between a point $r > R_M$ and R_M?

4.26 Show that the Kerr metric has $R_{\mu\nu} = 0$ (don't do it by hand!).

4.27 Take $r \gg R_M$ and $r \gg a$ and Taylor expand the Kerr metric from (4.98). Identify the leading order metric perturbation and compare with the slowly rotating sphere form from Example 3.3.

4.28 What spacetime does the Kerr metric describe if you take $a \to 0$? How about taking $R_M \to 0$?

4.29 Write out the geodesic Lagrangian for the Kerr metric with $\theta = \pi/2$ and $\dot{\theta} = 0$. Noting that the Lagrangian is independent of both t and ϕ, identify the two constants of integration and solve for $\dot{r}(\tau)^2$ in proper time parameterization. What is the effective potential? Characterize the orbits you can get from it.

4.30 Using the energy and angular momentum definitions from the previous problem, pick some interesting values for orbital motion in the Kerr geometry and solve the geodesic equation of motion numerically in coordinate time parameterization (take $a = R_M/2$).

4.31 Evaluate the scalar $R^{\mu\nu\alpha\beta}R_{\mu\nu\alpha\beta}$ in the Kerr geometry. What do you conclude about the event horizon and ring singularity ("true" or "coordinate")?

4.32 Taking the metric associated with the starting line element from (4.108), show that Einstein's equation in vacuum, $R_{\mu\nu} = 0$, gives back $\nabla^2 a = 0$ as in (4.112) and a pair of equations that give the derivatives of $b(s, z)$ in terms of $a(s, z)$. Show that once you have chosen a harmonic $a(s, z)$, the equations for $b(s, z)$ are solvable and find their solution in the case $a(s, z) = \alpha/\sqrt{s^2 + z^2}$ for constant α.

5 Gravitational Waves and Radiation

In this chapter, we will study a very different type of vacuum solution to Einstein's equation: radiation. Because gravitational radiation has a tiny effect, at least for terrestrial observations, the linearized approach from Section 3.6.1 will be our primary tool. In that setting, the physical effects are clear, since we have an "actual" field on a flat background. That field can act on particles in the usual tidal way that we expect from gravity. Throughout this chapter, we will assume a flat Minkowski background with Cartesian coordinates, so all derivative operators will be the familiar partials ∂_μ.[1]

We'll start off by developing plane wave solutions to the field equations, and in particular, pin down the gauge degrees of freedom so that we can physically interpret the metric perturbation itself. Because that process of gauge fixing shares much in common with electromagnetic radiation, we'll review electromagnetic plane wave potentials with an eye towards making the four-potential's direction match the polarization vector of the electric field. That allows us to focus just on electromagnetic potentials and make physical predictions based on them without reference to the gauge-invariant electric and magnetic fields. This is good on the gravitational side, since the metric is really the main object of study, and making its entries physically relevant is important for quantitative prediction.

After thinking about the waves themselves, we'll work out their effect on particle motion. We will establish that you need to measure relative separations of more than one particle to detect gravitational radiation. We'll consider some simple mechanisms that could in principle measure the presence of a gravitational wave and outline the LIGO experiment that has recently made multiple detections and led to the Nobel Prize for its primary scientists in 2017 [2].

From detection we'll return to the question of sourcing. Here, again, the theory of electricity and magnetism provides a vector model that is useful in understanding both what is similar and what is different when it comes to gravitational sources of radiation. We'll review the production of radiation in E&M, focusing on the role of conservation to constrain the form of that radiation, and develop both dipole and (electric) quadrupole sourced radiation fields. That same story can be told, with similar techniques and approximations, to show how gravitational radiation arises from moving mass and energy.

---

[1] Of course, you can elect to do everything more generally using $\partial_\mu \to D_\mu$.

5.1 Review of Plane Waves in E&M

Monochromatic plane waves are vacuum solutions to Maxwell's equation, with well-defined wavelength and frequency, related by the speed c of the radiation in vacuum. In Lorenz gauge, where $\partial_\mu A^\mu = 0$, the source-free Maxwell's equation is $\Box A^\mu = 0$. The form of the plane wave solution is

$$A^\mu = P^\mu e^{ik^\nu x_\nu} = P^\mu e^{i(-k^0 ct + \mathbf{k}\cdot\mathbf{r})}, \tag{5.1}$$

with the useful $3 + 1$ decomposition into time t and spatial $\mathbf{r}$ shown. We are using the convenient exponential form of the plane wave, but before any physical calculation, we would pick out either the real or imaginary part of the solution in (5.1). Taking one derivative, $\partial_\rho A^\mu = ik_\rho A^\mu$, and introducing another ∂^σ to get the d'Alembertian, we learn that to solve Maxwell's equation, we must have $k_\rho k^\rho = 0$: the four-vector k^μ is null (lightlike). We also have the Lorenz gauge that needs to be enforced in the solution and that one reads $k_\mu P^\mu = 0$. So we should augment the solution in (5.1) with its requirements: $k^\mu k_\mu = 0 = k_\mu P^\mu$.

It helps to have a concrete example in mind as we look at the implications of the constraints imposed by the field equation and gauge condition. Let's take the electromagnetic wave to be traveling in the $\hat{\mathbf{z}}$ direction, so that only k^0 and k^z are nonzero in k^μ. Let $k^0 \equiv \omega/c$. Then from $k^\mu k_\mu = 0$ we have $k_z = \pm\omega/c$ with the $\pm$ signs telling us whether the wave is traveling in the $+\hat{\mathbf{z}}$ or $-\hat{\mathbf{z}}$ direction.[2] Taking the plus sign (positive $\hat{\mathbf{z}}$ direction wave), from the gauge condition we learn that $P^z = P^0$. Using $V/c = A^0$ and recalling that the spatial part of A^μ constitutes the magnetic vector potential $\mathbf{A}$, the electric and magnetic fields are

$$\begin{aligned}
\mathbf{E} &= -\nabla V - \frac{\partial \mathbf{A}}{\partial t} = i\omega e^{i(-\omega t + \omega z/c)}\left(P^x\hat{\mathbf{x}} + P^y\hat{\mathbf{y}}\right), \\
\mathbf{B} &= \nabla \times \mathbf{A} = i\frac{\omega}{c}e^{i(-\omega t + \omega z/c)}\left(-P^y\hat{\mathbf{x}} + P^x\hat{\mathbf{y}}\right).
\end{aligned} \tag{5.2}$$

There are only two degrees of freedom in these gauge-independent fields, P^x and P^y, suggesting that P^0 is irrelevant and could be set to zero. That would make the P^μ from the four-potential identical to the polarization vector, the vector that gives the direction of the electric field. But how could we have established that P^0 is "removable" prior to computing the fields themselves?

Going back to Lorenz gauge in this plane wave context, we ask: Is it possible to perform a gauge transformation of solution (5.1) that leaves it in Lorenz gauge (hence allowing it to remain a solution)? Take $A_\mu \to \tilde{A}_\mu \equiv A_\mu + \psi_{,\mu}$ for some scalar ψ. Then the divergence becomes

$$\partial^\mu \tilde{A}_\mu = \partial^\mu(A_\mu + \psi_{,\mu}) = \partial^\mu A_\mu + \Box\psi = \Box\psi \tag{5.3}$$

---

[2] Throughout the chapter, I'll often use the coordinate vector labels 0, x, y, and z as indices to highlight the different directions present in the solution.

since $\partial^\mu A_\mu = 0$ from the start. So any ψ with $\Box\psi = 0$ gives back a four-potential that is still in Lorenz gauge. We can guarantee that ψ solves this equation by taking it to be

$$\psi = \psi_0 e^{ik^\nu x_\nu} \tag{5.4}$$

for some constant ψ_0. Then we clearly have $\Box\psi = 0$ since $k^\nu k_\nu = 0$ already. The potential with this additional gauge fixing is of the general form

$$\tilde{A}_\mu \equiv A_\mu + \psi_{,\mu} = (P_\mu + i\psi_0 k_\mu)e^{ik^\nu x_\nu}. \tag{5.5}$$

The coefficient for $\mu = 0$ is $P_0 + i\psi_0 k_0$, and unlike the other elements of k^μ, we know from $k^\mu k_\mu = -(k^0)^2 + \mathbf{k}\cdot\mathbf{k} = 0$ that $k^0 \neq 0$ (unless $k^\mu = 0$ for all four components). So we can pick $\psi_0 = -iP_0/k_0$ to get $\tilde{A}^0 = 0$. In this new gauge, then, we have eliminated the P^0 degree of freedom entirely (the process stops here, of course, as you can check in Problem 5.2).

Our theoretical plane wave potential now has the same number of degrees of freedom as the example fields in (5.2). That solution, with all gauge choices in place, reads

$$A^\mu \doteq \begin{pmatrix} 0 \\ P^x \\ P^y \\ 0 \end{pmatrix} e^{i(-\omega t + \omega z/c)} \tag{5.6}$$

and leads to the electric and magnetic fields in (5.2).

In the general case of a plane wave that is traveling in the $\hat{\mathbf{k}}$ direction, we can use a projection operator to isolate the portion of the vector P^j that is orthogonal to $\hat{\mathbf{k}}$. Given a unit vector $\hat{\mathbf{n}}$ with components n^j (for $j = 1, 2, 3$),[3] define the projection operator by

$$N^i{}_j \equiv \delta^i_j - n^i n_j. \tag{5.7}$$

This acts on a vector with components v^j to produce a new vector that is perpendicular to $\hat{\mathbf{n}}$,

$$N^i{}_j v^j = v^i - n^i\left(v^j n_j\right). \tag{5.8}$$

The operation is more familiar when you think of it in vector form. The piece of $\mathbf{v}$ that is perpendicular to $\hat{\mathbf{n}}$ is just all of $\mathbf{v}$ minus the bit that points in the $\hat{\mathbf{n}}$ direction, $\mathbf{v} - \hat{\mathbf{n}}(\mathbf{v}\cdot\hat{\mathbf{n}})$, and the component form of this equation is precisely (5.8).

Now generalize to motion in the $\hat{\mathbf{k}}$ direction (our example above had $\hat{\mathbf{k}} = \hat{\mathbf{z}}$) with polarization in the $\hat{\mathbf{P}}$ direction. The algorithm for generating a plane wave solution with polarization vector $\mathbf{P}$ that is the spatial portion of P^μ and is orthogonal to $\mathbf{k}$ is as follows: Start by setting $P^0 = 0$, which we know can be done. Then take the general spatial polarization components with $P^1 = P^x$, $P^2 = P^y$, and $P^3 = P^z$ and project out the portion lying in the $\hat{\mathbf{k}}$ direction

---

[3] In general, we use Roman indices to refer to the spatial 1, 2, 3 dimensions.

using the projection operator from (5.7) with $\hat{\mathbf{n}} = \hat{\mathbf{k}}$ to get $\bar{P}^i = N^i{}_j P^j$. The plane wave solution is then

$$
A^\mu \doteq \begin{pmatrix} 0 \\ \bar{P}^1 \\ \bar{P}^2 \\ \bar{P}^3 \end{pmatrix} e^{-i\omega(t - \hat{\mathbf{k}}\cdot\mathbf{r}/c)}. \tag{5.9}
$$

5.1.1 Detection of Electromagnetic Plane Waves

Plane waves have both electric and magnetic fields and so will cause charges to move. The presence of the radiation can be inferred from the motion of the charges themselves or from the radiation that those charges generate (see Section 5.5). To see the effect of the radiation on a charge q, take a plane wave solution like (5.2); we'll take the sine piece of the exponential (our choice – remember that the physical fields are the real or imaginary pieces of solutions like (5.2)) and pick polarization in the $\hat{\mathbf{x}}$ direction, so that the relevant fields are

$$
\begin{aligned}
\mathbf{E} &= E_0 \sin\big(\omega(t - z/c)\big)\hat{\mathbf{x}}, \\
\mathbf{B} &= \frac{E_0}{c} \sin\big(\omega(t - z/c)\big)\hat{\mathbf{y}}.
\end{aligned} \tag{5.10}
$$

The equation of motion for a particle of mass m and charge q that starts from rest at the origin is

$$
m\ddot{\mathbf{r}}(t) = q\mathbf{E} + q\dot{\mathbf{r}} \times \mathbf{B} \text{ with } \mathbf{r}(0) = 0 \text{ and } \dot{\mathbf{r}}(0) = 0. \tag{5.11}
$$

Working out the component equations, we have

$$
\begin{aligned}
m\ddot{x}(t) &= qE_0\left(1 - \frac{\dot{z}}{c}\right) \sin\big(\omega(t - z/c)\big), \\
m\ddot{y}(t) &= 0, \\
m\ddot{z}(t) &= qE_0\frac{\dot{x}}{c} \sin\big(\omega(t - z/c)\big).
\end{aligned} \tag{5.12}
$$

We are using nonrelativistic equations of motion here, and we'll assume that the speed of the charge is much less than c, so that we can drop the $\dot{x}/c$ and $\dot{z}/c$ terms; the magnetic force will be small compared to the electric force. The motion of the particle occurs only in the $\hat{\mathbf{x}}$ direction, with $y(t) = z(t) = 0$, solutions of the approximate equations of motion that are consistent with the initial conditions. Then the equation of motion in the $\hat{\mathbf{x}}$ direction becomes

$$
m\ddot{x}(t) = qE_0 \sin(\omega t) \longrightarrow x(t) = \frac{qE_0}{m\omega}t - \frac{qE_0}{m\omega^2} \sin(\omega t), \tag{5.13}
$$

and you should check under what conditions our assumption that $\dot{x}/c \ll 1$ holds in Problem 5.4. The motion of the charge is oscillatory, like the fields themselves, and occurs with the same frequency. We could detect this type of motion from the current that would be generated by a collection of charges moving in this way, in an antenna, for example.

5.2 Plane Waves for Gravitational Radiation

We have already worked on the weak field equations of motion for general relativity. For a flat Minkowski metric, $\eta_{\mu\nu}$ (in Cartesian coordinates), and a perturbation field $h_{\mu\nu}$ that "sits on top" of the flat background, the linearized Einstein equation is (3.97). We have already discussed gauge choices for $h_{\mu\nu}$, so we can pick a gauge in which $\partial^\mu h_{\mu\nu} = 0$.[4] Then the field equation with source $T_{\alpha\mu} = 0$ is

$$\frac{1}{2}(-\Box h_{\alpha\mu} - \partial_\alpha\partial_\mu h + \eta_{\mu\alpha}\Box h) = 0. \tag{5.14}$$

If we multiply through by $\eta^{\mu\alpha}$ (in four dimensions), then we get $\Box h = 0$. So we can encapsulate the content of the field equation in this gauge:

$$\Box h = 0, \qquad \Box h_{\alpha\mu} + \partial_\alpha\partial_\mu h = 0, \qquad \partial^\alpha h_{\alpha\mu} = 0. \tag{5.15}$$

Start with a plane wave form analogous to (5.1),

$$h_{\mu\nu} = P_{\mu\nu}e^{ik_\rho x^\rho}, \tag{5.16}$$

where the "polarization tensor" $P_{\mu\nu}$ is a symmetric set of coefficients, the second-rank analogue of P_μ for vectors. As with electromagnetic plane waves in exponential form, we will have to pick the real or imaginary part of this $h_{\mu\nu}$ before making predictions with it. From the first equation in (5.15) we have $Pk^\mu k_\mu = 0$, which tells us that either $P = 0$ (the trace $P \equiv P^\mu_\mu$) or $k^\mu k_\mu = 0$. The second equation is

$$k^\rho k_\rho P_{\alpha\mu} + k_\alpha k_\mu P = 0, \tag{5.17}$$

and we can evaluate our two options from the first equation. If $P = 0$, then we know that $k^\rho k_\rho = 0$. Similarly, if we take $k^\rho k_\rho = 0$, then $P = 0$; both must be true in either case. So far, then, $P_{\mu\nu}$ is traceless, and k^μ is a null vector, as it was in E&M (gravitational waves, like electromagnetic ones, travel at the speed of light).

The gauge condition relates $P_{\mu\nu}$ and k^μ:

$$\partial^\alpha h_{\alpha\mu} = 0 \longrightarrow ik^\alpha P_{\alpha\mu} = 0, \tag{5.18}$$

and we can use this equation to set four of the components of $P_{\alpha\mu}$. Is there additional gauge fixing we can perform? Yes, we have unresolved freedom that must be fixed before we can physically interpret the components of $P_{\alpha\mu}$. Take $\tilde{h}_{\mu\nu} = h_{\mu\nu} + v_{\mu,\nu} + v_{\nu,\mu}$ with $\partial^\mu v_\mu = 0$ (our choice here, you can always add the gradient of a scalar to the vector portion to establish that you

---

[4] This does not look like the DeDonder gauge choice that we made in (3.100). That's okay, we can pick the divergence of $h_{\mu\nu}$ to be whatever we like. As it will turn out in a moment, this choice is equivalent to the one in (3.100) in this vacuum setting.

can pick the divergence of v_μ to be zero – that's Lorenz gauge in the case of the electromagnetic vector A_μ). Then

$$\partial^\mu \tilde{h}_{\mu\nu} = \partial^\mu h_{\mu\nu} + \Box v_\nu = \Box v_\nu. \tag{5.19}$$

To remain in this gauge, we must have $\Box v_\nu = 0$, a set of four equations. Take $v_\nu = V_\nu e^{ik^\rho x_\rho}$ for constant V_μ to ensure that v_ν is harmonic. The field $\tilde{h}_{\mu\nu}$ has the components

$$\tilde{h}_{\mu\nu} = (P_{\mu\nu} + ik_\nu V_\mu + ik_\mu V_\nu)e^{ik^\rho x_\rho}. \tag{5.20}$$

We will choose our four V_ν so as to eliminate four of these terms. The natural choice here are the components $\tilde{h}_{0\nu}$ (just as we got rid of A_0 for the four-potential of E&M). To get these four components to be zero, we need

$$P_{0\nu} + ik_\nu V_0 + ik_0 V_\nu = 0. \tag{5.21}$$

Is this set solvable? Taking $\nu = 0$, we have $P_{00} + 2ik_0 V_0 = 0$, and this can be solved for V_0 since k_0 is not zero (unless $k_\mu = 0$, as discussed above). Then we just need to set

$$V_\nu = -\frac{P_{0\nu} - (P_{00}/(2k_0))k_\nu}{ik_0}, \tag{5.22}$$

which, again, we are guaranteed, exists.

Remember that the original $h_{\mu\nu}$ must be traceless, and we need to ensure that this is true for $\tilde{h}_{\mu\nu}$ as well. The trace is

$$\tilde{h} \equiv \eta^{\mu\nu}\tilde{h}_{\mu\nu} = P + 2ik^\nu V_\nu = P - \frac{1}{ik_0}\left(k^\nu P_{0\nu}\right), \tag{5.23}$$

where we have already used $k^\nu k_\nu = 0$. Since $k^\alpha P_{\alpha\mu} = 0$ from the original gauge choice in (5.18) and $P = 0$, we have $\tilde{h} = 0$ automatically, by construction.

We have now shown that we can achieve the "transverse-traceless" (or "TT") gauge for plane waves. All together, we can take

$$\boxed{h_{\mu\nu} = P_{\mu\nu}e^{ik^\rho x_\rho} \text{ with } P_{0\nu} = 0, \qquad k^\nu P_{\mu\nu} = 0, \text{ and } P \equiv P_{\mu\nu}\eta^{\mu\nu} = 0.}$$
$$\tag{5.24}$$

The traceless part is self-explanatory and is a requirement of the field equation in the original gauge. Transverse means that the polarization tensor has spatial components (the only ones left after setting $P_{0\nu} = 0$) that are orthogonal to the direction of wave propagation, $k^\nu P_{\mu\nu} = 0 \rightarrow k^j P_{jk} = 0$ with j and k running over just the spatial indices. Because the field is traceless, the field equations and gauge conditions are identical if we use the "trace-reversed" form $H_{\mu\nu} = h_{\mu\nu} - \eta_{\mu\nu}h/2$, so that we really are in a special case of DeDonder gauge here.

Returning to the specific setup from Section 5.1, for a z-directed plane wave, $k^0 = \omega/c = k^z$. The polarization tensor has $P_{0\nu} = 0$, and then $k^z P_{zk} = 0$ for each value of k means that $P_{zx} = P_{zy} = P_{zz} = 0$. We are left with P_{xx}, P_{xy}, and P_{yy}. The traceless requirement is $P_{xx} + P_{yy} = 0$, and we can take

$P_{yy} = -P_{xx}$. The fully gauge fixed solution here, displayed as in the E&M version (5.6), is

$$
h_{\mu\nu} \doteq \begin{pmatrix} 0 & 0 & 0 & 0 \\ 0 & P_{xx} & P_{xy} & 0 \\ 0 & P_{xy} & -P_{xx} & 0 \\ 0 & 0 & 0 & 0 \end{pmatrix} e^{i(-\omega t + \omega z/c)}. \tag{5.25}
$$

Given that there exists a TT gauge choice for any plane wave and that choice effects the components of the polarization tensor $P_{\mu\nu}$, we'd like a quick way of generating the TT gauge for a plane wave with arbitrary propagation direction $\hat{\mathbf{k}}$, like the transverse plane wave solutions for A^μ from (5.9). This can be done by setting $P_{0\nu}^{\ast} = 0$ for starters and then projecting out the components of P_{jk} that are in the direction of the wave. To do this, we'll again use the projection operator from (5.7) with $\hat{\mathbf{n}} = \hat{\mathbf{k}}$ acting on the polarization tensor P^{jk} to get the portion perpendicular to $\hat{\mathbf{k}}$. We must apply the operator twice, one for each index,

$$
\bar{P}^{jk} \equiv N^j{}_\ell N^k{}_m P^{\ell m} = P^{jk} - \left(n^k n_m P^{jm} + n^j n_m P^{mk} \right) + n^j n^k n_\ell n_m P^{\ell m}. \tag{5.26}
$$

It is clear from the form of $\bar{P}^{jk}$ that $n_k \bar{P}^{jk} = 0$ and $n_j \bar{P}^{jk} = 0$. But the new transverse tensor is not traceless. Subtracting off the trace, let

$$
\bar{P}^{jk} \equiv N^j{}_\ell N^k{}_m P^{\ell m} - \frac{1}{2} N^{jk} N_{\ell m} P^{\ell m}, \tag{5.27}
$$

where we are using the spatial metric η_{ij} (just the identity, since we are in Cartesian coordinates) to raise and lower indices. Taking the trace of this expression shows that

$$
\bar{P}^j{}_j = N^j{}_\ell N_{jm} P^{\ell m} - \frac{1}{2} N^j{}_j N_{\ell m} P^{\ell m} = 0 \tag{5.28}
$$

using $N^j{}_\ell N_{jm} P^{\ell m} = N_{\ell m} P^{\ell m}$ and $N^j{}_j = 2$ as you will establish in Problem 5.5. Since $N^{jk} n_j = 0$ and $N^{jk} n_k = 0$, the $\bar{P}^{jk}$ in (5.27) remains transverse. Taking $\hat{\mathbf{n}} = \hat{\mathbf{z}}$, you can verify that the projection operator returns the matrix shown in (5.25) for a z-directed plane wave in Problem 5.6. For a general $\hat{\mathbf{k}}$, we get the spatial piece of the polarization tensor from (5.27), and then the plane wave solution is

$$
h_{\mu\nu}^{\mathrm{TT}} = \bar{P}_{\mu\nu} e^{-i\omega(t - \hat{\mathbf{k}} \cdot \mathbf{r}/c)}. \tag{5.29}
$$

Other solutions to the wave equation can be built from these plane waves, and for each mode, we can use the projection in (5.7) to ensure that mode by mode, the waveform is in TT gauge, and hence the sum is as well.

Example 5.1 (Geodesics for Plane Waves). Take $g_{\mu\nu} = \eta_{\mu\nu} + h_{\mu\nu}$ for the TT-gauge plane wave metric perturbation in (5.25). The geodesic Lagrangian is (in proper time parameterization, but omitting the τ-dependence to save space)

$$
\begin{aligned}
L = \frac{m}{2} \Big[&-c^2 \dot{t}^2 + \dot{z}^2 + \dot{x}^2 \big(1 + e^{-i\omega(t-z/c)} P_{xx} \big) \\
&+ \dot{y}^2 \big(1 - e^{-i\omega(t-z/c)} P_{xx} \big) + 2\dot{x}\dot{y} e^{-i\omega(t-z/c)} \Big].
\end{aligned}
\tag{5.30}
$$

Since the Lagrangian does not depend on x or y, the momentum associated with these coordinates is constant,

$$
p_x = \frac{\partial L}{\partial \dot{x}}, \qquad p_y = \frac{\partial L}{\partial \dot{y}},
\tag{5.31}
$$

and we can extract $\dot{x}$ and $\dot{y}$ from this coupled pair,

$$
\begin{aligned}
\dot{x} &= \frac{1}{m} \frac{e^{2i\omega t} p_x - e^{i\omega(t+z/c)}(p_x P_{xx} + p_y P_{xy})}{e^{2i\omega t} - e^{2i\omega z/c}(P_{xx}^2 + P_{xy}^2)}, \\
\dot{y} &= \frac{1}{m} \frac{e^{2i\omega t} p_y - e^{i\omega(t+z/c)}(-p_x P_{xy} + p_y P_{xx})}{e^{2i\omega t} - e^{2i\omega z/c}(P_{xx}^2 + P_{xy}^2)}.
\end{aligned}
\tag{5.32}
$$

Imagine starting a particle off from rest at $z = 0$. Then we have $p_x = 0$ and $p_y = 0$ initially, and for all proper time, since these are constants of motion. So the particle does not move; it remains at rest. The z equation of motion is $\ddot{z} = 0$, and then $z = 0$ for all time, too. Evidently, a gravitational plane wave has no effect on a single particle. This result is not so surprising given what we know about general relativity. At any given point, the gravitational field can be removed entirely, leaving a Minkowski spacetime with no effective forcing. To see the effects of the wave, we must compare two particles traveling along two different geodesics, just as we did in Section 3.3. It is the deviation of two geodesics that is detectable.

5.3 Plane Wave Geodesic Deviation

We'll use the deviation equation from (3.63),

$$
\frac{D^2 s^\alpha}{D\tau^2} = R^\alpha{}_{\sigma\nu\mu} \dot{x}^\sigma \dot{x}^\nu s^\mu,
\tag{5.33}
$$

where $\dot{x}^\sigma$ is the tangent to a geodesic, and s^μ is the difference vector between that geodesic and a nearby one. The linearized Riemann tensor, appropriate for use with our plane wave solution from (5.24), is given in (3.93).

From Example 5.1, a single geodesic for a particle initially at rest has motion only in the temporal direction, so that $\dot{x}^\mu(\tau) = c\dot{t}(\tau)\delta_0^\mu$. Then only the $\sigma = \nu = 0$ elements contribute in (5.33), and in proper time parameterization, we have $\dot{x}^\mu \dot{x}_\mu = -c^2$, which tells us that $t(\tau) = \tau$, so $\dot{x}^0 = c$. The deviation equation can be written using $\tau \to t$ and $D \to d$ (the background is flat, and we'll use Cartesian coordinates),

$$\frac{d^2s^\alpha}{dt^2} = c^2 R^\alpha{}_{00\mu}s^\mu, \tag{5.34}$$

or, using the linearized form for the Riemann tensor,

$$\frac{d^2s^\alpha}{dt^2} = \frac{1}{2}c^2\eta^{\alpha\rho}[h_{\rho\mu,00} - h_{0\mu,\rho0} - h_{\rho0,0\mu} + h_{00,\mu\rho}]s^\mu. \tag{5.35}$$

In TT gauge, $h_{0\mu} = 0$ for all μ, so the deviation equation simplifies:

$$\frac{d^2s^\alpha}{dt^2} = \frac{1}{2}\ddot{h}^\alpha{}_\mu s^\mu, \tag{5.36}$$

where we have used the Minkowski metric to raise the index on $h_{\rho\mu}$ and introduced dots to refer to t derivatives (the 0 derivative is really with respect to ct, so the factors of c cancel).

For $\alpha = 0$, we get $\frac{d^2s^0}{dt^2} = 0$, and we can take the temporal separation to be zero (synchronize the clocks between the two stationary geodesics at $t = 0$). Then we have a purely spatial equation, and we'll use Roman indices that go from $1 \to 3$ to capture these,

$$\frac{d^2s^i}{dt^2} = \frac{1}{2}\ddot{h}^i{}_j s^j. \tag{5.37}$$

Remember that the metric perturbation $h_{\mu\nu}$ has magnitude governed by the polarization entries $P_{\mu\nu}$, and these are assumed to be small dimensionless constants. On the right-hand side of the geodesic deviation equation, then, we have quantities that are small, of order h. Take $s^i = d^i + \epsilon k^i(t)$ where d^i is a constant initial separation, and $\epsilon k^i(t)$ is a small perturbation on top of that constant separation with dimensionless $\epsilon \ll 1$. Putting this into (5.37), we have an equation in small parameters:

$$\epsilon\frac{d^2k^i(t)}{dt^2} = \frac{1}{2}\ddot{h}^i{}_j d^j + \underbrace{\frac{1}{2}c^2\epsilon\ddot{h}^i{}_j k^j(t)}_{\sim\epsilon h}. \tag{5.38}$$

The term on the left is of order ϵ, and the first term on the right is of order h, which we'll take to be of the same magnitude as ϵ. The second term on the right is then of size ϵh, an order of magnitude smaller than the rest.[5] Our final approximation, dropping the ϵh term on the right, gives us an equation that can be solved by integration

$$\epsilon\frac{d^2k^i(t)}{dt^2} = \frac{1}{2}\ddot{h}^i{}_j d^j \longrightarrow \epsilon k^i(t) = \frac{1}{2}h^i{}_j d^j, \tag{5.39}$$

and the deviation is

$$s^i(t) = d^i + \frac{1}{2}h^i{}_j d^j. \tag{5.40}$$

---

[5] Of course, this assumes that the frequency ω is "small."

Using the general plane wave form, we can write this in terms of the polarization tensor,

$$s^i(t) = d^i + \frac{1}{2} P^i_j d^j e^{i(-k^0 ct + \mathbf{k} \cdot \mathbf{r})}. \tag{5.41}$$

Example 5.2 (TT Gauge Plane Waves Traveling in the z Direction). Let's take the imaginary part of the z-directed plane waves from (5.25), and consider nearby geodesics that are in the xy plane with $z = 0$. Then we can write the time-dependent solution in (5.41) in vector form, with $\mathbf{d} = d^x \hat{\mathbf{x}} + d^y \hat{\mathbf{y}}$,

$$\mathbf{s}(t) = \mathbf{d} + \left[\left(d^x P_{xx} + d^y P_{xy} \right) \hat{\mathbf{x}} + \left(d^x P_{xy} - d^y P_{xx} \right) \hat{\mathbf{y}} \right] \sin(\omega t) \tag{5.42}$$

with initial displacement $\mathbf{d}$. For a pair of masses initially separated by $\mathbf{d}$, the solution above tells us that the masses oscillate relative to one another with frequency ω. The two polarization directions are interesting and have distinct effects on particle motion, so we'll think about them one at a time.

Take $P_{xy} = 0$, and set the initial displacement of the masses in the x direction, $d^x = d$, $d^y = 0$. Then

$$\mathbf{s}(t) = d \left(1 + P_{xx} \sin(\omega t) \right) \hat{\mathbf{x}}. \tag{5.43}$$

The separation in time starts at d and initially grows; the masses get further apart. If we instead take the initial separation to be in the y direction, with $d^x = 0$ and $d^y = d$, then the solution is

$$\mathbf{s}(t) = d \left(1 - P_{xx} \sin(\omega t) \right) \hat{\mathbf{x}}. \tag{5.44}$$

Here the initial separation of d decreases for $t > 0$, opposite the behavior for the $\hat{\mathbf{x}}$ separation. If you had a pair of masses separated horizontally (in x) and another pair separated vertically (in y), then the horizontal separation would initially increase while the vertical separation decreased, and vice versa once the argument of the sine function hits $\pi/2$. There is stretching in one direction with squeezing in the other, typical of the tidal forces associated with gravity. We have set the pairs of masses in a + pattern, and that pattern remains with increasing/decreasing arm length. For this reason, the polarization mode defined by P_{xx} is called the "plus polarization."

To see the effect of P_{xy}, set $P_{xx} = 0$. This time, we'll rotate the initial separations so that the first is along a 45° line with $\mathbf{d} = d(\hat{\mathbf{x}} + \hat{\mathbf{y}})/\sqrt{2}$. Then separation distance, as a function of time, is

$$\mathbf{s}(t) = \frac{d}{\sqrt{2}} (\hat{\mathbf{x}} + \hat{\mathbf{y}}) \left(1 + P_{xy} \sin(\omega t) \right). \tag{5.45}$$

Again, the initial separation grows for small $t > 0$. The orthogonal initial condition is the 45° line with negative slope, $\mathbf{d} = d(-\hat{\mathbf{x}} + \hat{\mathbf{y}})/\sqrt{2}$, with

$$\mathbf{s}(t) = \frac{d}{\sqrt{2}} (-\hat{\mathbf{x}} + \hat{\mathbf{y}}) \left(1 - P_{xy} \sin(\omega t) \right). \tag{5.46}$$

This separation is initially shrinking. The solutions in (5.45) and (5.46) form another tidal pair – when one separation vector is increasing, the other is decreasing. Since these are now tilted by forty five degrees, the polarization associated with P_{xy} is called the "cross polarization."

5.3.1 Detection of Gravitational Plane Waves

The previous section suggests a potential method for the detection of gravitational waves by observing the relative motion of a set of "test masses." If you had a flexible ring of masses in a plane orthogonal to an incident gravitational plane wave, then you could see the relative deformation of the ring with its characteristic tidal stretch/squeeze pairings. We can get away with a simpler setup involving just two masses, placed along perpendicular lines as shown in Figure 5.1. The relative motion of the pair of masses can be detected directly using a Michelson interferometer. Referring to Figure 5.1, two masses that are themselves mirrors are placed a distance d from the origin in both the $\hat{\mathbf{x}}$ and $\hat{\mathbf{y}}$ directions. If a laser beam is split and sent down each arm, then the beams reflect off of the mirrors and return to the origin. There they interfere constructively since they have gone the same distance. Suppose a gravitational plane wave comes along; because of the tidal nature of the relative motion, one arm of the interferometer will be extended by Δd, and the other will shrink by Δd, giving a total shift in relative path length of $2\Delta d$ (for a plus-polarized wave) – the pattern will oscillate in time with frequency given by the frequency of the plane wave. The difference Δd will cause a shift in phase for the beams returning from each arm, and the previously constructive interference will change, allowing Δd to be determined. Referring to the solution in (5.43), for example, if you knew the temporal evolution of Δd, then you could determine both the magnitude P_{xx} of the plane wave and its frequency ω.

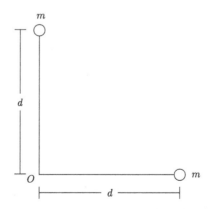

Fig. 5.1 An interferometer with arms of length d. The mirrors at the ends of the arms have mass m and serve as the "test masses" for detecting relative separations.

The LIGO (Laser Interferometer Gravitational-Wave Observatory) detector is a much more sophisticated version of the Michelson interferometer described here. We'll turn next to the production of gravitational radiation. That discussion will allow us to predict the basic size of metric perturbation that we expect to detect with an instrument like LIGO. You can read about some of the early ideas that led to LIGO in [53] and explore the LIGO website [1] to find out about more about its history and recent detections (the first of which is in [2]).

There were experiments designed to detect gravitational waves prior to LIGO. In the 1960s, Weber built "resonant bars," large cylinders made out of aluminum that were tuned to specific gravitational frequency signatures. The idea was to observe gravitational waves by looking at excitations of the bar set up by the tidal stretching of the waves [52]. This approach ultimately failed to detect waves, but was an early and interesting proposed detection mechanism.

5.4 Radiation Setup

There are two important results we need to develop the theories of both electromagnetic and gravitational radiation: Expressions of conservation and the Green's function for the sourced wave equation (i.e., a time-dependent point source solution). These are independent of the particular application and represent useful starting points for almost any theory that has conserved quantities and a PDE governing its fields (which covers most physical theories). We'll start by extracting relations for the temporal dependence of monopole, dipole, and quadrupole moments directly from a local statement of conservation and then move on to the Green's function before returning to carefully define radiation in the electromagnetic setting and compute the dominant terms.

5.4.1 Conservation Laws

We'll work in the language of charge and charge conservation, but you can easily substitute "energy" and "momentum" to apply the results from this section to gravity, as we will eventually do. To begin with, we'll develop relations for point charges, which we then generalize to arbitrary distributions of charge and current.

Let's start by thinking about the monopole and dipole moments of a point charge q that moves along some prescribed trajectory $\mathbf{w}(t)$, a vector that points from the origin to the location of the particle at time t. The charge density here is $\rho(\mathbf{r}, t) = q\delta^3(\mathbf{r} - \mathbf{w}(t))$, with current density $\mathbf{J}(\mathbf{r}, t) = \rho(\mathbf{r}, t)\dot{\mathbf{w}}(t)$. The monopole moment of the distribution is, integrating over all space,

$$Q \equiv \int \rho(\mathbf{r}, t)\, d\tau = q \tag{5.47}$$

and is time-independent, $\dot{Q} = 0$, a reminder that the charge q cannot just disappear – this is the statement of global charge conservation.

The dipole moment is

$$\mathbf{p}(t) \equiv \int \rho(\mathbf{r}, t)\mathbf{r}\, d\tau = q\mathbf{w}(t), \tag{5.48}$$

which is to be expected. The derivative of the dipole moment $\dot{\mathbf{p}}(t) = q\dot{\mathbf{w}}(t)$. The expression $q\dot{\mathbf{w}}(t)$ is just the integral of the current density, so we can relate the temporal derivative of the dipole moment to the integral, over all space, of the current density:

$$\frac{d\mathbf{p}(t)}{dt} = \int \mathbf{J}\, d\tau. \tag{5.49}$$

A similar expression can be generated for the magnetic dipole moment of the particle's motion. We'll use the familiar expression for the magnetic dipole moment $\mathbf{m} = I\mathbf{a}$ for current I and area vector $\mathbf{a}$. If we think of the charge's motion over a time interval dt, then the current $I = \lambda v$ can be built out of the one-dimensional charge density $\lambda = q/(\dot{w}(t)dt)$, with $v = \dot{w}$, so that $I = q/dt$. The area swept out over the time interval is $\mathbf{a} = \mathbf{w}(t) \times \dot{\mathbf{w}}(t)dt/2$, just half the area of the parallelogram with sides $\mathbf{w}(t)$ and $\dot{\mathbf{w}}(t)dt$. Then

$$\mathbf{m}(t) = I\mathbf{a} = \frac{1}{2}q\mathbf{w}(t) \times \dot{\mathbf{w}}(t). \tag{5.50}$$

Once again, we can write the right-hand side of this expression in terms of an integral involving $\mathbf{J}$:

$$\mathbf{m}(t) = \frac{1}{2}q\mathbf{w}(t) \times \dot{\mathbf{w}}(t) = \frac{1}{2}\int \mathbf{r} \times \mathbf{J}\, d\tau, \tag{5.51}$$

where the delta function inside $\mathbf{J}$ takes care of the integration for us automatically. The time-derivative of the magnetic dipole moment is then

$$\frac{d\mathbf{m}(t)}{dt} = \frac{1}{2}\int \mathbf{r} \times \dot{\mathbf{J}}\, d\tau. \tag{5.52}$$

We can keep going (last one, I promise!), the electric quadrupole moment of this distribution is, switching to index notation and using x^i to denote the ith element of $\mathbf{r}$,

$$P^{ik}(t) \equiv \int \rho x^i x^k\, d\tau = qw^i(t)w^k(t) \tag{5.53}$$

with derivative

$$\dot{P}^{ik}(t) = q\left(\dot{w}^i(t)w^k(t) + w^i(t)\dot{w}^k(t)\right). \tag{5.54}$$

The time derivative of the quadrupole moment can be expressed as the integral,

$$\frac{dP^{ik}(t)}{dt} = \int \left(J^i x^k + J^k x^i\right) d\tau, \tag{5.55}$$

again using the delta functions inside $\mathbf{J}$ to perform the integration, recovering the right-hand side of (5.54).

Continuum Form

It was easy to generate the expressions in (5.47), (5.49), and (5.55) for the point charge, and appealing to superposition, we can build up any continuum distribution we like from individual point charges, so we expect these equations to hold quite generally. There is another way to establish these results directly from the statement of local charge conservation,

$$\frac{\partial \rho}{\partial t} = -\nabla \cdot \mathbf{J} \tag{5.56}$$

for a continuous distribution ρ and current density $\mathbf{J}$, assuming that the source is finite and bounded.

To start, integrate both sides of (5.56) over all space:

$$\frac{d}{dt} \int \rho \, d\tau = - \int \nabla \cdot \mathbf{J} \, d\tau = - \oint \mathbf{J} \cdot d\mathbf{a} = 0, \tag{5.57}$$

where the surface integral on the right vanishes since (by assumption) $\mathbf{J} \to 0$ as $r \to \infty$. This is the global statement of charge conservation, $\dot{Q} = 0$.

To get the analogous (electric) dipole expression, we can multiply the right-hand side of (5.56) by x^i (again, the ith component of $\mathbf{r}$) and then integrate to get

$$\frac{d}{dt} \int \rho x^i \, d\tau = - \int \frac{\partial J^\ell}{\partial x^\ell} x^i \, d\tau. \tag{5.58}$$

On the left, we have the time derivative of the dipole moment of the distribution. This time, we cannot use the divergence theorem directly to convert the right-hand side into a surface integral. Instead, we can generate an appropriate integration-by-parts formula by using the divergence theorem on an integrand of the form $\partial_\ell (J^\ell x^i)$:

$$\int \frac{\partial}{\partial x^\ell} (J^\ell x^i) \, d\tau = \oint J^\ell x^i \, da_\ell = 0, \tag{5.59}$$

where we have again evaluated the integrand at spatial infinity, where we assume that $J^\ell x^i$ vanishes. Using the product rule on the integral on the left, we can expand the result:

$$\int \frac{\partial J^\ell}{\partial x^\ell} x^i \, d\tau + \int J^\ell \frac{\partial x^i}{\partial x^\ell} \, d\tau = 0. \tag{5.60}$$

The derivative of the coordinates with respect to themselves yields a Kronecker delta as usual, $\frac{\partial x^i}{\partial x^\ell} = \delta^i_\ell$, and using that delta in the sum, we end up with the identity

$$\int \frac{\partial J^\ell}{\partial x^\ell} x^i \, d\tau = - \int J^i \, d\tau. \tag{5.61}$$

The integral on the left of this expression appears in (5.58), which becomes

$$\frac{d}{dt} \int \rho x^i \, d\tau = \int J^i \, d\tau \qquad (5.62)$$

or, back in vector notation,

$$\frac{d\mathbf{p}}{dt} = \int \mathbf{J} \, d\tau, \qquad (5.63)$$

that is, the temporal change in the dipole moment of the charge distribution is equal to the integral of the current density, just as we established in (5.49).

Moving on to the electric quadrupole target, this time we'll multiply both sides of (5.56) by $x^i x^k$ to get the time derivative of the quadrupole on the left, then integrate over all space as usual:

$$\frac{d}{dt} \int \rho x^i x^k \, d\tau = - \int \frac{\partial J^\ell}{\partial x^\ell} x^i x^k \, d\tau. \qquad (5.64)$$

Focusing on the right-hand side, we again want to develop an integration-by-parts formula to flip the divergence off of the current density and onto the $x^i x^k$ terms, yielding a spray of Kronecker deltas. To that end, consider the divergence $\partial_\ell(J^\ell x^i x^k)$: when integrated, this divergence becomes an integral of $J^\ell x^i x^j$ over the surface at infinity, which evaluates to zero for bounded current. Then using the product rule, we have

$$\int \frac{\partial J^\ell}{\partial x^\ell} x^i x^k \, d\tau + \int J^\ell \frac{\partial}{\partial x^\ell}\left(x^i x^k\right) d\tau$$
$$= \int \frac{\partial J^\ell}{\partial x^\ell} x^i x^k \, d\tau + \int J^\ell \left(\delta^i_\ell x^k + \delta^k_\ell x^i\right) d\tau. \qquad (5.65)$$

Using the deltas and setting the whole expression to zero, we learn that

$$\int \frac{\partial J^\ell}{\partial x^\ell} x^i x^k \, d\tau = - \int \left(J^i x^k + J^k x^i\right) d\tau, \qquad (5.66)$$

and using this result in (5.64) with P^{ik} defined by (5.53),

$$\frac{dP^{ik}}{dt} = \int \left(J^i x^k + J^k x^i\right) d\tau, \qquad (5.67)$$

as we found in (5.55).

5.4.2 Green's Function Solution

We'll review the radiation associated with the electric field of a moving particle from E&M. What is radiation? The defining answer is: The portion of the electromagnetic field (or potential) that carries energy to infinity. Extracting the portion of the field that carries this energy will take a while, but the set of approximations we end up with are useful in shifting the definition over to gravitational radiation, since we will basically identify analogous components and use those to relate the sources of radiation to their fields.

The Poynting vector $\mathbf{S} = \mathbf{E} \times \mathbf{B}/\mu_0$ tells us how energy flows in E&M. For time-varying fields, especially oscillatory ones, it is natural to compute the time-average $\langle \mathbf{S} \rangle$ of the Poynting vector, giving us a power per unit area. The power carried by the electromagnetic field can be integrated over a sphere of radius R,

$$P = \oint \langle \mathbf{S} \rangle \cdot d\mathbf{a}, \tag{5.68}$$

and for radiation, this power must be the same for spheres of all radii including $R \to \infty$. So any electromagnetic field that has $\mathbf{S} \sim 1/r^2 \hat{\mathbf{r}}$ is a "radiation" field. For the most part, we want electric and magnetic fields that each contribute a factor of $1/r$, and this is where we first run into trouble. Far away from any localized, *stationary* distribution of charge, the electric and magnetic fields each fall off like $1/r^2$, leading to $S \sim 1/r^4$. It must be that when we make the sources accelerate, a new contribution to the electric and magnetic fields emerges.

We'll work with the potentials (both the electromagnetic A^μ and, later, the gravitational $h_{\mu\nu}$). Maxwell's equation for A^μ in the Lorenz gauge is

$$\Box A^\mu = -\mu_0 J^\mu \quad \text{with } \Box \equiv -\frac{1}{c^2}\frac{\partial^2}{\partial t^2} + \nabla^2. \tag{5.69}$$

Compare this with the linearized Einstein equation governing $h_{\mu\nu}$ in Lorenz gauge,

$$\Box H^{\mu\nu} = -\frac{16\pi G}{c^4} T^{\mu\nu} \tag{5.70}$$

with $H^{\mu\nu} \equiv h^{\mu\nu} - \eta^{\mu\nu} h/2$.

Both field equations (and many others) have multiple copies of a $(3 + 1)$-dimensional Poisson problem

$$\Box U(\mathbf{r}, t) = k\rho(\mathbf{r}, t), \tag{5.71}$$

where U is a component of the field, k is a constant (used to set units), and ρ is a provided source density of appropriate tensorial type. Using the Green's function approach that we first encountered in Section 1.2, we can solve for $U(\mathbf{r}, t)$ given any $\rho(\mathbf{r}, t)$. The idea is to solve for the field of a point source and then appeal to superposition to build up general solutions by integration. To see the method in action, suppose we had solved the problem posed by (5.71) for a source of the form $\rho = \delta(ct' - ct)\delta^3(\mathbf{r} - \mathbf{r}')$, that is, a source that flashed on and off instantaneously at time t' and location $\mathbf{r}'$. We'll omit the constant k, which we can add back in later, and call the solution G for "Green's function,"

$$\Box G(\mathbf{r}, t, \mathbf{r}', t') = \delta(ct' - ct)\delta^3(\mathbf{r} - \mathbf{r}'). \tag{5.72}$$

Now we can build U solving (5.71) given a more general $\rho(\mathbf{r}, t)$ by adding up the point source contributions,

$$U(\mathbf{r}, t) = k \int G(\mathbf{r}, t, \mathbf{r}', t')\rho(\mathbf{r}', t')d(ct')d\tau', \tag{5.73}$$

where the integral is over all spacetime. Acting on both sides of the above with $\Box$, which takes derivatives with respect to the unprimed t and $\mathbf{r}$ coordinates,

$$\Box U(\mathbf{r}, t) = k \int \rho(\mathbf{r}', t')(\Box G(\mathbf{r}, t, \mathbf{r}', t'))d(ct')d\tau' = k\rho(\mathbf{r}, t), \quad (5.74)$$

where we have used the four-dimensional delta function source for G to perform the integrals. So it is clear that U from (5.73) solves the Poisson PDE from (5.71) with the generally implicit boundary condition that $U \to 0$ as $\mathbf{r} \to \infty$.

To solve the PDE that defines the Green's function of interest from (5.72), we will work with the temporal Fourier transform of the Green's function. Here are the Fourier transform conventions we will use: Given a function of time and position (although we will omit the position dependence) $p(t)$, its Fourier transform is

$$\tilde{p}(f) = \int_{-\infty}^{\infty} e^{i2\pi f t} p(t)dt \quad (5.75)$$

with inverse Fourier transform

$$p(t) = \int_{-\infty}^{\infty} e^{-i2\pi f t} \tilde{p}(f)df. \quad (5.76)$$

For this pair to be self-consistent, we must have the following representation for the Dirac delta function (see Problem 5.8):

$$\delta(t - \bar{t}) = \int_{-\infty}^{\infty} e^{i2\pi f(t - \bar{t})}df. \quad (5.77)$$

All right, let's go back to (5.72). To simplify the problem, we'll start off by setting the point source to $t' = 0$, $\mathbf{r}' = 0$; we can always move it later. Then we have to solve

$$-\frac{1}{c^2}\frac{\partial^2 G}{\partial t^2} + \nabla^2 G = \delta(ct)\delta^3(\mathbf{r}) = \frac{1}{c}\delta(t)\delta^3(\mathbf{r}). \quad (5.78)$$

To isolate the Fourier transform of G, multiply both sides of this equation by $e^{i2\pi f t}$ and integrate over all time:

$$-\frac{1}{c^2}\int_{-\infty}^{\infty}\frac{\partial^2 G}{\partial t^2}e^{i2\pi f t}dt + \nabla^2 \tilde{G} = \frac{1}{c}\delta^3(\mathbf{r}), \quad (5.79)$$

where the Fourier transform $\tilde{G}$ of G is a function of f and the position variables $\mathbf{r}$. In the first term, we'll integrate by parts twice, assuming that the Green's function vanishes as $t \to \pm\infty$, which is a reasonable piece of the boundary condition for this maximally localized source. The integration by parts pulls down two factors of $i2\pi f$, which come outside the integral leaving us with the Fourier transform of G,

$$\left(\frac{2\pi f}{c}\right)^2 \tilde{G}(\mathbf{r}, f) + \nabla^2 \tilde{G}(\mathbf{r}, f) = \frac{1}{c}\delta^3(\mathbf{r}). \quad (5.80)$$

From the spherical symmetry of the source, we can impose a boundary condition at $r \to \infty$, demanding that $\tilde{G}(\mathbf{r}, f)$ itself be spherically symmetric, so that it depends only on r. If we do this and consider the PDE at all points except $r = 0$, the equation becomes

$$\left(\frac{2\pi f}{c}\right)^2 \tilde{G}(r, f) + \frac{1}{r}\frac{\partial^2}{\partial r^2}\left(r\tilde{G}(r, f)\right) = 0. \tag{5.81}$$

Multiplying by r and defining $K \equiv r\tilde{G}$ we obtain the familiar

$$\frac{\partial^2 K}{\partial r^2} = -\left(\frac{2\pi f}{c}\right)^2 K \longrightarrow \tilde{G} = \frac{Ae^{\pm i2\pi rf/c}}{r} \tag{5.82}$$

with constant of integration A. We have taken only one of the two solutions here, although we remain agnostic about which (plus or minus) we retain. To set A, we can appeal to the known point-source solution for the electrostatic potential. If you have $\rho = q\delta^3(\mathbf{r})$ in $\nabla^2 V = -\rho/\epsilon_0$ and call $k \equiv -q/\epsilon_0$, then the electric potential for a point charge at the origin, $V = q/(4\pi\epsilon_0 r)$, gives us the Green's function with $G = -1/(4\pi r)$. Since the function in (5.82) must reduce to this when $f = 0$, we learn that $A = -1/(4\pi c)$ (the c comes from the original $1/c\delta^3(\mathbf{r})$ in (5.80)).

So far we have the Fourier transform of the Green's function,

$$\tilde{G}(\mathbf{r}, f) = -\frac{e^{\pm i2\pi rf/c}}{4\pi cr}, \tag{5.83}$$

and we have to use the inverse Fourier transform to find $G(\mathbf{r}, t)$. Putting $\tilde{G}(\mathbf{r}, f)$ into (5.76),

$$G(\mathbf{r}, t) = -\int_{-\infty}^{\infty}\frac{e^{\pm i2\pi rf/c}}{4\pi cr}e^{-i2\pi ft}df = -\frac{1}{4\pi cr}\int_{-\infty}^{\infty}e^{-i2\pi f(t\mp r/c)}df. \tag{5.84}$$

Using identity (5.77), we can evaluate the integral, and the Green's function is

$$G(\mathbf{r}, t) = -\frac{\delta(t \pm r/c)}{4\pi rc} = -\frac{\delta(ct \pm r)}{4\pi r}. \tag{5.85}$$

Now the role of the $\pm$ is clear. If the source flashes on and off at the origin at time $t = 0$, then the effect of that source shows up at a location a distance r away at time $t = \mp r/c$, arriving both *before* (minus sign) and after (plus sign) the signal occurs. This is the usual advanced vs. retarded time, and we'll take the plus sign to ensure the physically relevant causal choice, $t = r/c > 0$.

Moving the source from $0 \to t'$ and its location from $0 \to \mathbf{r}'$, we are just shifting axes, so the Green's function for an arbitrary location is

$$\boxed{G\left(\mathbf{r}, t, \mathbf{r}', t'\right) = -\frac{\delta(c(t - t') - |\mathbf{r} - \mathbf{r}'|)}{4\pi|\mathbf{r} - \mathbf{r}'|}.} \tag{5.86}$$

The source for this four-dimensional Green's function problem is not a realistic one. The point source we used had $\rho(\mathbf{r}, t) = q\delta(ct)\delta^3(\mathbf{r})$, similar to the sourcing on the right-hand side of (5.72). As a point source, it is fine to spatially localize to the origin using $\delta^3(\mathbf{r})$, but the additional temporal delta violates charge conservation. You can't have a real charge blink instantaneously in and out of existence as the $\delta(ct)$ implies. So the Green's function source is not physically realizable. That's okay, and the delta in the numerator of (5.86) reminds us of this fact and instructs us to use G only under the integral sign as in (5.73). Still, there is a nice physical interpretation for the form of the Green's function. If you throw a rock into a pond, then a circular ripple emanates from the rock's entry point traveling radially outwards at a (roughly) constant speed (set by the medium) and decreasing in magnitude as it gets further from its point of generation. Similarly, there is a sphere of influence represented by the numerator of (5.86), which moves radially away from the charge's instantaneous "blink" at speed c. The falloff with distance, as $1/|\mathbf{r} - \mathbf{r}'|$ reminds us of the effect of a point charge. So we can interpret (5.86) roughly as the spherical wave that emanates from a point source that turns on and off at time t' and falls off as we expect for a point source.

5.5 Electromagnetic Radiation

Radiation is defined to be the portion of a field that carries energy all the way out to infinity. In electricity and magnetism, the energy transport properties are encoded in the Poynting vector $\mathbf{S} \equiv \frac{1}{\mu_0}\mathbf{E} \times \mathbf{B}$ with units of power per unit area. Then the total power passing through a sphere of radius r is

$$P(r) = \oint_{S(r)} \mathbf{S} \cdot d\mathbf{a} = \int_0^{2\pi} \int_0^{\pi} S_r r^2 \sin\theta d\theta d\phi, \qquad (5.87)$$

where S_r is the $\hat{\mathbf{r}}$ component of the Poynting vector. In order for $P(r)$ to be nonzero as $r \to \infty$, which *defines* the radiation fields, $S_r \sim 1/r^2$ is required.[6] To get the radial component of the Poynting vector to fall off like $1/r^2$, we need the associated electric and magnetic fields to fall off like $1/r$ each. You might try to cook up an electric field that falls off like $1/r^2$ (easy to do) with a magnetic field that is uniform, but uniform magnetic fields only occur for infinite distributions of current and are therefore unavailable.

We already know something about the radiation fields of E&M: spherically symmetric distributions of charge do not radiate. Aside from the fact that even spherically symmetric distributions moving about a fixed center do not generate magnetic fields, we also know that the electric field outside of a spherically symmetric distribution of charge with total charge Q goes like $\mathbf{E} = Q/(4\pi\epsilon_0 r^2)\hat{\mathbf{r}}$, which does *not* fall off like $1/r$. Do we have *any* examples of electric and magnetic fields, generated by finite distributions, that lead to fields with $1/r$ dependence? Not in electrostatics. Let's probe the Green's function solution to the potential to see if we can isolate $\mathbf{E}$ and $\mathbf{B}$ fields with

---

[6] Anything that falls off slower would indicate that the total power through the sphere is *increasing* as you go away from the source, a delightful scenario if one could realize it!

$1/r$ dependence and ones that have a nontrivial Poynting vector in the radial direction (so that $\mathbf{E}$ and $\mathbf{B}$ point in the $\hat{\boldsymbol{\theta}}$ and $\hat{\boldsymbol{\phi}}$ directions).

The integral solution of the potential field is four copies of (5.73) with appropriate units in place,

$$A^\mu(\mathbf{r}, t) = \mu_0 \int J^\mu(\mathbf{r}', t') \frac{\delta(c(t - t') - |\mathbf{r} - \mathbf{r}'|)}{4\pi |\mathbf{r} - \mathbf{r}'|} d(ct') \, d\tau'. \qquad (5.88)$$

The integral is over all time and space, and we can perform the temporal integration using the delta function from the Green's function. That integral enforces $t' = t - |\mathbf{r} - \mathbf{r}'|/c$, leaving us with the purely spatial integral solution that will be our starting point,

$$A^\mu(\mathbf{r}, t) = \frac{\mu_0}{4\pi} \int \frac{J^\mu(\mathbf{r}', t - |\mathbf{r} - \mathbf{r}'|/c)}{|\mathbf{r} - \mathbf{r}'|} \, d\tau'. \qquad (5.89)$$

We will take derivatives of elements of A^μ to obtain the electric and magnetic fields far away from the source, so we will focus on the pieces of A^μ that already go like $1/r$; those are the best candidates for obtaining the target electric and magnetic fields. Our tool will be the approximation to $|\mathbf{r} - \mathbf{r}'|$, where the spatial dependence lives. Since $\mathbf{r}'$ is bounded by the finite extent of the source, we know that it is possible to go far enough away from the distribution to have $r \gg r'$. Remember that we want the "large r" behavior to identify the fields that generate power that makes it out to infinity (and beyond). Then expanding $|\mathbf{r} - \mathbf{r}'|$ in $r'/r \ll 1$, we have

$$|\mathbf{r} - \mathbf{r}'| = r\sqrt{1 - \frac{2\hat{\mathbf{r}} \cdot \mathbf{r}'}{r} + \left(\frac{r'}{r}\right)^2} \approx r - \hat{\mathbf{r}} \cdot \mathbf{r}'. \qquad (5.90)$$

We will use this approximation twice, once in the denominator of the integrand from (5.89) and once in the evaluation of the current,

$$A^\mu(\mathbf{r}, t) \approx \frac{\mu_0}{4\pi} \int \frac{J^\mu(\mathbf{r}', t - r/c + \hat{\mathbf{r}} \cdot \mathbf{r}'/c)}{r(1 - \hat{\mathbf{r}} \cdot \mathbf{r}'/r)} \, d\tau'. \qquad (5.91)$$

In the numerator, we will expand the current density in the small quantity $\hat{\mathbf{r}} \cdot \mathbf{r}'/r$, whereas in the denominator, it suffices to retain just the r piece (other contributions lead to terms in the potential that already go like $1/r^2$, as you will show in Problem 5.14),

$$A^\mu(\mathbf{r}, t) \approx \frac{\mu_0}{4\pi r} \int \left[J^\mu(\mathbf{r}', t - r/c) + \frac{\hat{\mathbf{r}} \cdot \mathbf{r}'}{c} \dot{J}^\mu(\mathbf{r}', t - r/c) \right] d\tau'. \qquad (5.92)$$

We can write the two contributions separately as

$$A^\mu(\mathbf{r}, t) \approx \frac{\mu_0}{4\pi r} \left[\int J^\mu(\mathbf{r}', t - r/c) \, d\tau' + \frac{\hat{\mathbf{r}}}{c} \cdot \int \mathbf{r}' \dot{J}^\mu(\mathbf{r}', t - r/c) \, d\tau' \right], \qquad (5.93)$$

and, finally, using x^k to refer to the kth element of $\mathbf{r}$ (with $\hat{x}^k$ the kth element of $\hat{\mathbf{r}}$ and x'^k the kth element of $\mathbf{r}'$) and defining $t_r \equiv t - r/c$ (just to make the notation easier to keep track of), we have the indexed form

$$A^\mu(\mathbf{r}, t) \approx \frac{\mu_0}{4\pi r} \left[\int J^\mu(\mathbf{r}', t_r) \, d\tau' + \frac{\hat{x}^k}{c} \int x'^k \dot{J}^\mu(\mathbf{r}', t_r) \, d\tau' \right]. \qquad (5.94)$$

Keep in mind that the summed k index here is spatial, and we are working in Cartesian coordinates, so the up-down placement doesn't change the value of objects like x^k. We'll retain our repeated index summation convention, but at times relax the up-down pairing requirement in favor of visual clarity.

5.5.1 Electric Dipole Radiation

Since "monopole" sources (spherically symmetric or point-like distributions) do not lead to energy transfer out to infinity, electromagnetic radiation is dominated by the electric dipole contribution (see [21] for electric and magnetic dipole radiation and [26, 55] for a discussion of electric quadrupole radiation). Starting with the $\mu = 0$ term, with $A^0 = V/c$ and $J^0 = \rho c$, the zero component of (5.94) is

$$V(\mathbf{r}, t) = \frac{\mu_0 c^2}{4\pi r}\left[\int \rho(\mathbf{r}', t_r)\, d\tau' + \frac{\hat{x}^k}{c}\int x'^k \dot{\rho}(\mathbf{r}', t_r)\, d\tau'\right]. \quad (5.95)$$

The first term is precisely the Coulomb potential $Q/(4\pi\epsilon_0 r)$, which will not contribute to radiation since its gradient shows up in $\mathbf{E}$, and that gradient will deliver a $1/r^2$ to the electric field, not part of the radiation piece. The integral in the second term is the time-derivative of the dipole moment, so the relevant "radiation" electric potential, written in vector form, is

$$V_{\text{rad}}(\mathbf{r}, t) = \frac{\mu_0 c}{4\pi r}\hat{\mathbf{r}} \cdot \dot{\mathbf{p}}(t_r) \qquad \text{with} \qquad \dot{\mathbf{p}}(t_r) \equiv \left.\frac{d\mathbf{p}(t)}{dt}\right|_{t=t_r \equiv t-r/c}. \quad (5.96)$$

Moving on to the spatial components, to begin with, we need the first term from (5.94) (the remaining term we will pick up in a moment), which, written in vector form, is

$$\mathbf{A}_{\text{rad}}(\mathbf{r}, t) = \frac{\mu_0}{4\pi r}\int \mathbf{J}(\mathbf{r}', t_r)\, d\tau'. \quad (5.97)$$

From (5.49), the right-hand-side integral is precisely $\dot{\mathbf{p}}(t_r)$, so the magnetic vector potential reads

$$\mathbf{A}_{\text{rad}}(\mathbf{r}, t) = \frac{\mu_0 \dot{\mathbf{p}}(t_r)}{4\pi r}. \quad (5.98)$$

Let's compute the electric and magnetic fields associated with V_{rad} and $\mathbf{A}_{\text{rad}}$. The magnetic field is

$$\mathbf{B} = \nabla \times \mathbf{A}_{\text{rad}} = \frac{\mu_0}{4\pi}\left(\frac{1}{r}\nabla \times \dot{\mathbf{p}}(t - r/c) - \dot{\mathbf{p}}(t - r/c) \times \nabla\left(\frac{1}{r}\right)\right). \quad (5.99)$$

The second term will not contribute to radiation since the gradient of $1/r$ will again return $1/r^2$ as in the electric case. So we only need to worry about the curl of $\dot{\mathbf{p}}(t - r/c)$, which depends on coordinates through its argument. As you will establish explicitly in Problem 5.15, the relevant curl is

$$\nabla \times \dot{\mathbf{p}}(t - r/c) = -\frac{1}{c}\hat{\mathbf{r}} \times \ddot{\mathbf{p}}(t - r/c), \quad (5.100)$$

giving the radiation piece of the magnetic field,

$$\mathbf{B}_{\text{rad}} = -\frac{\mu_0}{4\pi cr}\hat{\mathbf{r}} \times \ddot{\mathbf{p}}(t_r). \tag{5.101}$$

For the electric field, we need both the time derivative of $\mathbf{A}_{\text{rad}}$, which just puts another dot on the $\dot{\mathbf{p}}$, and the gradient of V_{rad},

$$\nabla V_{\text{rad}} = \frac{\mu_0 c}{4\pi r}\nabla\big(\hat{\mathbf{r}} \cdot \dot{\mathbf{p}}(t_r)\big) + \frac{\mu_0 c}{4\pi}\hat{\mathbf{r}} \cdot \dot{\mathbf{p}}(t_r)\nabla\Big(\frac{1}{r}\Big). \tag{5.102}$$

As with the curl of the magnetic vector potential in (5.99), the second term of ∇V_{rad} will lead to $1/r^2$ contributions to the electric field that are not part of the radiation. The first term can be simplified using a gradient product rule:

$$\nabla\big(\hat{\mathbf{r}} \cdot \dot{\mathbf{p}}(t_r)\big) = \hat{\mathbf{r}} \times \big(\nabla \times \dot{\mathbf{p}}(t_r)\big) + \dot{\mathbf{p}}(t_r) \times (\nabla \times \hat{\mathbf{r}}) + (\hat{\mathbf{r}} \cdot \nabla)\dot{\mathbf{p}}(t_r) + \big(\dot{\mathbf{p}}(t_r) \cdot \nabla\big)\hat{\mathbf{r}}. \tag{5.103}$$

The curl of $\hat{\mathbf{r}}$ is zero, as always, and the term that involves the derivatives of $\hat{\mathbf{r}}$ will produce a factor of $1/r$ that will combine with the $1/r$ out front in (5.102), rendering it irrelevant to the radiation fields. We know how to write the curl of $\dot{\mathbf{p}}(t_r)$ from (5.100), and the final relevant derivative appearing in (5.103) can be simplified,

$$(\hat{\mathbf{r}} \cdot \nabla)\dot{\mathbf{p}}(t_r) = -\frac{1}{c}\ddot{\mathbf{p}}(t_r). \tag{5.104}$$

Putting the pieces back together, we have

$$\nabla V_{\text{rad}} = -\frac{\mu_0}{4\pi r}\big(\hat{\mathbf{r}} \times \big(\hat{\mathbf{r}} \times \ddot{\mathbf{p}}(t_r)\big) + \ddot{\mathbf{p}}(t_r)\big). \tag{5.105}$$

The radiation portion of the electric and magnetic fields, responsible for carrying energy away from the source to infinity, are

$$\mathbf{E}_{\text{rad}} = \frac{\mu_0}{4\pi r}\big(\hat{\mathbf{r}} \times \big(\hat{\mathbf{r}} \times \ddot{\mathbf{p}}(t_r)\big) + \ddot{\mathbf{p}}(t_r) - \ddot{\mathbf{p}}(t_r)\big) = \frac{\mu_0}{4\pi r}\big(\hat{\mathbf{r}} \times \big(\hat{\mathbf{r}} \times \ddot{\mathbf{p}}(t_r)\big)\big),$$
$$\mathbf{B}_{\text{rad}} = -\frac{\mu_0}{4\pi cr}\hat{\mathbf{r}} \times \ddot{\mathbf{p}}(t_r). \tag{5.106}$$

It is useful to note that, for radiation, the electric field can be written in terms of the magnetic field,

$$\boxed{\mathbf{E}_{\text{rad}} = -c\hat{\mathbf{r}} \times \mathbf{B}_{\text{rad}},} \tag{5.107}$$

which is true even for sources beyond the dipole one.

The Poynting vector for the radiation fields, which started all of this off, is

$$\mathbf{S}_{\text{rad}} = \frac{1}{\mu_0}\mathbf{E}_{\text{rad}} \times \mathbf{B}_{\text{rad}} = \frac{1}{\mu_0 c}\mathbf{B}_{\text{rad}} \times (\hat{\mathbf{r}} \times \mathbf{B}_{\text{rad}}) = \frac{c}{\mu_0}B_{\text{rad}}^2\hat{\mathbf{r}}, \tag{5.108}$$

and, finally, the total power radiated, as collected by a sphere of radius r, from (5.87), is

$$P(r,t) = \oint_{S(r)} \mathbf{S}_{\text{rad}} \cdot d\mathbf{a} = \frac{c}{\mu_0}\int_0^{2\pi}\int_0^{\pi} B_{\text{rad}}^2 r^2 \sin\theta d\theta d\phi. \tag{5.109}$$

Let's put in our expression for $\mathbf{B}_{\text{rad}}$ from (5.106),

$$P_{\text{dip}}(t) = \frac{\mu_0 \ddot{p}(t_r)^2}{8\pi c} \int_0^\pi \sin^2 \alpha \sin\theta d\theta, \tag{5.110}$$

where α is the angle between the dipole moment direction and the field point, coming from the magnitude of $\hat{\mathbf{r}} \times \ddot{\mathbf{p}}(t_r)$. If we align the dipole moment with the $\hat{\mathbf{z}}$ axis, so that $\alpha = \theta$, then

$$P_{\text{dip}}(t) = \frac{\mu_0 \ddot{p}(t_r)^2}{6\pi c}. \tag{5.111}$$

As advertised (or rather, by construction), the power is independent of r, except through the temporal dependence t_r, which has to do with when the power "arrives" (as opposed to how the magnitude of the power changes). You can build quite a bit of intuition for electric dipole radiation by imagining point particles moving along bounded trajectories, like $\mathbf{w}(t) = d_0 \sin(\omega t)\hat{\mathbf{z}}$, with associated dipole moment $\mathbf{p}(t) = q\mathbf{w}(t)$, and then computing the electric and magnetic fields. Try out Problem 5.16 and Problem 5.17 to see what the radiation fields look like and how the power depends on the motion of the charges.

Example 5.3 (Gauge Choice). The solutions in (5.96) and (5.98) represent a wave that travels in the radial direction, that's what $t_r = t - r/c$ tells us. So there should be a "transverse" gauge in which the direction of $\mathbf{A}$ is perpendicular to the $\hat{\mathbf{r}}$ vector. We can use the projection operator from (5.7) with $\hat{\mathbf{n}} = \hat{\mathbf{r}}$ to find the polarization components in this transverse gauge. If you work out that projection in Problem 5.20, you'll get

$$\begin{aligned} \mathbf{A}(\mathbf{r}, t) = \frac{\mu_0}{4\pi r} \Big[\big(\cos\theta \big(\dot{p}^x(t_r) \cos\phi + \dot{p}^y(t_r) \sin\phi \big) - \dot{p}^z(t_r) \sin\theta \big) \hat{\boldsymbol{\theta}} \\ + \big(\dot{p}^y(t_r) \cos\phi - \dot{p}^x(t_r) \sin\phi \big) \hat{\boldsymbol{\phi}} \Big]. \end{aligned} \tag{5.112}$$

We could also obtain the components here directly using the transformation from Cartesian to spherical coordinates as in Problem 5.21. Those components are with respect to the natural coordinate basis in (2.64), not the unit vectors found in (5.112).

Example 5.4 (Electromagnetic Dipole Oscillator). Suppose you have a charge moving up and down along the z axis with frequency ω. The charge density is $\rho(\mathbf{r}, t) = q\delta(x)\delta(y)\delta(z - d\cos(\omega t))$ with dipole moment

$$\mathbf{p}(t) = \int \rho(\mathbf{r}', t)\mathbf{r}' d\tau' = qd\cos(\omega t)\hat{\mathbf{z}}. \tag{5.113}$$

The magnetic vector potential from (5.112) is, in spherical coordinates (and basis vectors),

$$\mathbf{A}(\mathbf{r}, t) = \frac{\mu_0 q d\omega}{4\pi r} \sin\big(\omega(t - r/c)\big) \sin\theta \hat{\boldsymbol{\theta}}. \tag{5.114}$$

The electric and magnetic fields are

$$\mathbf{E}(\mathbf{r}, t) = -\frac{\partial \mathbf{A}}{\partial t} = -\frac{\mu_0 q d \omega^2}{4\pi} \frac{\sin \theta}{r} \cos(\omega(t - r/c))\hat{\boldsymbol{\theta}},$$

$$\mathbf{B}(\mathbf{r}, t) = \nabla \times \mathbf{A} = -\frac{\mu_0 q d \omega^2}{4\pi c} \frac{\sin \theta}{r} \cos(\omega(t - r/c))\hat{\boldsymbol{\phi}}. \tag{5.115}$$

Notice that the fields oscillate with the same frequency as the source and will cause the charges that they effect to also oscillate at frequency ω as we saw in Section 5.1.1.

The radiated power comes from the Poynting vector $\mathbf{S} = \frac{1}{\mu_0}\mathbf{E} \times \mathbf{B}$, which is

$$\mathbf{S} = \frac{\mu_0 q^2 d^2 \omega^4}{16\pi^2 c} \frac{\sin^2 \theta}{r^2} \cos^2(\omega(t - r/c))\hat{\mathbf{r}}, \tag{5.116}$$

so that the power falls off like $1/r^2$ (again, that's the whole point of our assumptions) and is radially directed. Time-averaging over one full cycle just throws in a factor of $1/2$ and removes the oscillatory dependence: let $T \equiv \omega/(2\pi)$; then the time average of the Poynting vector is

$$\langle \mathbf{S} \rangle \equiv \frac{1}{T} \int_{t_0}^{t_0 + T} \mathbf{S}(t)dt = \frac{\mu_0 q^2 d^2 \omega^4}{32\pi^2 c} \frac{\sin^2 \theta}{r^2}\hat{\mathbf{r}}. \tag{5.117}$$

The magnitude of the time averaged power is called the "intensity," $I \equiv |\langle \mathbf{S} \rangle|$.

The total power radiated by the oscillator is

$$P = \oint \langle \mathbf{S} \rangle \cdot d\mathbf{a} = \frac{\mu_0 (q d \omega^2)^2}{12\pi c}, \tag{5.118}$$

where we have used a sphere of radius R as the surface of integration. This result agrees with (5.111), which was developed for an arbitrary dipole moment.

5.5.2 Electric Quadrupole Radiation

Charge conservation means that there is no electric monopole radiation, so the first radiating term, given a source, is its dipole, as we have just established. In general relativity, both energy and momentum are conserved, eliminating monopole and dipole radiation, so in the case of gravity, the first source term that contributes to radiation is the quadrupole. For reasons of comparison, then, we will generate the relevant expressions for electric quadrupole radiation.

Quadrupole radiation comes from the second term in (5.94), which also includes magnetic dipole radiation. It is clear from the general form of radiation fields in (5.107) that everything (here meaning $\mathbf{E}_{\mathrm{rad}}$, $\mathbf{S}_{\mathrm{rad}}$, and P_{rad}) can be written in terms of the magnetic field, and then, since the magnetic field depends only on $\mathbf{A}$, that our focus should be on the spatial components ($\mu = 1, 2, 3$) of the second term in (5.94). Writing out those terms using Roman indices, for the second integral, our starting point is

$$A^i(\mathbf{r}, t) = \frac{\mu_0}{4\pi rc} \hat{x}^k \int \left[\frac{1}{2}(x'^k J^i + x'^i J^k) + \frac{1}{2}(x'^k J^i - x'^i J^k) \right] d\tau', \quad (5.119)$$

where we have broken $x'^k J^i$ up into its symmetric and antisymmetric pieces in the integrand.

For the antisymmetric piece, we can write, in vector notation,

$$\begin{aligned}
\mathbf{A}_{\text{md}}(\mathbf{r}, t) &\equiv \frac{\mu_0}{8\pi rc} \int \left((\hat{\mathbf{r}} \cdot \mathbf{r}')\mathbf{J} - (\hat{\mathbf{r}} \cdot \mathbf{J})\mathbf{r}' \right) d\tau' \\
&= \frac{\mu_0}{4\pi rc} \frac{1}{2} \int \hat{\mathbf{r}} \times (\mathbf{J} \times \mathbf{r}') \, d\tau'.
\end{aligned} \quad (5.120)$$

Referring to (5.52), this part of the magnetic vector potential can be written as

$$\mathbf{A}_{\text{md}}(\mathbf{r}, t) = -\frac{\mu_0}{4\pi rc} \hat{\mathbf{r}} \times \dot{\mathbf{m}}(t_r), \quad (5.121)$$

and it is clear that this vector potential is associated with magnetic dipole radiation (you should work through Problem 5.24 to parallel the electric dipole radiation discussion from the previous section).

We want the electric quadrupole piece, so we'll focus on the symmetric bit of (5.119), defining

$$A^i_{\text{quad}}(\mathbf{r}, t) \equiv \frac{\mu_0}{4\pi rc} \hat{x}^k \frac{1}{2} \int (x'^k J^i + x'^i J^k) \, d\tau'. \quad (5.122)$$

From (5.67), the integral here is just the second derivative of the electric quadrupole moment, so that

$$A^i_{\text{quad}}(\mathbf{r}, t) = \frac{\mu_0}{8\pi rc} \hat{x}^k \ddot{P}^{ik}(t_r) \quad (5.123)$$

is the relevant piece of the vector potential.

Since we are interested in the radiated power, we can use (5.108) written in terms of the vector potential, as you will show in Problem 5.25,

$$S^i_{\text{rad}} = \frac{c}{\mu_0} \left(\partial^j A^\ell \partial^j A^\ell - \partial^j A^\ell \partial^\ell A^j \right) \hat{x}^i, \quad (5.124)$$

where, as always, we drop any terms that lead to dependence beyond $1/r^2$. Thinking about the derivatives of A^i, it is clear that whenever a derivative operator hits $1/r$ or $\hat{x}^k$, it will generate additional factors of $1/r$, so we only really need to worry about when the derivative acts on $\ddot{P}^{ik}(t_r)$:

$$\left(\partial_j A^i \right)_{\text{rad}} = \frac{\mu_0}{8\pi rc} \hat{x}^k \partial_j \ddot{P}^{ik}(t_r) = \frac{\mu_0}{8\pi rc} \hat{x}^k \dddot{P}^{ik}(t_r) \left(-\frac{1}{c} \hat{x}_j \right). \quad (5.125)$$

Then the Poynting vector components become

$$S^i_{\text{rad}} = \frac{c}{\mu_0} \hat{x}^i \left(\frac{\mu_0}{8\pi rc^2} \right)^2 \left[\hat{x}^k \hat{x}^\ell \dddot{P}^{jk}(t_r) \dddot{P}^{j\ell}(t_r) - \hat{x}^k \hat{x}^\ell \hat{x}^m \hat{x}^n \dddot{P}^{k\ell}(t_r) \dddot{P}^{mn}(t_r) \right]. \quad (5.126)$$

The Poynting vector clearly "points" in the $\hat{\mathbf{r}}$ direction (note the presence of $\hat{x}^i$ on the right of S^i_{rad}), so the whole vector is directed radially. To find the total power, we have to integrate the spray of unit vectors,

$$
P_{\text{rad}}(r) = \frac{\mu_0}{64\pi^2 c^3} \left[\dddot{P}^{jk}(t_r) \dddot{P}^{j\ell}(t_r) \int_0^{2\pi} \int_0^{\pi} \hat{x}^k \hat{x}^\ell \sin\theta d\theta d\phi \right.
$$
$$
\left. - \dddot{P}^{k\ell}(t_r) \dddot{P}^{mn}(t_r) \int_0^{2\pi} \int_0^{\pi} \hat{x}^k \hat{x}^\ell \hat{x}^m \hat{x}^n \sin\theta d\theta d\phi \right]. \tag{5.127}
$$

Let's focus on the integrals of the basis vectors. The tensor $\hat{x}^k \hat{x}^\ell$ has components given by

$$
\hat{x}^k \hat{x}^\ell \doteq \frac{1}{x^2 + y^2 + z^2} \begin{pmatrix} x^2 & xy & xz \\ xy & y^2 & yz \\ xz & yz & z^2 \end{pmatrix} \tag{5.128}
$$

or, written in the spherical coordinates of interest,

$$
\hat{x}^k \hat{x}^\ell \doteq \begin{pmatrix} \sin^2\theta \cos^2\phi & \sin^2\theta \sin\phi \cos\phi & \sin\theta \cos\theta \cos\phi \\ \sin^2\theta \sin\phi \cos\phi & \sin^2\theta \sin^2\phi & \sin\theta \cos\theta \sin\phi \\ \sin\theta \cos\theta \cos\phi & \sin\theta \cos\theta \sin\phi & \cos^2\theta \end{pmatrix}. \tag{5.129}
$$

Every off-diagonal term in this tableaux has ϕ dependence that will vanish when integrating in ϕ from $0 \to 2\pi$,

$$
\int_0^{2\pi} \hat{x}^k \hat{x}^\ell \sin\theta d\phi \doteq \pi \begin{pmatrix} \sin^3\theta & 0 & 0 \\ 0 & \sin^3\theta & 0 \\ 0 & 0 & 2\sin\theta \cos^2\theta \end{pmatrix}. \tag{5.130}
$$

Integrating this result with respect to θ, all nonzero terms pick up a factor of $4/3$, so that

$$
\int_0^{2\pi} \int_0^{\pi} \hat{x}^k \hat{x}^\ell \sin\theta d\theta d\phi = \frac{4\pi}{3} \delta^{k\ell}. \tag{5.131}
$$

To compute the integral of the quartic integrand $\hat{x}^k \hat{x}^\ell \hat{x}^m \hat{x}^n$, we'll avoid brute force and demonstrate a more general method based on the index symmetries of the integral. The integrand is symmetric under the interchange of any index pair, so the result of integration must be as well. In terms of building blocks, there are no preferred directions, because of the integration over the entire sphere, and the radial unit vector is also unavailable for the same reason. Referring to the discussion in Section 4.1, the only tensor available for building the integral is the Kronecker delta. Since the delta takes two indices, and the integrand has four, we must have pairs of deltas, and the indices must be appropriately symmetrized. So far, then, we have

$$
\int_0^{2\pi} \int_0^{\pi} \hat{x}^k \hat{x}^\ell \hat{x}^m \hat{x}^n \sin\theta d\theta d\phi = \alpha \left(\delta^{k\ell} \delta^{mn} + \delta^{km} \delta^{\ell n} + \delta^{kn} \delta^{\ell m} \right), \tag{5.132}
$$

where α is a constant. To set that constant, we just need to pick indices to isolate a single term, let $k = \ell = 1$ and $m = n = 2$. Using the basis vectors written in spherical coordinates, this choice has the integral

$$\int_0^{2\pi} \int_0^{\pi} (\sin\theta\cos\phi)^2 (\sin\theta\sin\phi)^2 \sin\theta \, d\theta \, d\phi = \frac{4\pi}{15}, \tag{5.133}$$

so that $\alpha = 4\pi/15$.

With the integrals in place, the total power is

$$
\begin{aligned}
P_{\text{rad}} &= \frac{\mu_0}{64\pi^2 c^3} \left(\frac{4\pi}{3} \dddot{P}^{ik}(t_r) \dddot{P}^{ik}(t_r) \right. \\
&\quad \left. - \frac{4\pi}{15} \left(2\dddot{P}^{km}(t_r) \dddot{P}^{km}(t_r) + \dddot{P}^{mm}(t_r) \dddot{P}^{nn}(t_r) \right) \right) \\
&= \frac{\mu_0}{80\pi c^3} \left(\dddot{P}^{ik}(t_r) \dddot{P}^{ik}(t_r) - \frac{1}{3} \dddot{P}^{mm}(t_r) \dddot{P}^{nn}(t_r) \right).
\end{aligned}
\tag{5.134}
$$

If we define the traceless quadrupole moment by

$$Q^{ij} \equiv P^{ij} - \frac{1}{3}\delta^{ij} P^{mm}, \tag{5.135}$$

then

$$Q^{ij} Q^{ij} = P^{ij} P^{ij} - \frac{1}{3} P^{mm} P^{nn}, \tag{5.136}$$

so that the power can be written compactly as

$$P_{\text{rad}} = \frac{\mu_0}{80\pi c^3} \dddot{Q}^{ij}(t_r) \dddot{Q}^{ij}(t_r). \tag{5.137}$$

In many cases, the total power is also time-averaged over, say, one full cycle of the source motion, so we'll often see

$$\boxed{P_{\text{rad}} = \frac{\mu_0}{80\pi c^3} \langle \dddot{Q}^{ij}(t_r) \dddot{Q}^{ij}(t_r) \rangle.} \tag{5.138}$$

You can compare the power radiated by the quadrupole versus the dipole for an oscillating charge in Problem 5.29.

5.6 Gravitational Radiation

Electromagnetic radiation is defined to be the portion of the electric and magnetic fields that carry energy out to infinity. To employ the definition, we needed to be able to evaluate the gauge-independent fields themselves to compute the (also gauge-independent) Poynting vector, which tells us about the energy flow. On the gravity side, we primarily work with small metric perturbations, analogous to the electromagnetic vector potential, without computing the gauge-independent fields that show up in, for example, the geodesic equation of motion. Fortunately, the electromagnetic case gave us a way to focus on the components of A^μ, and those lessons can be carried over directly to gravity.

We learned that the potential (here the metric perturbation $H^{\mu\nu}$) should (1) fall off like $1/r$ so that the gauge-independent fields, which are obtained from derivatives of the potential, fall off at most like $1/r$, and (2) be time dependent – dependence on $t - r/c$ ensures that derivatives can be taken without necessarily introducing additional factors of $1/r$. These rules are based on the observation that the electromagnetic Poynting vector for radiation should have $1/r^2$ dependence, and any potential with smaller than $1/r$ dependence leads to a Poynting vector that falls off faster than $1/r^2$. While we won't compute the analogue of the Poynting vector for gravity directly, the same basic rules apply. Our model, then, is a form for $H^{\mu\nu}$ that mimics the radiation potential in (5.98) or (5.123).

We have done much of the work relevant to gravitational radiation already, in schematic form. The primary difference between electromagnetic and gravitational radiation, structurally, is the expanded set of conservation laws present in gravity. Instead of the single charge-current pair from (5.56), we have a set of four conservation statements

$$\frac{1}{c}\frac{\partial T^{0\nu}}{\partial t} + \frac{\partial T^{k\nu}}{\partial x^k} = 0. \tag{5.139}$$

Then the three main results from Section 5.4.1, namely the charge (5.47), dipole (5.49), and quadrupole (5.55) derivatives are updated, under the mapping $\rho \to T^{0\nu}/c$, $J^k \to T^{k\nu}$, to read:

$$\frac{1}{c}\frac{d}{dt}\int T^{0\nu}\,d\tau = 0,$$
$$\frac{1}{c}\frac{d}{dt}\int T^{0\nu}x^i\,d\tau = \int T^{\nu i}\,d\tau, \tag{5.140}$$
$$\frac{1}{c}\frac{d}{dt}\int T^{0\nu}x^ix^j\,d\tau = \int \left(T^{\nu i}x^k + T^{\nu k}x^i\right)d\tau,$$

where the integrals are over all space, as always, and we assume that the source distributions are finite and bounded. Each one of these relations is four separate statements, which we will break up using $\nu = 0$ and then $\nu = 1, 2, 3$ for the spatial pieces (these will be denoted using a Roman letter, as usual). To set the multipole notation that is relevant to gravity and that already appears in (5.140), define the "energy monopole" to be

$$E \equiv \int T^{00}\,d\tau, \tag{5.141}$$

where the integral is over all space. The "energy dipole" moment is

$$P^i \equiv \int T^{00}x^i\,d\tau, \tag{5.142}$$

and the energy quadrupole is

$$P^{ij} \equiv \int T^{00}x^ix^j\,d\tau. \tag{5.143}$$

These are the source moments that replace the electric Q, $\mathbf{p}$, and P^{ij}, with energy density T^{00} appearing instead of charge density ρ.

From the first equation in (5.140), with $\nu = 0$, we learn that

$$\frac{d}{dt} \int T^{00} \, d\tau = 0, \qquad (5.144)$$

which tells us that total energy is conserved. Setting $\nu = i$ for spatial $i = 1, 2, 3$, we obtain

$$\frac{d}{dt} \int T^{0i} \, d\tau = 0, \qquad (5.145)$$

and the global source momentum is conserved.

Moving on to the nonzero derivatives, from the second equation in (5.140) with $\nu = 0$,

$$\frac{1}{c} \frac{d}{dt} \int T^{00} x^i \, d\tau = \int T^{0i} \, d\tau, \qquad (5.146)$$

where the left-hand side looks like the time derivative of the "energy dipole moment." For $\nu = j$, we get

$$\frac{1}{c} \frac{d}{dt} \int T^{0j} x^i \, d\tau = \int T^{ji} \, d\tau \qquad (5.147)$$

with the time derivative of a "momentum dipole moment" on the left.

Finally, it turns out that we will only need the $\nu = 0$ component of the third equation in (5.140),

$$\frac{1}{c} \frac{d}{dt} \int T^{00} x^i x^j \, d\tau = \int \left(T^{0i} x^j + T^{0j} x^i \right) d\tau. \qquad (5.148)$$

The left-hand side represents the time derivative of the energy quadrupole moment, which will end up being the dominant source of gravitational radiation. This is the first available term, since the energy "monopole" (total energy) is constant via (5.144), and the energy "dipole" from (5.146) has

$$\frac{1}{c^2} \frac{d^2}{dt^2} \int T^{00} x^i \, d\tau = \frac{1}{c} \frac{d}{dt} \int T^{0i} \, d\tau = 0 \qquad (5.149)$$

using (5.145). Technically, this equation tells us only that the energy dipole is at most linear in time, but growing linearly with time violates our source assumptions, so in fact, the energy dipole is time independent.

These expressions allow us to relate the integral of the spatial components of the stress tensor to the time derivatives of the quadrupole moment of the energy density. Take the time derivative of (5.148) and use (5.147) on the right-hand side to get

$$\frac{1}{c^2} \frac{d^2}{dt^2} \int T^{00} x^i x^j \, d\tau = \frac{1}{c} \frac{d}{dt} \int \left(T^{0i} x^j + T^{0j} x^i \right) d\tau = 2 \int T^{ij} \, d\tau. \quad (5.150)$$

Rearranging a little, we have

$$\int T^{ij} \, d\tau = \frac{1}{2c^2} \frac{d^2}{dt^2} \int T^{00} x^i x^j \, d\tau. \qquad (5.151)$$

This equation will be the main result we need to connect the metric perturbation to its source.

We understand the source relationships enforced by conservation, so let's turn to the solution of

$$\Box H^{\mu\nu} = -\frac{16\pi G}{c^4} T^{\mu\nu}, \tag{5.152}$$

the field equation governing a metric perturbation with the source in place from (3.101). The Green's function solution in (5.73), applied to our tensor source with appropriate units, is

$$H^{\mu\nu}(\mathbf{r}, t) = \frac{16\pi G}{c^4} \int T^{\mu\nu}(\mathbf{r}', t') \frac{\delta(c(t - t') - |\mathbf{r} - \mathbf{r}'|)}{4\pi |\mathbf{r} - \mathbf{r}'|} d(ct') d\tau'. \tag{5.153}$$

We can again perform the t' integration using the delta function, leaving

$$H^{\mu\nu}(\mathbf{r}, t) = \frac{16\pi G}{c^4} \int \frac{T^{\mu\nu}(\mathbf{r}', t - |\mathbf{r} - \mathbf{r}'|/c)}{4\pi |\mathbf{r} - \mathbf{r}'|} d\tau'. \tag{5.154}$$

If we make the same assumptions about the source, that it is localized near the origin and that the field points of interest are far from the origin, then we can use the approximation from (5.90), but in this case, it suffices to keep just the leading order term $|\mathbf{r} - \mathbf{r}'| \approx r$,

$$H^{\mu\nu}(\mathbf{r}, t) \approx \frac{4G}{c^4 r} \int T^{\mu\nu}(\mathbf{r}', t - r/c) d\tau'. \tag{5.155}$$

The $\mu = \nu = 0$ component of the solution involves the integral of T^{00}, which is a constant in time by (5.144). Similarly, the H^{0i} component is constant by (5.149) (together with the bounded source assumption). Since these elements of the metric perturbation are constant in time, they cannot contribute to gravitational radiation any more than the constant, in time, electric potential $V = Q/(4\pi \epsilon_0 r)$ can contribute to electromagnetic radiation. We now see how monopole- and dipole-sourced gravitational radiation is excluded, structurally, by conservation of energy and momentum.

That leaves us with the spatial components of the perturbation. These depend on the integral of T^{ij}, which we can relate directly to the second derivative of the (energy) quadrupole moment using (5.151),

$$H^{ij}(\mathbf{r}, t) = \frac{4G}{c^4 r} \int T^{ij}(\mathbf{r}', t_r) d\tau' = \frac{2G}{c^6 r} \frac{d^2}{dt^2} \int T^{00} x^i x^j d\tau \tag{5.156}$$

with $t_r \equiv t - r/c$. Written in terms of the electric quadrupole moment defined in (5.143), this solution becomes the simple

$$\boxed{H^{ij}(\mathbf{r}, t) = \frac{2G}{c^6 r} \ddot{P}^{ij}(t_r).} \tag{5.157}$$

While we developed this far field solution using the sources, the components of the metric perturbation actually solve the wave equation $\Box H^{ij}(\mathbf{r}, t) = 0$, since at the field point, there *is* no source, one of our early assumptions. Then we know that the solution can be put into TT gauge by appropriate gauge choice. In that gauge, the trace of $H^{\mu\nu}$ is zero, and we have $H^{\mu\nu} = h^{\mu\nu}$.

We'll start by making the solution in (5.157) traceless and then use the projection approach to achieve transversality (while preserving tracelessness). Subtracting off the trace of the quadrupole tensor, as in the electromagnetic case (5.135), let

$$Q^{mn}(t) \equiv P^{mn} - \frac{1}{3} P \eta^{mn} = \int T^{00} \left(x'^m x'^n - \frac{1}{3} x'^j x'^j \eta^{mn} \right) d\tau'. \quad (5.158)$$

If we use this instead of P^{mn} in (5.157), then $h^{mn} = H^{mn}$ since these differ only by a trace. Take

$$h^{mn}(\mathbf{r}, t) \equiv \frac{2G}{c^6 r} \ddot{Q}^{mn}(t - r/c), \quad (5.159)$$

and then use $\hat{\mathbf{n}} = \hat{\mathbf{r}}$ to define the projector from (5.27). You can show in Problem 5.30 that the TT projection of Q^{mn} is identical to that of P^{mn}. Using that projection, we have

$$\left[h^{mn}(\mathbf{r}, t) \right]^{TT} = \frac{2G}{c^6 r} \left(N^m_j N^n_k - \frac{1}{2} N^{mn} N_{jk} \right) \ddot{Q}^{jk}(t - r/c). \quad (5.160)$$

If we carry out the projection and then transform to spherical coordinates, then the TT solution to the wave equation for spherical waves has nonzero components

$$\begin{aligned}
h^{22}(\mathbf{r}, t) &= \frac{2G}{c^6 r} \frac{1}{8r^2} \left(\left(\left(\ddot{P}^{xx}(t_r) - \ddot{P}^{yy}(t_r) \right) \cos(2\phi) \right.\right. \\
&\quad \left. + 2 \ddot{P}^{xy}(t_r) \sin(2\phi) \right) \left(3 + \cos(2\theta) \right) \\
&\quad - 2 \left(\ddot{P}^{xx}(t_r) + \ddot{P}^{yy}(t_r) - 2 \ddot{P}^{zz}(t_r) \right) \sin^2 \theta \\
&\quad \left. - 4 \left(\ddot{P}^{xz}(t_r) \cos \phi + \ddot{P}^{yz}(t_r) \sin \phi \right) \sin(2\theta) \right), \\
h^{23}(\mathbf{r}, t) &= \frac{2G}{c^6 r} \frac{\cot \theta \cos \phi}{r^2} \left(\ddot{P}^{xy}(t_r) \cos \phi \right. \\
&\quad - \sin \phi \left(\ddot{P}^{xx}(t_r) - \ddot{P}^{yy}(t_r) + \ddot{P}^{xy}(t_r) \tan \phi \right) \\
&\quad \left. + \left(\ddot{P}^{xz}(t_r) \tan \phi - \ddot{P}^{yz}(t_r) \right) \tan \theta \right) = h^{32}(\mathbf{r}, t), \\
h^{33}(\mathbf{r}, t) &= \frac{2G}{c^6 r} \frac{1}{4r^2} \left(\ddot{P}^{xx}(t_r) + \ddot{P}^{yy}(t_r) - 2 \ddot{P}^{zz}(t_r) \right. \\
&\quad + 4 \cot \theta \left(\ddot{P}^{xz}(t_r) \cos \phi + \ddot{P}^{yz}(t_r) \sin \phi \right) \\
&\quad - \left(\left(\ddot{P}^{xx}(t_r) - \ddot{P}^{yy}(t_r) \right) \cos(2\phi) \right. \\
&\quad \left. \left. + 2 \ddot{P}^{xy}(t_r) \sin(2\phi) \right) \left(-1 + \frac{2}{\sin^2 \theta} \right) \right),
\end{aligned} \quad (5.161)$$

with $t_r \equiv t - r/c$ as usual. The additional factors of r out front, which would seem to violate our rule to keep only $1/r$ terms, come from the transformation to spherical coordinates. If we wrote out these terms using unit basis vectors like $\hat{\boldsymbol{\theta}}$ and $\hat{\boldsymbol{\phi}}$ (the only two that contribute), then those additional factors would go away.

Example 5.5 (Circular Motion in the xy Plane). Suppose you have a point mass m moving in a circle of radius R at $z = 0$ with angular frequency ω. The energy density associated with the object is

$$T^{00} = mc^2 \delta(z) \delta(x - R\cos(\omega t)) \delta(y - R\sin(\omega t)). \quad (5.162)$$

The traceless quadrupole moment can be computed from (5.158),

$$Q^{mn}(t) \doteq mc^2 R^2 \begin{pmatrix} \cos^2(\omega t) - \frac{1}{3} & \cos(\omega t)\sin(\omega t) & 0 \\ \cos(\omega t)\sin(\omega t) & \sin^2(\omega t) - \frac{1}{3} & 0 \\ 0 & 0 & -\frac{1}{3} \end{pmatrix}. \quad (5.163)$$

Then the metric perturbation at large distances, from (5.159) is

$$h^{mn}(r, t) = \frac{4Gm\omega^2 R^2}{c^4 r} \begin{pmatrix} -\cos(2\omega t_r) & -\sin(2\omega t_r) & 0 \\ -\sin(2\omega t_r) & \cos(2\omega t_r) & 0 \\ 0 & 0 & 0 \end{pmatrix} \quad (5.164)$$

with $t_r \equiv t - r/c$ as usual. Notable are its size, set by the dimensionless constant $Gm\omega^2 R^2/(c^4 r)$ out front, and its temporal evolution: The oscillation of the metric components is at a frequency that is twice the source frequency.

We can run the quadrupole moment tensor through (5.161) to get

$$\begin{aligned} &\left[h^{mn}(\mathbf{r}, t) \right]^{\text{TT}} \\ &= -\frac{4Gm\omega^2 R^2}{c^4 r^3} \\ &\quad \times \begin{pmatrix} 0 & 0 & 0 \\ 0 & \frac{1}{4}(3 + \cos(2\theta))\cos(2\psi) & -\cot\theta\sin(2\psi) \\ 0 & -\cot\theta\sin(2\psi) & -\frac{1}{4\sin^2\theta}(3 + \cos(2\theta))\cos(2\psi) \end{pmatrix}, \\ &\psi \equiv \phi - \omega(t - r/c), \end{aligned} \quad (5.165)$$

and write the result in a manner reminiscent of vectors like $\mathbf{v} = v^j \mathbf{e}_j$ using $\mathbb{H} = h^{mn}\mathbf{e}_m\mathbf{e}_n$, and then expressing $\mathbf{e}_m\mathbf{e}_n$ in terms of the unit spherical basis vectors from (2.64), we have the "dyadic"[7]

$$\begin{aligned} \mathbb{H} = -\frac{4Gm\omega^2 R^2}{c^4 r}\Big[&\frac{1}{4}(3 + \cos(2\theta))\cos(2\psi)(\hat{\theta}\hat{\theta} - \hat{\phi}\hat{\phi}) \\ &- \cos\theta\sin(2\psi)(\hat{\phi}\hat{\theta} + \hat{\theta}\hat{\phi})\Big]. \end{aligned} \quad (5.166)$$

The "matrix" $\mathbb{H}$ has elements that are associated with the spherical basis vectors, so that $H^{\theta\theta}$, for example, is the component sitting in front of $\hat{\theta}\hat{\theta}$, just as E^θ is the component of the electric field sitting in front of $\hat{\theta}$. In this form, you can clearly see a spherical representation of our two polarization modes.

Some mechanism is sustaining the circular motion here, and the obvious choice is gravitational. Let M be the mass of a spherical source at the origin.

---

[7] The dyadic representation carries the same information as the matrix form – the basis vectors tell you which row (first basis vector) and column (second) of the matrix to associate with the component.

Then using Newtonian gravity to relate ω to the source and test particle parameters,

$$\frac{mv^2}{R} = \frac{GMm}{R^2} \longrightarrow \omega^2 = \frac{GM}{R^3}. \qquad (5.167)$$

The magnitude of the metric perturbation from (5.164) is then

$$\frac{4Gm\omega^2 R^2}{c^4 r} = \frac{2Gm}{c^2} \frac{2GM}{c^2} \frac{1}{Rr} \equiv \frac{R_M R_m}{Rr}, \qquad (5.168)$$

where R_M and R_m are the Schwarzschild radii of the source and test masses, respectively. Since we are using the radiation expression for the solution, we must have $r \gg R$. As an example, to get a feel for the size of the perturbation, take the Earth's orbit about the Sun. If we assume a circular orbit of radius $R = 1.5 \times 10^8$ km, observed from a distance of $r = 100R$, with the appropriate masses, then the size of the metric perturbation for the system will be

$$\frac{R_M R_m}{Rr} \approx 1 \times 10^{-23}. \qquad (5.169)$$

How sensitive would an instrument like LIGO have to be in order to detect a metric perturbation of this size? Referring back to Section 5.3.1 and the relative separation expression in (5.44), the detector needs to be able to measure displacements on the order of hd, where h is the size of the metric perturbation (P^{xx} from (5.44)), and d is the distance separating the masses. For LIGO, $d = 4$ km, so the displacements of the test masses will be on the order of

$$\Delta d \sim hd \approx \times 10^{-20} \, \text{m}. \qquad (5.170)$$

Compare this sensitivity with the 5×10^{-11} m that is the Bohr radius of hydrogen and it is clear that LIGO needs to be able to measure incredibly small displacements. The model here is artificial and uses familiar objects. In fact, the gravitational radiation that LIGO has detected comes from much larger sources that are also much further away. Still, you can see why, for many years, the ability of LIGO to detect gravitational waves was met with some skepticism.

5.6.1 Power Radiated Away

Gravitational radiation extracts energy from its sources just as electromagnetic radiation does. We computed the total power radiated by electric quadrupole radiation in (5.138), and we can use our update prescription $\mu_0 \to 4\pi G/c^2$ to generate the equivalent expression for gravity. Remember, though, that this formal replacement is associated with massive, as opposed to energetic, sources of Newtonian gravity. Since our source $T^{\mu\nu}$ has units of energy density, and those units are inherited (together with length-squared) by the quadrupole moment, we pick up an additional factor of $1/c^4$ in going from (5.138) to its gravitational form:

$$P_{\text{rad}} = \frac{G}{20c^9}\langle \dddot{Q}^{ij} \dddot{Q}^{ij} \rangle. \tag{5.171}$$

As usual, with electromagnetic predictions, the numerical factor $1/20$ out front is not quite right; it turns out that

$$\boxed{P_{\text{rad}} = \frac{G}{5c^9}\langle \dddot{Q}^{ij} \dddot{Q}^{ij} \rangle} \tag{5.172}$$

is the correct expression for gravitational radiation (see [35]).

If we use the quadrupole moment associated with a circular-motion source from (5.163), then the total radiated power is

$$P_{\text{rad}} = \frac{32 G m^2 R^4 \omega^6}{5c^5}. \tag{5.173}$$

Using the expression for ω^2 from (5.167), appropriate for uniform circular motion caused by Newtonian gravity,

$$P_{\text{rad}} = \frac{32 G^4 m^2 M^3}{5c^5 R^5}. \tag{5.174}$$

We have leaned on our E&M expressions and experience here, and for good reason. While it is possible to take a fully linearized view of gravitational radiation, and use field theoretic techniques to generate its stress tensor, then develop radiation from there, it is much more difficult to assign energy (and all the rest) to the full gravitational field, the metric, of general relativity. Physically, the problem is that at any point in spacetime, we can reduce that field to the Minkowski metric representing flat spacetime. At a point, then, there is no gravitational field, and it is hard to associate energy content with that nothing. A more structural observation is that energy densities that we know, like the $u = 1/2\epsilon_0 \nabla V \cdot \nabla V$ of electrostatics, depend on the derivative of the field. Think of the two types of metric derivative that we have seen: $g_{\mu\nu;\alpha}$ and $g_{\mu\nu,\alpha}$. As candidates for generating a stress tensor, both of these are flawed. The first is automatically zero in the spacetimes relevant to general relativity, with their metric connections. The second is a nontensorial building block, and it's hard to know how to combine these nontensors to form a viable stress tensor. There exist ways to develop relatively consistent stress-energy "pseudotensors" (see [35] and [31]), but in the end, these return the total radiation result we have used in this section.

The radiation power loss due to gravitational wave generation is responsible for the inspiral of orbiting bodies, just as electromagnetic radiation causes inspiral of orbiting charges. The 1993 Nobel Prize was awarded to Hulse and Taylor for their observation of a pulsar-neutron star binary [25], spiraling in towards each other due to gravitational radiation power loss.

Exercises

5.1   For a four-vector potential of the form $A^\mu = P^\mu e^{i(-\omega t + \omega z/c)}$ with $P^z = P^0$ and $P^x = P^y = 0$, what are the electric and magnetic fields? What does this tell you about P^z and P^0?

5.2   Suppose you have a plane wave potential $A^\mu = P^\mu e^{ik_\nu x^\nu}$ that is in Lorenz gauge, $\partial_\mu A^\mu = 0$ with $P^0 = 0$. Show that you cannot introduce a function ϕ with $A_\mu \to A_\mu + \phi_{,\mu}$, while remaining in Lorenz gauge (and with no 0 component), to set other components of P^μ to zero.

5.3   A particle with mass m and charge q sits at rest at the origin. A plane wave electromagnetic field $A^\mu = P^\mu e^{-i\omega(t-z/c)}$ with $P^x = P$ the only nonzero component comes by. Taking the real part of the resulting electromagnetic fields, write the equation of motion for the charge and solve assuming that the speed of the (now moving) charge is much less than c.

5.4   A particle of mass m and charge q is at rest at the origin. An electromagnetic plane wave of the form (5.10) comes by. We assumed that $\dot{x}(t)/c \ll 1$ in working out the solution to the equations of motion from (5.12), what constraint does this condition put on the solution from (5.13)? More specifically, what is the relation between the plane wave's frequency ω and the rest of the parameters in the problem?

5.5   Show that for the projection operator defined in (5.7), we have $N^j{}_j = 2$ and $N^j{}_\ell N_{jm} P^{\ell m} = N_{\ell m} P^{\ell m}$.

5.6   Using the projection operator from (5.7) with $\hat{\mathbf{n}} = \hat{\mathbf{z}}$, show that starting from a full polarization tensor

$$P^{jk} \doteq \begin{pmatrix} P^{xx} & P^{xy} & P^{xz} \\ P^{xy} & P^{yy} & P^{yz} \\ P^{xz} & P^{yz} & P^{zz} \end{pmatrix}, \tag{5.175}$$

the projection in (5.27) recovers the form of the polarization tensor found in (5.25).

5.7   Taking $\bar{h}_{\alpha\mu} \equiv h_{\alpha\mu} - (1/2)\eta_{\alpha\mu}h$ with $h \equiv h_{\mu\nu}\eta^{\mu\nu}$ (and $\eta_{\mu\nu}$ is the Minkowski metric written in Cartesian coordinates), what does the Lorenz gauge condition $\partial^\mu h_{\mu\nu} = (1/2)\partial_\nu h$ give for the divergence of $\bar{h}_{\alpha\mu}$?

5.8   Take the Fourier transform $\tilde{p}(f)$ from (5.75) and put it into (5.76). Show that to recover $p(t)$ from the integrals on the right (watch out for integration variable name clashing!), (5.77) must hold.

5.9   The Levi-Civita "symbol" (it's not a tensor) is defined, in three dimensions, to be

$$\epsilon^{ijk} = \begin{cases} 1 & \text{if } i, j, k \text{ is an even permutation of } 1, 2, 3, \\ -1 & \text{if } i, j, k \text{ is an odd permutation of } 1, 2, 3, \\ 0 & \text{else.} \end{cases} \tag{5.176}$$

From this definition: (1) Show that the curl of a vector field has elements that can be written (in Cartesian coordinates) $(\nabla \times \mathbf{A})^i = \epsilon^{ijk}\partial_j A_k$ (index placement can be up or down here, since all elements are spatial), and (2) The sum of two symbols with two indices contracted: $\epsilon^{ijk}\epsilon_{ij\ell} = 2\delta^k_\ell$ and with one index contracted, $\epsilon^{ijk}\epsilon_{imn} = \delta^j_m\delta^k_n - \delta^j_n\delta^k_m$.

5.10 Evaluate the negative gradient of the potential in (5.96), keeping in mind that $t_r \equiv t - r/c$, so itself has position dependence. Isolate the $1/r$ piece of the electric field that comes from this potential.

5.11 Starting from A^μ with $A^0 = V/c$ with V from (5.96) and $\mathbf{A}$ from (5.98), make a gauge transformation to get $A^0 = 0$. What is the resulting $\mathbf{A}$? Check that the electric and magnetic radiation fields you get from this new vector potential match the ones you get from the original set.

5.12 For a gravitational plane wave propagating in the $\hat{\mathbf{x}}$ direction, write the TT gauge metric perturbation.

5.13 Show that the linearized Riemann tensor $R_{\alpha\mu\beta\nu}$ is unchanged under $h_{\mu\nu} \rightarrow h_{\mu\nu} + v_{\mu,\nu} + v_{\nu,\mu}$, that is, under a gauge transformation. Show that for a metric perturbation in TT gauge,

$$R_{j0k0} = -\frac{1}{2}h^{TT}_{jk,00}. \tag{5.177}$$

5.14 Show that the expansion of the integrand from (5.89) using the approximation for $|\mathbf{r} - \mathbf{r}'|$ from (5.90) yields the integrand in (5.92) through first order in $1/r$ (i.e., corrective terms go like $1/r^2$ and hence do not contribute to radiation).

5.15 Compute the curl of the time-derivative of the dipole moment, evaluated at $t - r/c$, to verify (5.100).

5.16 For an electric dipole moment of the form $\mathbf{p}(t) = p_0 \cos(\omega t)\hat{\mathbf{z}}$, write out the radiation portion of the electric and magnetic fields, find the total radiated power using (5.110), and time average that expression – the time average of an oscillatory function $f(t)$ with period T is

$$\langle f \rangle \equiv \frac{1}{T}\int_0^T f(t)dt. \tag{5.178}$$

5.17 Find the electric and magnetic dipole radiation fields for a charge q moving with position vector

$$\mathbf{w}(t) = a\cos(\omega t)\hat{\mathbf{x}} + b\sin(\omega t)\hat{\mathbf{y}} \tag{5.179}$$

for constants $a > b$ and frequency ω. Evaluate the time-averaged dipole power. Which radiates more, a particle moving in a circular trajectory of radius a or a particle moving along a line of length $2a$ (each traversed with constant angular frequency ω)?

5.18 In the classical view of hydrogen, the electron orbits the proton in a circular orbit of radius $a \approx 0.5$ Å (the Bohr radius). Assuming nonrelativistic, quasi-uniform circular motion, how much energy is lost per cycle to dipole radiation? How long will it take for this energy loss mechanism to cause the electron to fall into the nucleus?

5.19 In (5.157), we are given an expression that should solve the wave equation at points away from the source. Check that it does.

5.20 Show that the solution in (5.98), when projected using (5.7) with $n^j = x^j/r$ and written in terms of the spherical basis vectors, gives (5.112).

5.21 Redo Problem 5.20, but this time, transform the components A^i to spherical coordinates to get $\bar{A}^i$ and then form the vector $\bar{\mathbf{A}} = \bar{A}^1 \mathbf{e}_r + \bar{A}^2 \mathbf{e}_\theta + \bar{A}^3 \mathbf{e}_\phi$ written in terms of $\hat{\mathbf{r}}$, $\hat{\boldsymbol{\theta}}$, and $\hat{\boldsymbol{\phi}}$. Now the projection amounts to throwing away the $\hat{\mathbf{r}}$ component. Show that this procedure recovers the vector you got in Problem 5.20.

5.22 In Newtonian gravity, where the sources are massive (as opposed to being all forms of energy), we can compute the spatial quadrupole moment from the mass density ρ:

$$P^{ij} = \int_{\text{all space}} \rho x^i x^j \, d\tau. \tag{5.180}$$

Find P^{ij} and $Q^{ij} \equiv P^{ij} - (1/3)P\eta^{ij}$ (where P is the trace of P^{ij}, and η^{ij} is one if $i = j$, zero otherwise) for a uniform sphere of total mass M and radius R.

5.23 For a stationary Newtonian (i.e., mass only) source with (reduced) quadrupole moment Q^{ij}, find the form (up to numerical constants) of the gravitational potential generated by the quadrupole source using only units, together with the building blocks: G (the gravitational constant), the vector $\mathbf{r}$, and its magnitude, assuming that you are "away" from the source, in vacuum. Verify that this potential satisfies Laplace's equation.

5.24 From (5.121), find the radiation piece of the electric and magnetic fields, analogous to (5.106). Does the relation in (5.107) hold? Using these radiation fields, find the Poynting vector for magnetic dipole radiation and compute the total power radiated.

5.25 In developing the Poynting vector for radiation, we used the identity

$$(\nabla \times \mathbf{A}) \cdot (\nabla \times \mathbf{A}) = \partial^j A^\ell \partial^j A^\ell - \partial^j A^\ell \partial^\ell A^j. \tag{5.181}$$

Show that it is true (you might want to use the Levi-Civita symbol from Problem 5.9).

5.26 A neutral ring of current carrying wire with current $I(t) = I_0 \cos(\omega t)$ and radius R sits in the xy plane. Find the radiation components of the associated electric and magnetic fields and compute the total power radiated (use your result from Problem 5.24) and its time-average.

5.27 Evaluate the integrals

$$\int_0^{2\pi} \int_0^\pi \hat{x}^k \sin\theta \, d\theta \, d\phi \quad \text{and} \quad \int_0^{2\pi} \int_0^\pi \hat{x}^k \hat{x}^\ell \hat{x}^j \sin\theta \, d\theta \, d\phi, \tag{5.182}$$

where $\hat{x}^k$ is the kth component of the unit vector $\hat{\mathbf{r}}$.

5.28 For a charge q moving in uniform circular motion with $\mathbf{w}(t) = R \times (\cos(\omega t)\hat{\mathbf{x}} + \sin(\omega t)\hat{\mathbf{y}})$, compute the quadrupole moment tensor $P^{ij}(t)$. Using this, find the quadrupole radiation vector potential $\mathbf{A}_{\text{rad}}$ and associated radiation electric and magnetic fields.

5.29 Compute the total power radiated by the quadrupole contribution for the charge in the previous problem. Compare that with the power radiated in the dipole case for the same setup from Problem 5.17.

5.30 Show that using the projection from (5.27) with direction $\hat{\mathbf{n}} = \hat{\mathbf{r}}$, the projection of $P^{mn}(t)$ from (5.143) and the traceless form $Q^{mn}(t)$ defined in (5.158) are the same.

5.31 Using the Earth's nearly circular orbit about the sun as an example, find the dimensionless coefficient governing the magnitude of the metric perturbation generated by the Earth in (5.164) as measured at $r = R$.

5.32 You move a $M = 1\,\text{kg}$ mass back and forth over $d = 1\,\text{m}$ with a period of $T = 4\,\text{s}$, where $x(t) = (d/2)\sin(2\pi t/T)$ is the mass location along the $\hat{\mathbf{x}}$ axis as shown in Figure 5.2. What is the maximum magnitude of the radiative gravitational perturbation, $h_{\mu\nu}$, as measured from $1\,\text{km}$ away (write the expression clearly before putting in the numbers)? What frequency (or frequencies) f (as opposed to angular frequencies) does this radiation signal contain?

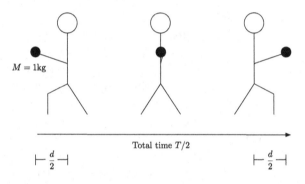

$M = 1\text{kg}$

Total time $T/2$

$\vdash \frac{d}{2} \dashv$ $\vdash \frac{d}{2} \dashv$

Fig. 5.2   Figure for Problem 5.32.

5.33 How long would it take for the Earth to fall into the Sun due to energy lost to gravitational radiation? (this is similar to Problem 5.18; see Example 5.5 for the model).

5.34 We have studied the linearized form of gravitational waves, but there are also solutions to the full set of Einstein's equations that look wave-like [4]. Starting from the line element

$$ds^2 = -c^2 dt^2 + dz^2 + L(u)^2 \left(e^{2\beta(u)}dx^2 + e^{-2\beta(u)}dy^2\right) \quad (5.183)$$

with $u \equiv ct - z$ and unknown functions $L(u)$ and $\beta(u)$, transform to coordinates $u = ct - z$, $v = ct + z$. Extract the associated metric, which will have functional dependence only on $u = ct - z$, associated with wave motion in the z direction. What does the vacuum form of Einstein's equation tell you about the relation between $L(u)$ and $\beta(u)$?

6 Gravitational Sources

Vacuum solutions in general relativity provide an accessible starting point for understanding the theory and building intuition about motion in different geometries generated by distributions of energy that are "over there," away from a test particle's location. That's reasonable for empty space, and we have thought about all sorts of physics in that context. In this chapter, we take up solutions that are inside material sources, at least setting up and discussing, if not solving, the equations that come from various models of "fluids," a relatively generic term that here refers to any continuous collection of matter. This chapter presents problems that are similar to studying the electric and magnetic fields "in material."

We'll start by developing the classical (prerelativistic, both general and special) equations that govern flowing mass, a subset of the full Navier–Stokes equations. That is an interesting and somewhat involved task that will require us to think about how volumes containing a continuum of "particles" (or energy) respond to forces, both external and internal. We will start with a version of Newton's second law that can be applied to fluids described by density, velocity, and pressure (so-called "perfect fluids"). With the relevant equations in hand, we can study the Newtonian gravitational field set up by materials for later comparison with the relativistic version. In particular, we'll review the static solution for spherically symmetric sources (the interior of "stars") in Newtonian gravity. Then we'll develop the form of the sources' stress tensor so that we can study the same problem in general relativity.

Finally, we will turn from specifying sources and finding fields to the inverse problem: Given a metric, what source generated it? This problem is similar to being given an electrostatic field and finding the charge density ρ that generated it. In that electrostatic case, we just turn Gauss' law around, $\rho = \epsilon_0 \nabla \cdot \mathbf{E}$, and a similar approach will work to identify the stress tensor that gives rise to a metric. Cosmology, the study of the evolution of the universe due to the mass inside of it, is another example of working inside a "material," and an example of the inverse procedure: The Robertson–Walker[1] metric is constructed using a set of target properties that represent assumptions about the form of our universe. These assumptions then constrain the source distributions. There are more speculative examples, and we'll close the chapter by discussing two

---

[1] Known also as the Friedmann–Lemaître metric, the Friedmann–Robertson–Walker metric, or Friedmann–Lemaître–Robertson–Walker metric. I will use the shortened "Roberston–Walker metric" throughout.

famous cases of "engineering" a metric with desired properties and then characterizing the source needed to produce them: the Ellis wormhole connection between two asymptotically flat spacetimes and the Alcubierre warpdrive.

6.1 Dynamics and Continuum Distributions

Given a set of n particles with masses $\{m_i\}_{i=1}^n$, positions $\{\mathbf{r}_i(t)\}_{i=1}^n$, and velocities $\{\mathbf{v}_i(t) \equiv \dot{\mathbf{r}}_i(t)\}_{i=1}^n$ that interact under the influence of a pairwise force (like Newtonian gravity or electrostatic forcing) $\mathbf{H}(\mathbf{x}, \mathbf{y})$ (where $\mathbf{x}$ and $\mathbf{y}$ are particle locations that determine the force magnitude and direction) and an externally applied force of some sort, a function of position, $\mathbf{F}(\mathbf{r})$, Newton's second law reads

$$ m \frac{d\mathbf{v}_i(t)}{dt} = \mathbf{F}\big(\mathbf{r}_i(t)\big) + \sum_{j=1 \neq i}^n \mathbf{H}\big(\mathbf{r}_i(t), \mathbf{r}_j(t)\big) \qquad \text{for } i = 1 \rightarrow n. \qquad (6.1) $$

As the number of particles gets large, we can take a continuum view of the configuration, where instead of discrete particles moving around, we keep track of the mass density $\rho(\mathbf{r}, t)$ at each point in space and time, together with a velocity field $\mathbf{v}(\mathbf{r}, t)$ that tells us, at time t, the velocity of any mass at location $\mathbf{r}$.[2] How should we update the description above to refer to these new, continuum variables?

To answer the question, we need to understand the continuum description of the motion of a specific parcel of mass. In (6.1) the index i tells us which particle is moving around, but when we move to mass specified by $\rho(\mathbf{r}, t)$, all we know is how much mass density is at a point $\mathbf{r}$ at time t, not whence that mass came. To track the motion of a patch of mass and hence develop a version of (6.1) that we can use in the continuum case, we need to introduce some additional ideas and notation. This material is covered in fluid mechanics texts like [24, 34].

6.1.1 Material Derivative

Suppose we have a particle (a packet of mass), within the continuum, that starts at $\mathbf{x}$ at time $t = 0$. We denote by $\mathbf{w}(\mathbf{x}, t)$ the vector that points from the origin to the particle's location at time t. The tangent to the curve swept out by the particle, referring to Figure 6.1, is what we typically call velocity, and

---

[2] The velocity field is independent of the presence of mass at a particular location, that is, you can have $\mathbf{v}(\mathbf{r}, t)$ at locations and times that have $\rho(\mathbf{r}, t) = 0$.

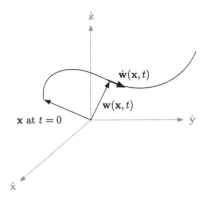

Fig. 6.1 The vector $\mathbf{w}(\mathbf{x}, t)$ tells us the location, at time t, of the mass that was at $\mathbf{x}$ at time $t = 0$. The temporal derivative of $\mathbf{w}$ gives us the velocity, tangent to the curve at time t.

so we'll associate it with the velocity field $\mathbf{v}(\mathbf{r}, t)$ defined above to enforce this connection:

$$\mathbf{v}\big(\mathbf{w}(\mathbf{x}, t), t\big) = \frac{\partial \mathbf{w}(\mathbf{x}, t)}{\partial t}. \tag{6.2}$$

Now consider a generic function of time and space $u(\mathbf{r}, t)$; this function assigns some physical quantity to the point $\mathbf{r}$ at time t. The value of u at a particle that is moving along the curve $\mathbf{w}(\mathbf{x}, t)$ is $u(\mathbf{w}(\mathbf{x}, t), t)$. How does the value of u change as the mass packet moves along? To track the change of u, felt by the particle, in time, we have to consider both the explicit temporal dependence of u (its value at any point changes in time), but we also have to account for the motion of the particle through u's spatially varying landscape. The relevant derivative, then, is

$$\frac{d}{dt} u\big(\mathbf{w}(\mathbf{x}, t), t\big) = \frac{\partial u}{\partial t} + \frac{\partial \mathbf{w}}{\partial t} \cdot \nabla u = \frac{\partial u}{\partial t} + \mathbf{v} \cdot \nabla u, \tag{6.3}$$

and when we use this construct in the context of the function $u(\mathbf{r}, t)$, it is called the "material" or "convective" derivative and is denoted

$$\frac{D}{Dt} u(\mathbf{r}, t) \equiv \frac{\partial u}{\partial t} + \mathbf{v} \cdot \nabla u. \tag{6.4}$$

As an example, the change in velocity, as a function of time, is

$$\frac{D\mathbf{v}}{Dt} \equiv \frac{\partial \mathbf{v}}{\partial t} + (\mathbf{v} \cdot \nabla)\mathbf{v}, \tag{6.5}$$

a candidate for the time derivative of particle velocity found on the left-hand side in a continuum version of (6.1).

Because the continuum description of mass provided by the mass density $\rho(\mathbf{r}, t)$ assigns zero mass to individual points, it will be easier to develop an integral analogue of Newton's second law and from that obtain the local form. If we considered any finite volume containing our n discrete particles, only the "external" force $\mathbf{F}$ in (6.1) would contribute. Newton's third law ensures that the pairwise interaction force $\mathbf{H}$ cancels in the closed domain. The same is true in the continuum case, except that we have to be careful with the boundary of the domain, where there can be uncanceled force pairs with particles outside the domain. There is no real discrete version of that story, since it is "highly unlikely" (mathematicians have fancier ways to describe this notion) that a discrete particle will fall on any continuous boundary. Let's start by thinking about how integrals of densities like $\rho(\mathbf{r}, t)$ behave.

6.1.2 Integral Domains

Imagine a continuum of particles (a "fluid") traveling along trajectories given by $\mathbf{w}(\mathbf{x}, t)$, where, again, $\mathbf{x}$ refers to the starting location of the particles at time $t = 0$. Pick a domain Ω_0 enclosing some subset of the fluid particles at $t = 0$. Imagine that domain evolving in time to enclose those "same" particles as they move through space. At time t, let Ω_t be the set of locations $\mathbf{w}(\mathbf{x}, t)$ for $\mathbf{x} \in \Omega_0$, the initial domain. Suppose we have a physical variable described by the density field $\rho(\mathbf{r}, t)$ (that could be mass density, charge density, energy density, whatever you like), and we'd like to know the value of the integral of $\rho(\mathbf{r}, t)$ for the particles in Ω_t. That quantity is given by

$$M(t) \equiv \int_{\Omega_t} \rho(\mathbf{r}, t)\, d\tau, \tag{6.6}$$

where $d\tau$ is the infinitesimal volume element.

Question: How does the total amount of "stuff" (mass, charge, energy, etc.) associated with that initial domain Ω_0 change as time goes on? From (6.6) it is clear that there are two contributions to the evolution: (1) the density $\rho(\mathbf{r}, t)$ is time dependent, and (2) the domain Ω_t is time dependent. The *answer* to the question is

$$\frac{d}{dt} \int_{\Omega_t} \rho(\mathbf{r}, t)\, d\tau = \int_{\Omega_t} \frac{\partial \rho(\mathbf{r}, t)}{\partial t}\, d\tau + \oint_{\partial \Omega_t} \rho(\mathbf{r}, t)\mathbf{v}(\mathbf{r}, t) \cdot d\mathbf{a}, \tag{6.7}$$

where the first term on the right keeps track of the temporal change in $\rho(\mathbf{r}, t)$ over the original Ω_t domain, and the second, surface integral, accounts for the change in the domain.

To understand the surface term, refer to Figure 6.2. The integration domain changes from Ω_t to $\Omega_{t+\Delta t}$ because particles on the boundary of Ω_t move with velocity $\mathbf{v}$ over a time interval Δt, so that the change in the value of the integral due to the domain motion (only) is

$$\int_{\Omega_{t+\Delta t}} \rho(\mathbf{r}, t)\, d\tau - \int_{\Omega_t} \rho(\mathbf{r}, t)\, d\tau \approx \oint_{\partial \Omega_t} \rho(\mathbf{r}, t)\big(\mathbf{v}(\mathbf{r}, t)\Delta t\big) \cdot d\mathbf{a}, \tag{6.8}$$

and taking $\Delta t \to 0$ gives

$$\lim_{\Delta t \to 0} \frac{1}{\Delta t} \left[\int_{\Omega_{t+\Delta t}} \rho(\mathbf{r}, t)\, d\tau - \int_{\Omega_t} \rho(\mathbf{r}, t)\, d\tau \right] = \oint_{\partial \Omega_t} \rho(\mathbf{r}, t)\mathbf{v}(\mathbf{r}, t) \cdot d\mathbf{a}. \tag{6.9}$$

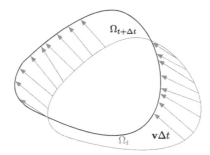

Fig. 6.2 Two domains, Ω_t (gray) and $\Omega_{t+\Delta t}$ (black) with the boundary of $\Omega_{t+\Delta t}$ defined to be the boundary of Ω_t as it has moved with velocity $\mathbf{v}$ (a function of position) in time Δt. The vector arrows give the motion $\mathbf{v}\Delta t$ of the points from $\partial\Omega_t$ to define $\partial\Omega_{t+\Delta t}$.

The difference in the shape of the volume integrals is accounted for by integrating over the original surface. An example with simplified geometry and velocity field $\mathbf{v}$ helps to establish the interpretation of the second term.

Example 6.1 (Change in $M(t)$). To be concrete, yet keep things simple, we'll take a cubical domain of side length ℓ that moves with constant speed v (the speed of the particles within the box) along the $\hat{\mathbf{y}}$ axis as shown in Figure 6.3. We'll start at $t = 0$ with the domain Ω_0 and consider a small temporal interval Δt that, together with $\mathbf{v}(\mathbf{r}, t)$, gives us the new domain $\Omega_{\Delta t}$.

We are interested in the integral of $\rho(\mathbf{r}, t)$ over the two domains, that is, we would like to compare

$$M_0 \equiv \int_{\Omega_0} \rho(\mathbf{r}, 0)\, d\tau \tag{6.10}$$

with

$$M_{\Delta t} \equiv \int_{\Omega_{\Delta t}} \rho(\mathbf{r}, 0 + \Delta t)\, d\tau \tag{6.11}$$

to see how the quantity $M(t)$ has changed for a collection of particles. The value of M changes both because of the change in the domain and because of the change in ρ itself during the time interval Δt.

In terms of our chosen Ω_0 and $\Omega_{\Delta t}$, we have

$$M_0 = \int_0^\ell \int_0^\ell \int_0^\ell \rho(x, y, z, 0) dx\, dy\, dz \tag{6.12}$$

and

$$M_{\Delta t} = \int_0^\ell \int_{v\Delta t}^{\ell+v\Delta t} \int_0^\ell \rho(x, y, z, \Delta t) dx\, dy\, dz. \tag{6.13}$$

We will Taylor expand in Δt. In preparation for that, consider a single-variable function $u(y)$; an integral with infinitesimal limits can be evaluated and expanded (as you will show in Problem 6.1),

$$\int_\epsilon^{\ell+\epsilon} u(y)dy = \int_0^\ell u(y)dy + \epsilon\big(u(\ell) - u(0)\big). \tag{6.14}$$

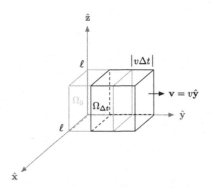

Fig. 6.3 The two domains Ω_0 and $\Omega_{\Delta t}$.

Using this relation in the expression for $M_{\Delta t}$ gives

$$M_{\Delta t} \approx \int_0^\ell \int_0^\ell \int_0^\ell \rho(x, y, z, \Delta t) dx dy dz$$
$$+ v\Delta t \int_0^\ell \int_0^\ell \rho(x, \ell, z, \Delta t) dx dz - v\Delta t \int_0^\ell \int_0^\ell \rho(x, 0, z, \Delta t) dx dz.$$
$$(6.15)$$

There are two "surface" terms here: One at $y = \ell$ with a plus sign and one at $y = 0$ that shows up with a minus sign. Noting the direction of $\mathbf{v}$, we can write those terms together as an integral over the boundary of the original domain using the dot product of $\mathbf{v}$ with the area element:

$$v\Delta t \int_0^\ell \int_0^\ell \rho(x, \ell, \Delta t) dx dz - v\Delta t \int_0^\ell \int_0^\ell \rho(x, 0, z, \Delta t) dx dz$$
$$= \Delta t \oint_{\partial\Omega_0} \rho(\mathbf{r}, \Delta t) \mathbf{v} \cdot d\mathbf{a}.$$
$$(6.16)$$

Then

$$M_{\Delta t} = \int_{\Omega_0} \rho(\mathbf{r}, \Delta t) \, d\tau + \Delta t \oint_{\partial\Omega_0} \rho(\mathbf{r}, \Delta t) \mathbf{v} \cdot d\mathbf{a}, \qquad (6.17)$$

and we Taylor expand the explicit t dependence in $\rho(\mathbf{r}, \Delta t)$, noting that in the surface term, we will drop the Δt^2 piece, so that

$$M_{\Delta t} = \int_{\Omega_0} \rho(\mathbf{r}, 0) \, d\tau + \Delta t \int_{\Omega_0} \left.\frac{\partial \rho(\mathbf{r}, t)}{\partial t}\right|_{t=0} d\tau + \Delta t \oint_{\partial\Omega_0} \rho(\mathbf{r}, 0) \mathbf{v} \cdot d\mathbf{a}.$$
$$(6.18)$$

The first term is M_0, and then

$$\left.\frac{dM}{dt}\right|_{t=0} \equiv \lim_{\Delta t \to 0} \frac{M_{\Delta t} - M_0}{\Delta t} = \int_{\Omega_0} \left.\frac{\partial \rho(\mathbf{r}, t)}{\partial t}\right|_{t=0} d\tau + \oint_{\partial\Omega_0} \rho(\mathbf{r}, 0) \mathbf{v}(\mathbf{r}, 0) \cdot d\mathbf{a},$$
$$(6.19)$$

and while we have developed this expression for $t = 0$ and $t = \Delta t$, it holds for any $t \to t + \Delta t$ interval, recovering (6.7).

The expression in (6.7) can be written entirely in terms of a volume integral over Ω_t, without reference to its boundary using the divergence theorem,

$$\frac{d}{dt} \int_{\Omega_t} \rho(\mathbf{r}, t) \, d\tau = \int_{\Omega_t} \left[\frac{\partial \rho(\mathbf{r}, t)}{\partial t} + \nabla \cdot \left(\rho(\mathbf{r}, t) \mathbf{v} \right) \right] d\tau. \tag{6.20}$$

Writing out the divergence using the product rule,

$$\frac{d}{dt} \int_{\Omega_t} \rho(\mathbf{r}, t) \, d\tau = \int_{\Omega_t} \left[\frac{\partial \rho(\mathbf{r}, t)}{\partial t} + (\mathbf{v} \cdot \nabla)\rho(\mathbf{r}, t) + \rho(\mathbf{r}, t)\nabla \cdot \mathbf{v}(\mathbf{r}, t) \right] d\tau, \tag{6.21}$$

it becomes clear how the generalization to vector-valued integrals must go. For a vector field $\mathbf{z}(\mathbf{r}, t)$, we apply three copies of (6.21), written in vector form:

$$\frac{d}{dt} \int_{\Omega_t} \mathbf{z}(\mathbf{r}, t) \, d\tau = \int_{\Omega_t} \left[\frac{\partial \mathbf{z}(\mathbf{r}, t)}{\partial t} + (\mathbf{v} \cdot \nabla)\mathbf{z}(\mathbf{r}, t) + \mathbf{z}(\mathbf{r}, t)\nabla \cdot \mathbf{z}(\mathbf{r}, t) \right] d\tau. \tag{6.22}$$

Example 6.2 (Conservation of Mass). The total mass contained in a volume represents an integral of interest, and its time-variation should be sensible from a conservation of mass point of view. Using the moving domain Ω_t, mass conservation reads

$$\frac{d}{dt} \int_{\Omega_t} \rho(\mathbf{r}, t) \, d\tau = 0 \tag{6.23}$$

for density $\rho(\mathbf{r}, t)$. No mass enters or leaves the domain Ω_t, because the domain itself is moving so as to capture all the mass. According to (6.20), we have

$$\frac{d}{dt} \int_{\Omega_t} \rho(\mathbf{r}, t) \, d\tau = \int_{\Omega_t} \left[\frac{\partial \rho}{\partial t} + \nabla \cdot (\rho \mathbf{v}) \right] d\tau = 0, \tag{6.24}$$

and for this to be zero for any starting domain, Ω_0 leading to some later Ω_t, we recover the usual local statement of mass conservation,

$$\frac{\partial \rho}{\partial t} + \nabla \cdot (\rho \mathbf{v}) = 0. \tag{6.25}$$

6.1.3 Newton's Second Law

For a moving domain Ω_t containing mass with density $\rho(\mathbf{r}, t)$ and velocity field $\mathbf{v}(\mathbf{r}, t)$, the total momentum in Ω_t at time t is

$$\mathbf{P}(t) = \int_{\Omega_t} \rho(\mathbf{r}, t)\mathbf{v}(\mathbf{r}, t) \, d\tau. \tag{6.26}$$

The forces that act on the mass contained within Ω_t come in two varieties as in (6.1): "external forces" ($\mathbf{F}$ in (6.1)) and "internal forces" that cancel in pairs ($\mathbf{H}$ in (6.1), pressure is an example). We will specify the external "force density" (force per unit volume) $\mathbf{f}(\mathbf{r}, t)$; then the contribution to the change in momentum that comes from these forces acting throughout Ω_t is

$$\int_{\Omega_t} \mathbf{f}(\mathbf{r}, t)\, d\tau. \tag{6.27}$$

The "internal" forces will cancel when integrated over the volume Ω, except at the boundary. Let $S^{ij}(\mathbf{r}, t)$ be a function that gives the ith component of the force per unit area acting on a surface with normal in the jth direction; then we can write Newton's second law, in spatial component form, as

$$\frac{d P^i(t)}{dt} = \int_{\Omega_t} f^i(\mathbf{r}, t)\, d\tau + \oint_{\partial\Omega_t} S^{ij}(\mathbf{r}, t)\, da_j. \tag{6.28}$$

Again using the divergence theorem to rewrite the surface term as a divergence integrated over Ω_t,

$$\frac{d P^i(t)}{dt} = \int_{\Omega_t} \left[f^i(\mathbf{r}, t) + \partial_j S^{ij}(\mathbf{r}, t) \right] d\tau. \tag{6.29}$$

Applying (6.22) on the left, we have

$$\int_{\Omega_t} \left[\frac{\partial(\rho v^i)}{\partial t} + (\mathbf{v} \cdot \nabla)(\rho v^i) + \rho v^i \nabla \cdot \mathbf{v} \right] d\tau = \int_{\Omega_t} \left[f^i(\mathbf{r}, t) + \partial_j S^{ij}(\mathbf{r}, t) \right] d\tau. \tag{6.30}$$

The local form of Newton's second law, obtained by removing the integrals, is

$$\frac{\partial(\rho v^i)}{\partial t} + (\mathbf{v} \cdot \nabla)(\rho v^i) + \rho v^i \nabla \cdot \mathbf{v} = f^i + \partial_j S^{ij}. \tag{6.31}$$

To this we add the usual conservation of mass equation from above,

$$\frac{\partial \rho}{\partial t} + \nabla \cdot (\rho \mathbf{v}) = 0. \tag{6.32}$$

From (6.31), using the product rule for temporal derivatives and replacing $\frac{\partial \rho}{\partial t}$ with $-\nabla \cdot (\rho \mathbf{v})$, we also have

$$\rho \frac{\partial v^i}{\partial t} + \rho(\mathbf{v} \cdot \nabla) v^i = f^i + \partial_j S^{ij}$$
$$\frac{\partial \rho}{\partial t} = -\nabla \cdot (\rho \mathbf{v}), \tag{6.33}$$

where I include the conservation of mass equation as a reminder (that is necessary in getting from (6.31) to the current form). Finally, referring to the "material" derivative definition in (6.4), extended to vectors, we can write the set

$$\rho \frac{D v^i}{Dt} = f^i + \partial_j S^{ij},$$
$$\frac{\partial \rho}{\partial t} = -\nabla \cdot (\rho \mathbf{v}), \tag{6.34}$$

which looks about as much like Newton's second law as we could ask for. In this continuum setting, the first equation is called the "Cauchy momentum equation." When specific assumptions about the form of S^{ij} are introduced, it becomes one of the equations of the "Navier–Stokes" set, which also includes mass conservation, the second equation in (6.34).

6.1.4 Perfect Fluid

For a fixed spatial domain Ω, the total force on the particles in the domain at time t is obtained by integrating the right-hand side of (6.34) and using the divergence theorem

$$F_\Omega^i = \int_\Omega f^i(\mathbf{r}, t)\, d\tau + \oint_{\partial\Omega} S^{ij}(\mathbf{r}, t)\, da_j. \tag{6.35}$$

The surface term depends on the structure of the pairwise interactions we assume to exist between the material elements within Ω. We can make some simplifying assumptions about the form of those interactions.

Surfaces forces that are perpendicular to the surface normal are called "viscous" forces and have components that act along the surface. In a "perfect fluid," the assumption is that there are no viscous forces (the "inviscid" assumption), so that at any surface, $S^{ij} da_j \parallel da^i$. Under this assumption, the tensor S^{ij} must be diagonal. To see this implication, first consider a surface normal in the $\hat{\mathbf{x}}$ direction, $\hat{\mathbf{n}} = \hat{\mathbf{x}}$. Then the sum that gives the force direction, $S^{ij} n_j$, must have $S^{yx} = S^{zx} = 0$ in order to have component only in the $\hat{\mathbf{x}}$ direction. Similarly, for a surface normal in the $\hat{\mathbf{y}}$ direction, we require $S^{xy} = S^{zy} = 0$, and for normal in the $\hat{\mathbf{z}}$ direction, $S^{xz} = S^{yz} = 0$. That leaves us with nonzero components S^{xx}, S^{yy}, and S^{zz}. Finally, taking $\hat{\mathbf{n}} = n_x\hat{\mathbf{x}} + n_y\hat{\mathbf{y}} + n_z\hat{\mathbf{z}}$, the force at the surface is

$$\mathbf{F}_{\partial\Omega} = da\left[S^{xx} n_x\hat{\mathbf{x}} + S^{yy} n_y\hat{\mathbf{y}} + S^{zz} n_z\hat{\mathbf{z}}\right]. \tag{6.36}$$

By requiring that $\mathbf{F}_{\partial\Omega} = \alpha\hat{\mathbf{n}}$ we see that not only must S^{ij} be diagonal, but also the diagonal components must be equal: $S^{xx} = S^{yy} = S^{zz}$. The "perfect fluid" model, then, has $S^{ij} = -p\delta^{ij}$, where p has dimension of force per unit area, a pressure. The sign here is to make p play the role of an energy density in the equations of motion, as shown below.

With this simplifying assumption in place, the term $\partial_j S^{ij}$ appearing in (6.35) becomes

$$\partial_j S^{ij} = -\partial^i p, \tag{6.37}$$

and the continuum version of Newton's second law, written in local form, is

$$\boxed{\begin{aligned} \rho\frac{D\mathbf{v}}{Dt} &= \mathbf{f} - \nabla p, \\ \frac{\partial\rho}{\partial t} &= -\nabla \cdot (\rho\mathbf{v}). \end{aligned}} \tag{6.38}$$

This special case forms part of "Euler's equations" describing inviscid flow, itself a special case of the Navier–Stokes equations governing more general fluids.

The set of equations (6.38) is incomplete: there are four unknowns $\{\rho, \mathbf{v}\}$, yet five variables $\{\rho, \mathbf{v}, p\}$, once the external forces have been specified. We need to connect the pressure to other variables, and that connection is provided by a model of particle interaction. In general, an "equation of state" is a relation between thermodynamic variables. The ideal gas law is an example, but there are many others. For our set of variables, the relevant equation of state is a relation between the density and pressure variables. Such relations can be very complicated, but many simple ones take the form $p = A\rho^q$ for constants A and q. In general, this type of relation is called a "polytropic" equation of state (for astrophysically relevant motivation and examples, see [8] and [56]; a nice discussion of polytropic equations of state more generally can be found in [49]). Two simple examples are the noninteracting dust, which has $p = 0$, and the relation between electromagnetic field energy density and pressure (skip ahead and try Problem 7.16 if you want to see how it arises) with $p = \rho/3$. Different values for A and q are relevant for other model interactions.

6.2 Newtonian Gravity and Static, Interacting Dust

Let's take the external force in (6.38) to be gravitational. For a distribution of mass with density $\rho(\mathbf{r}, t)$, the gravitational force density is $\mathbf{f} = \rho(\mathbf{r}, t)\mathbf{g}(\mathbf{r}, t)$ with $\mathbf{g}(\mathbf{r}, t)$ the Newtonian gravitational field, itself sourced by the mass density in the usual way from (1.16).[3] We have a combined set of equations of motion (from (6.38)) together with the field equation for $\mathbf{g}$, which can also be written in terms of its potential φ:

$$\rho(\mathbf{r}, t)\left[\frac{\partial \mathbf{v}(\mathbf{r}, t)}{\partial t} + \big(\mathbf{v}(\mathbf{r}, t) \cdot \nabla\big)\mathbf{v}(\mathbf{r}, t)\right] = \rho(\mathbf{r}, t)\mathbf{g}(\mathbf{r}, t) - \nabla p(\mathbf{r}, t),$$

$$\frac{\partial \rho(\mathbf{r}, t)}{\partial t} = -\nabla \cdot \big(\rho(\mathbf{r}, t)\mathbf{v}(\mathbf{r}, t)\big), \tag{6.39}$$

$$\nabla \cdot \mathbf{g}(\mathbf{r}, t) = -4\pi G\rho(\mathbf{r}, t) \longrightarrow \nabla^2\varphi(\mathbf{r}, t) = 4\pi G\rho(\mathbf{r}, t).$$

There are five equations here, for the five unknowns ρ, $\mathbf{v}$, and φ. As mentioned above, the pressure $p(\mathbf{r}, t)$ must be stipulated, either through a general equation of state or through a simplified version in which p is a function of the density ρ: $p(\mathbf{r}, t) = w(\rho(\mathbf{r}, t))$ for some single-variable $w(u)$. We'll leave it unspecified for now and pick some common choices as necessary.

To start, we'll simplify the set (6.39) by specializing to a spherically symmetric distribution of mass density and velocity field. Inside a spherically symmetric body, we can assume that all functions depend only on r, the distance to the origin, and t, with the velocity field taking the form $\mathbf{v}(\mathbf{r}, t) = v(r, t)\hat{\mathbf{r}}$ and similarly spherically symmetric gravitational field $\mathbf{g}(\mathbf{r}, t) = g(r, t)\hat{\mathbf{r}}$. Then

---

[3] We are adding in the time dependence without introducing the full dynamics of Newtonian gravity here, but our fundamental applications will be to static problems, where we don't have to worry about time dependence.

equations (6.39) become

$$\frac{\partial v}{\partial t} + v\frac{\partial v}{\partial r} = g - \frac{1}{\rho}\frac{\partial p}{\partial r}, \qquad \frac{\partial \rho}{\partial t} = -\frac{\partial}{\partial r}(\rho v), \qquad \frac{1}{r^2}\frac{\partial}{\partial r}(r^2 g) = -4\pi G\rho.$$
(6.40)

Defining the mass enclosed by a sphere of radius r to be the integral

$$m(r,t) \equiv \int_{B(r)} \rho(\mathbf{r}')\, d\tau' = 4\pi \int_0^r \rho(r',t)r'^2 dr',$$
(6.41)

we can integrate the last equation in (6.40) to write $g(r,t)$ in terms of $m(r,t)$,

$$g(r,t) = -\frac{m(r,t)G}{r^2},$$
(6.42)

a manifestation of the shell theorem: the gravitational acceleration at r depends only on the total mass enclosed by a sphere of radius r.

Let's see if we can find a static solution to (6.40), for which there is no time dependence, and we take $v = 0$, no mass motion. In this type of solution, the gravitational force is balanced by the pressure force. The mass conservation equation, the second in (6.40), is automatically satisfied, the third equation from the set is solved by $g(r) = -m(r)G/r^2$, and the first equation becomes

$$\frac{1}{\rho(r)}\frac{dp(r)}{dr} = g(r) = -\frac{m(r)G}{r^2}.$$
(6.43)

Here we can see the need for some external relation to solve for $\rho(r)$, $p(r)$, and $g(r)$. Suppose we look for solutions that have uniform mass density $\rho(r) = \rho_0$. Then $m(r) = 4\pi r^3 \rho_0/3$, and we can solve (6.43) for $p(r)$,

$$p(r) = -\frac{2}{3}G\pi r^2 \rho_0^2 + p_c,$$
(6.44)

where p_c is the pressure at the center of the distribution. At the center of the sphere, there is no mass enclosed, and the pressure is p_c. As you move out from the center, the mass enclosed increases, whereas the pressure decreases, eventually reaching $p(R) = 0$ at the boundary of the sphere. Given a sphere of total mass M and radius R, we can identify the constants $\rho_0 = M/(4/3\pi R^3)$ and $p_c = 3GM^2/(8\pi R^4)$. We can also find the interior gravitational field and potential if desired. You can find static equilibrium solutions using different assumptions about the form of $\rho(r)$, as in Problem 6.2.

In more general static settings, where we do not simply demand that the density take on a specific form, the information contained in (6.43) is often given in terms of m and ρ rather than in terms of g and ρ:

$$\frac{1}{\rho(r)}\frac{dp(r)}{dr} = -\frac{m(r)G}{r^2}, \qquad \frac{dm(r)}{dr} = 4\pi\rho r^2,$$
(6.45)

where the derivative of $m(r)$ comes directly from its (integral) definition (6.41). These equations can be solved numerically once an equation of state is provided.

Example 6.3 (Numerical Solution of (6.45)). A typical equation of state is the "polytropic"

$$p(\rho) = \frac{p_c}{\rho_c^\gamma} \rho^\gamma, \tag{6.46}$$

where p_c is the pressure at $r = 0$, ρ_c is the density there (a value provided by the initial condition), and γ is a constant. Then the pressure equation from (6.45) becomes

$$\frac{dp}{dr} = \frac{dp}{d\rho}\frac{d\rho}{dr} = -\frac{Gm\rho}{r^2} \longrightarrow \frac{d\rho}{dr} = -\frac{Gm\rho^{2-\gamma}\rho_c^\gamma}{\gamma p_c r^2}, \tag{6.47}$$

and the pair of interest is now

$$\frac{d\rho}{dr} = -\frac{Gm\rho^{2-\gamma}\rho_c^\gamma}{\gamma p_c r^2}, \qquad \frac{dm}{dr} = 4\pi\rho r^2. \tag{6.48}$$

To solve this set of equations, we'll use the Runge–Kutta approach outlined in Appendix B. The vectors $\mathbf{f}$ and $\mathbf{G}$ from (B.1) that define the problem are

$$\mathbf{f}(r) \equiv \begin{pmatrix} \rho(r) \\ m(r) \end{pmatrix}, \qquad \mathbf{G}(r, \mathbf{f}(r)) = \begin{pmatrix} -\dfrac{Gf_2 f_1^{2-\gamma}\rho_c}{\gamma p_c r^2} \\ 4\pi f_1 r^2 \end{pmatrix}. \tag{6.49}$$

As our initial condition, take $\rho_c = 150\,000\,\text{kg/m}^3$ (roughly the density at the center of the sun) with $m_c = 0$ (no mass at the point $r = 0$) and $p_c =$

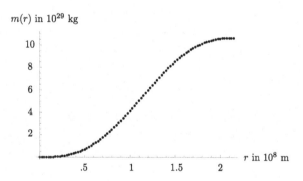

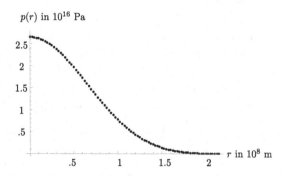

The mass and pressure as functions of r for a numerical solution of (6.45) with central density and pressure that are approximately those of the Sun.

26.5×10^{15} Pa (the pressure at the Sun's center). For the exponent in (6.46), we'll use $\gamma = 5/3$ (associated with a monoatomic ideal gas; see [8] for more on polytropic equations of state). The result of the numerical solution is shown in Figure 6.4. The solution ends when the approximation to pressure transitions from positive to negative, giving, to within the step size in radius, the approximate zero of the pressure that defines the edge of the mass distribution (a description of this type of approach to solving equations like (6.48) is given in [35]). That zero occurs at $R \approx 2.2 \times 10^8$ m, with a total mass enclosed, at that radius, of $M \approx 1.1 \times 10^{30}$ kg. These do not match the values for the Sun, with a radius of $R \approx 7 \times 10^8$ m and a total mass of $M \approx 2 \times 10^{30}$ kg, but the order of magnitude is correct. Try changing the value of γ to see if you can get a closer match in Problem 6.3.

6.3 Stress Tensor and Conservation

We'd like a general relativistic version of the set in (6.39). We naturally expect Einstein's equation to replace the field equation, the third relation in (6.39), but what becomes of the mass-conservation and force equations? As a related question, how should we encode the mass density and pressure in a particle stress tensor that can act as the source in Einstein's equation? We'll start from the expression developed in Section 3.4.2, at first with no pressure term, then cook up the correct extension to include pressure.

First, let's work in a flat space with Cartesian coordinates, and we'll use the u and $\frac{dx^\mu}{dt}$ form of the stress tensor from (3.68), that is, $T^{\mu\nu} = u \frac{dx^\mu}{dt} \frac{dx^\nu}{dt}$ (as opposed to the rest-energy u_0 and four-velocity η^μ form from (3.70)). The stress tensor conservation law $T^{\mu\nu}{}_{,\mu} = 0$ provides the set of four equations

$$\frac{1}{c}\frac{\partial u}{\partial t} + \nabla \cdot (u\mathbf{v}/c) = 0, \qquad \frac{1}{c}\frac{\partial(uv^j/c)}{\partial t} + \frac{\partial(uv^j v^i/c^2)}{\partial x^i} = 0, \qquad (6.50)$$

using $\frac{dx^0}{dt} = c$ to evaluate T^{00} and the spatial vector $\mathbf{v}$ for $v^j \equiv \frac{dx^j}{dt}$ (all Roman indices are spatial, and we'll relax our up-down requirement for visual clarity where necessary, since the coordinates are Cartesian).

The first equation is a statement of energy conservation, and if the energy in question is due only to mass, then we can write $u = \rho c^2$ and recover

$$\frac{\partial \rho}{\partial t} + \nabla \cdot (\rho \mathbf{v}) = 0 \qquad (6.51)$$

from the first equation in (6.50). The other three equations ($j = 1, 2, 3$ in the second set of (6.50)) represent momentum conservation for $u = \rho c^2$,

$$\frac{\partial(\rho v^j)}{\partial t} + \frac{\partial(\rho v^i v^j)}{\partial x^i} = 0. \qquad (6.52)$$

We can expand the expression (product rule),

$$\frac{\partial \rho}{\partial t} v^j + \rho \frac{\partial v^j}{\partial t} + \frac{\partial (\rho v^i)}{\partial x^i} v^j + \rho v^i \frac{\partial v^j}{\partial x^i} = 0, \tag{6.53}$$

and use the conservation statement from (6.51), then the first and third terms cancel, leaving

$$\rho \frac{D\mathbf{v}}{Dt} = 0 \tag{6.54}$$

in vector form, written in terms of the material derivative definition from (6.5). We have recovered a force-free version of Newton's second law and mass conservation from the stress tensor conservation statement.

Comparing (6.54) with the first equation in (6.38), there are two things missing: (1) external forcing and (2) internal interactions in the form of a pressure term. The external forcing for us will be gravity, of course, and that will be introduced naturally and automatically when we move from the flat space, Cartesian form of conservation $T^{\mu\nu}{}_{,\mu} = 0$ to the relativistic $T^{\mu\nu}{}_{:\mu} = 0$. For the pressure term, our goal is to obtain the term ∇p on the left-hand side of (6.54). The easiest way to introduce the gradient is in the position-derivative term in (6.52). If we could take $\rho v^i v^j \to \rho v^i v^j + p \delta^{ij}$, then the divergence appearing in (6.52) would read

$$\frac{\partial}{\partial x^i} \left(\rho v^i v^j + p \delta^{ij} \right) = \frac{\partial (\rho v^i v^j)}{\partial x^i} + \frac{\partial p}{\partial x^j}, \tag{6.55}$$

giving us precisely the missing term.

To construct the correct form of the correction term, let's work in the local rest frame of the source at a point with $\frac{d\bar{x}^0}{dt} = c$ and $\bar{\mathbf{v}} = 0$, where the stress tensor will take the target form (i.e., with the spatial pressure terms in place):

$$\bar{T}^{\mu\nu} \doteq \begin{pmatrix} \bar{u} & 0 & 0 & 0 \\ 0 & \bar{p} & 0 & 0 \\ 0 & 0 & \bar{p} & 0 \\ 0 & 0 & 0 & \bar{p} \end{pmatrix}. \tag{6.56}$$

To construct this object using special relativistic tensors, our building blocks are $\eta^{\mu\nu}$ (the Minkowski metric) and η^μ, the four-velocity,[4]

$$\bar{T}^{\mu\nu} = (\bar{u} + \bar{p}) \bar{\eta}^\mu \bar{\eta}^\nu / c^2 + \bar{p} \eta^{\mu\nu}, \tag{6.57}$$

where the first term puts a $\bar{p}$ in $\bar{T}^{00}$, and the second term adds in a $-\bar{p}$ in the same location. The pressure function already has the correct units to be an energy density, so we don't even need any additional factors of c. In general, then, not just in the local rest frame of the source at a point,

$$T^{\mu\nu} = (u + p) \eta^\mu \eta^\nu / c^2 + p \eta^{\mu\nu}. \tag{6.58}$$

This equation is the special relativistic stress tensor for an inviscid, perfect fluid. The general relativistic form can be obtained by replacing $\eta^{\mu\nu}$ with the

---

[4] There is no relation between $\eta^{\mu\nu}$, the traditional symbol for the Minkowski metric, and η^μ, the four-velocity of a chunk of fluid, despite the overloaded symbol η.

metric $g^{\mu\nu}$ and understanding the η^μ to be the four-velocity in the general sense, with $\eta^\mu g_{\mu\nu}\eta^\nu = -c^2$. Note that the energy u and pressure p are given in the pointwise instantaneous rest frame of the fluid (I've omitted the subscript from Section 3.4 to avoid confusion with uniform energy density solutions, so that the rest energy density $u_0 \to u$ from here on).

6.4 General Relativity and Static, Interacting Dust

We'll reproduce the Newtonian calculation from Section 6.2 in the general relativistic setting. Our goal is to find the metric inside a "star" or other massive body. Since we are thinking about actual material (as opposed to pure energy), we'll move the $1/c^2$ factor from (6.58) inside the parentheses and use $\rho = u/c^2$ to focus our attention on the mass density, writing the stress tensor as

$$T^{\mu\nu} = \left(\rho + \frac{p}{c^2}\right)\eta^\mu\eta^\nu + pg^{\mu\nu}. \tag{6.59}$$

The full set of equations that update the set from (6.39) in general relativity is

$$T^{\mu\nu}{}_{;\mu} = 0,$$
$$R^{\mu\nu} - \frac{1}{2}g^{\mu\nu}R = \frac{8\pi G}{c^4}T^{\mu\nu}, \tag{6.60}$$

where the first equation is really a set of four, combining energy conservation ($\nu = 0$) and momentum ($\nu = 1, 2, 3$) conservation. The latter provides the form of Newton's second law as described in Section 6.3, with gravitational interaction introduced by the covariant derivative, which makes the equation metric dependent. The second equation in (6.60) is the field equation of general relativity. Whereas the energy-momentum conservation equation tells us how matter responds to the metric, the Einstein equation tells us how the metric responds to matter. The two are nonlinearly coupled together.

Specializing to the spherically symmetric static limit from the start, all functions will depend only on the coordinate r,[5] and the four-velocity at any point has spatial components $\eta = 0$ with η^0 set by the proper time requirement: $\eta^\mu\eta_\mu = -c^2 = \eta^0 g_{00}\eta^0$, so that

$$\eta^0 = \sqrt{-\frac{c^2}{g_{00}}}, \qquad \eta_0 = \sqrt{-c^2 g_{00}}. \tag{6.61}$$

We'll use our spherically symmetric metric from Section 4.1, written in the traditional form for this type of nonvacuum setting,

$$ds^2 = -e^{2\Phi(r)}c^2dt^2 + e^{2\Psi(r)}dr^2 + r^2d\theta^2 + r^2\sin^2\theta d\phi^2 \tag{6.62}$$

---

[5] Technically, since we don't know what the metric is, we have no idea how to assign coordinates for spherical symmetry, but it all works out okay in this case.

for unknown functions $\Phi(r)$ and $\Psi(r)$. The stress tensor then has $T^{00} = \rho(r)(\eta^0)^2 = \rho(r)c^2 e^{-2\Phi(r)}$ with spatial components $T^{rr} = p(r)e^{-2\Psi(r)}$, $T^{\theta\theta} = p(r)/r^2$, and $T^{\phi\phi} = p(r)/(r^2 \sin^2\theta)$. Our goal is to find the functions $\Phi(r)$, $\Psi(r)$, $\rho(r)$, and $p(r)$. Our tool will be the set of equations in (6.60) (some of which will be redundant – try working them out explicitly using a symbolic algebra package in Problem 6.5) and an equation of state relating p to ρ.

From the equation of motion, the first of (6.60), we learn that

$$\frac{dp}{dr} = -\left(\rho + \frac{p}{c^2}\right)\frac{d(c^2\Phi)}{dr}, \tag{6.63}$$

which is the same as the first equation in (6.45) with $\rho \to \rho + p/c^2$ and $g \sim -m(r)G/r^2 \to -(c^2\Phi)'$ (because of the appearance of Φ as the argument of the exponential, it must be dimensionless, so to recover an object with the dimension of gravitational potential, there must be a factor of c^2); then $c^2\Phi$ is like the Newtonian gravitational potential φ. The 00 field equation reads

$$\frac{e^{2\Phi}}{r^2}\frac{d}{dr}\left(r\left(1 - e^{-2\Psi}\right)\right) = e^{2\Phi}\frac{8\pi G\rho}{c^2}, \tag{6.64}$$

suggesting that we define $m(r)$ by the equation

$$r\left(1 - e^{-2\Psi}\right) \equiv \frac{2Gm(r)}{c^2}. \tag{6.65}$$

Then (6.64) becomes

$$\frac{dm(r)}{dr} = 4\pi\rho r^2, \tag{6.66}$$

matching the mass equation from (6.45). With this alternative to Ψ, the g_{rr} component of the metric reads

$$g_{rr} = \frac{1}{1 - 2Gm(r)/(c^2 r)}, \tag{6.67}$$

which looks just like the Schwarzschild metric's radial component, with $m(r)$ playing the role of the central body mass M. I have omitted the r dependence of the familiar functions ρ, p, Φ, and Ψ but retain it in $m(r)$ to remind us that this symbol is a function of position (it will end up having its usual interpretation as the mass contained in a sphere that has radial coordinate r, as we shall see). Finally, from the rr component of Einstein's equation we get, for the function Φ,

$$\frac{d\Phi}{dr} = \frac{Gm(r) + 4\pi Gr^3(p/c^2)}{c^2 r^2 - 2Grm(r)}. \tag{6.68}$$

All together, the equations governing spherically symmetric hydrostatic equilibrium are (compare with (6.40))

$$\frac{dp}{dr} = -\left(\rho + \frac{p}{c^2}\right)\frac{d(c^2\Phi)}{dr}, \qquad \frac{dm(r)}{dr} = 4\pi\rho r^2,$$
$$\frac{d\Phi}{dr} = \frac{Gm(r) + 4\pi Gr^3(p/c^2)}{c^2 r^2 - 2Grm(r)}. \tag{6.69}$$

There are three equations here and three unknowns, once $p(\rho)$ is given. There are still two remaining field equations, associated with the $\theta\theta$ and $\phi\phi$ components, but as you will show in Problem 6.6, those are redundant if the equations in (6.69) are satisfied.

As a final observation, we can eliminate Φ from the set in (6.69) by replacing $\frac{d\Phi}{dr}$ in the first equation with its expression from the third, leading to the "Oppenheimer–Volkov" equation

$$\frac{dp}{dr} = -\left(\rho + \frac{p}{c^2}\right)\frac{Gm(r) + 4\pi Gr^3(p/c^2)}{r^2 - 2Grm(r)/c^2}, \qquad (6.70)$$

and this relation, together with the equation governing $m(r)$ and the equation of state $p(\rho)$, can be solved once the "initial" conditions at $r = 0$ are given: $m(0) = 0$, $\rho(0) = \rho_c$, $p(\rho_c) = p_c$. As in the Newtonian case, you start at $r = 0$ and step forward in r (numerically, say, as in Problem 6.7) until the pressure is zero, which defines the radius of the central body, call it R. Pressure is a continuous variable, so that once the pressure is zero, it remains zero (again, the defining feature here). The density function goes to zero at R, potentially discontinuously (i.e., it is not a good variable to use in finding the physical extent of the central body). Once you have the pressure, density, and mass, the functions Φ and Ψ are pinned down.

At $r = R$, we must match the external, Schwarzschild solution, so that $m(R) = M$ gives the total mass of the central body, and

$$e^{2\Phi(R)} = \left(1 - \frac{2GM}{c^2 R}\right) \qquad (6.71)$$

is the boundary condition for $\Phi(R)$, which can be reconstructed from the final equation in (6.69).

As one of the few closed-form solutions, we'll work out the static, uniform density (also known as incompressible) noninteracting fluid configuration here. This case, as with its Newtonian version from Section 6.2, has uniform $\rho(r) = \rho_c$. Then we can integrate to get $m(r) = (4/3)\pi r^3 \rho_c$ inside. The Oppenheimer–Volkov equation (6.70) can be solved using separation of variables, together with the initial condition $p(0) = p_c$:

$$p(r) = \frac{\rho_c c^2[-3c(p_c + \rho_c c^2) + (3p_c + \rho_c c^2)\sqrt{9c^2 - 24G\pi\rho_c r^2}]}{9c(p_c + \rho_c c^2) - (3p_c + \rho_c c^2)\sqrt{9c^2 - 24G\pi\rho_c r^2}}. \qquad (6.72)$$

From this expression for $p(r)$ the radius at which the pressure drops to zero, $p(R) = 0$, can be found:

$$R^2 = \frac{3p_c c^2(2p_c + \rho_c c^2)}{2\pi G\rho_c(3p_c + \rho_c c^2)^2}. \qquad (6.73)$$

Alternatively, we can leave R arbitrary and pick the central pressure p_c so that $P(R) = 0$ at a given R, inverting the above equation to find p_c,

$$p_c = \frac{\rho_c c^2(12\pi G\rho_c R^2 - 3c^2 + c\sqrt{9c^2 - 24\pi G\rho_c R^2})}{12(c^2 - 3\pi G\rho_c R^2)}. \qquad (6.74)$$

This central pressure goes to infinity as $3\pi G\rho_c R^2 \to c^2$. That implies a limit to the ratio of the mass "length" GM/c^2 and the radius R of the central body. Using $\rho_c = M/(4/3\pi R^3)$, the infinite value for p_c occurs when

$$\frac{GM}{c^2 R} = \frac{4}{9}. \tag{6.75}$$

Evidently, hydrostatic equilibrium for uniform density central bodies of total mass M constrains the radius to be

$$R > \frac{9GM}{4c^2}. \tag{6.76}$$

This result holds more generally, for nonuniform density, and is known as "Buchdahl's theorem" [6].

Finally, we'll write out the metric for the uniform density case – for a spherical body of mass M and radius R (satisfying (6.76)), the pressure is

$$p(r) = -3Mc^2\left(R^3 - R^3\sqrt{1 - \frac{2GMr^2}{c^2 R^3}}\sqrt{1 - \frac{2GM}{c^2 R}} + \frac{GM}{c^2}(r^2 - 3R^2)\right)$$
$$\Big/\left(4\pi R^3\left(4R^3 + \frac{MG}{c^2}(r^2 - 9R^2)\right)\right). \tag{6.77}$$

Working backwards, from $p(r)$ and $m(r)$, we can solve the third equation in (6.69) for $\Phi(r)$ subject to the boundary condition in (6.71), eventually returning the interior metric element

$$g_{00} = -e^{2\Phi(r)} = -\frac{5}{2} + \frac{1}{2}\frac{GM}{c^2}\left(\frac{r^2}{R^3} + \frac{9}{R}\right) + \frac{3}{2}\sqrt{1 - \frac{2GM}{c^2 R}}\sqrt{1 - \frac{2GMr^2}{c^2 R^3}} \tag{6.78}$$

with

$$g_{rr} = \left(1 - \frac{2GMr^2}{c^2 R^3}\right)^{-1} \tag{6.79}$$

from (6.67). The angular components are, as always, $g_{\theta\theta} = r^2$ and $g_{\phi\phi} = r^2\sin^2\theta$. As with any newly calculated metric, you should verify, in Problem 6.8, that it is not flat. This interior solution matches up with the Schwarzschild exterior smoothly at $r = R$. Schwarzschild worked out this interior solution for uniform density [45] around the same time that he computed the exterior Schwarzschild solution [44].

Using Einstein's equation to find the metric given some target density and pressure is one direction of inquiry. The procedure is similar to being given a distribution of charge and current in E&M. From the charge and current you find **E** and **B**, and you can use those to predict particle motion. But we can also use the field equations in the "inverse" direction, taking some target electric and magnetic fields and finding the charge and current distributions that lead to those fields. That same inversion is useful for Einstein's equation. The goal is to create metrics with certain target properties and then find distributions of matter and energy that generate them. We'll start off with the homogenous, isotropic Robertson–Walker metric, used in cosmology, and then use the same basic approach to define wormholes and the Alcubierre warp drive.

6.5 Robertson–Walker Metric

The Robertson–Walker metric is built to describe "cosmology," the long-term dynamical structure of the universe sourced by the energy inside of it (see, for example, [42]). The model that informs the metric is that of time-evolving spatial "slices," with the coordinate time used to parameterize the slices. In general, such a metric has line element of the form[6]

$$ds^2 = -c^2 dt^2 + R(t)^2 g_{ij} dx^i dx^j, \tag{6.80}$$

where g_{ij} is a static spatial metric, and $R(t)$ is the dimensionless "scale factor," to be set by solving Einstein's equation. In this setup, points in space retain their coordinate labels (the points are "comoving"), and the fundamental geometry of the space piece of the spacetime does not change (g_{ij} is constant in time). Beyond the relatively general construct in (6.80), the large-scale structure of the universe is, to good approximation, uniform (density, for example, is constant over very large length scales) and contains no "preferred direction." The uniform density leads to an assumption of "homogeneity," where all spatial points are equivalent. The lack of a preferred direction is called "isotropy." The assumption that the universe is uniform and isotropic is called the "cosmological principle" and guides the starting form of the metric.

Over small scales, of course, the universe is neither uniform nor isotropic. Sitting in a classroom, there are desks at some locations and not others, different points in the room have different amounts of mass associated with them, not a uniform density. There's also a clear direction in the room – objects fall down rather than up. One can imagine a situation in which there is a uniform density but a preferred direction, think of water flowing in a river. The density of water is basically uniform, but it is all moving in a specific direction. There are also configurations in physics that are isotropic but not uniform. The electric field of a point particle is an example, with field lines that spread out from the charge equally in all directions but emanate from a particular point, the charge location, and not other points (the field is isotropic about a point).

The Robertson–Walker metric has homogeneity and isotropy built into its spatial surfaces. Let's think about how to construct g_{ij} in (6.80) to support these two assumptions. Homogenous means that all points are "the same" (no preferred points), and isotropic means that all directions are "the same" (no preferred directions). Given a point in space, we can enforce isotropy by requiring spherical symmetry about that point (make it the origin), with its democratic $\hat{r}$ direction. To achieve homogeneity, we require that isotropy hold for every point in the space, that is, that any point can be chosen as the origin without

---

[6]   To be fully general, we should include a "cross-term" of the form $g_{0j} cdt dx^j$, but this can be removed by assuming that the spatial slices are themselves orthogonal to the direction of temporal increase.

changing any observable result. A coordinate-free way to enforce homogeneity and isotropy is to require that the Ricci curvature scalar be a constant: the same value at all points. Two familiar examples of spaces with constant Ricci scalar curvature are: (1) the flat three-dimensional piece of Minkowski spacetime with $R = 0$ and (2) the surface of a sphere of radius a with $R = 2/a^2$. In fact, there is one other important example in which the curvature is negative, as we shall see.

We will require, then, that g_{ij} define a space of "constant curvature." Start by picking a point to be the origin of the spatial surface. With respect to that point, the metric must be spherically symmetric, so we can take

$$g_{ij} \doteq \begin{pmatrix} f(r) & 0 & 0 \\ 0 & r^2 & 0 \\ 0 & 0 & r^2 \sin^2 \theta \end{pmatrix} \tag{6.81}$$

as in Section 4.1. The Ricci scalar for this purely spatial geometry is

$$R(r) = \frac{2(rf'(r) + f(r)^2 - f(r))}{r^2 f(r)^2}. \tag{6.82}$$

This Ricci scalar depends on the radial coordinate relative to our arbitrary origin. To ensure that any origin leads to the same curvature, we will demand that $R(r) = R_0$, a constant. Solving the resulting ODE for $f(r)$ gives

$$f(r) = \frac{1}{1 - \alpha/r - R_0 r^2/6}, \tag{6.83}$$

where α is a constant of integration. The homogeneity condition means that *no* scalar quantity made from the metric can have position dependence, else there is a coordinate-independent way to distinguish between different points. Computing the value of the Kretschmann scalar $R^{\mu\nu\rho\sigma} R_{\mu\nu\rho\sigma}$, as you will do in Problem 6.9, and demanding that it is independent of r gives $\alpha = 0$ in (6.83). Letting $k \equiv R_0/6$, the homogenous, isotropic metric with constant curvature is

$$g_{ij} \doteq \begin{pmatrix} \frac{1}{1-kr^2} & 0 & 0 \\ 0 & r^2 & 0 \\ 0 & 0 & r^2 \sin^2 \theta \end{pmatrix}. \tag{6.84}$$

This is one form of the spatial components of the Robertson–Walker metric. It can also be written in conformally flat form as you will show in Problem 6.11. The specific value of k can be adjusted by rescaling the radial coordinate, so it suffices to consider only the sign of k to identify the space.

Flat

The simplest case is $k = 0$, corresponding to a vanishing (three-)Ricci scalar. The spatial line element is

$$d\ell^2 = dr^2 + r^2 d\theta^2 + r^2 \sin^2 \theta d\phi^2, \tag{6.85}$$

which is just the usual Euclidean line element written in spherical coordinates. So the spatial metric is flat, and the full metric takes the form

$$ds^2 = -c^2dt^2 + R(t)^2[dr^2 + r^2d\theta^2 + r^2\sin^2\theta d\phi^2]. \tag{6.86}$$

Closed

For $k > 1$, the line element is

$$d\ell^2 = \frac{dr^2}{1 - kr^2} + r^2d\theta^2 + r^2\sin^2\theta d\phi^2, \tag{6.87}$$

and to eliminate the denominator (and its singularity), take $\sqrt{k}r = \sin\chi$; then

$$d\ell^2 = \frac{1}{k}[d\chi^2 + \sin^2\chi d\theta^2 + \sin^2\chi\sin^2\theta d\phi^2]. \tag{6.88}$$

This metric has the form of a three-sphere. In three-dimensional spherical coordinates, fixing the angle θ and radial coordinate r defines a circle of radius $r\sin\theta$. In a four-dimensional space, with line element

$$d\ell^2 = dx^2 + dy^2 + dz^2 + du^2 \tag{6.89}$$

for new spatial coordinate u, we can define a polar angle χ such that fixing the four-dimensional radial coordinate and χ leads to a sphere of radius $r\sin\chi$. The radial coordinate is defined by $r^2 = x^2 + y^2 + z^2 + u^2$, and then we can define three angles θ, χ, and ϕ by

$$x = r\sin\chi\sin\theta\cos\phi, \qquad y = r\sin\chi\sin\theta\sin\phi,$$
$$z = r\sin\chi\cos\theta, \qquad u = r\cos\chi. \tag{6.90}$$

In these coordinates, the line element (6.89) becomes

$$d\ell^2 = dx^2 + dy^2 + dz^2 + du^2 = dr^2 + r^2[d\chi^2 + \sin^2\chi d\theta^2 + \sin^2\chi\sin^2\theta d\phi^2]. \tag{6.91}$$

If we are constrained to live on a surface of fixed r, with $dr = 0$, then the resulting three-dimensional space has precisely the line element of (6.88).

It is interesting that the total volume of the space in (6.88) is finite. Remember that there is a factor $R(t)^2$ sitting out front in the full line element from (6.80); then the total volume of the space piece at time t is

$$V = \frac{R(t)^3}{k^{3/2}} \int_0^\pi \int_0^\pi \int_0^{2\pi} \sin^2\chi\sin\theta d\phi d\theta d\chi = \frac{2\pi^2 R(t)^3}{k^{3/2}}. \tag{6.92}$$

Open

Taking $k < 0$ now, the line element is

$$d\ell^2 = \frac{dr^2}{1 + |k|r^2} + r^2d\theta^2 + r^2\sin^2\theta d\phi^2, \tag{6.93}$$

suggesting a hyperbolic sine replacement to achieve the simplification of the closed case. Let $\sqrt{k}r = \sinh\eta$, leading to

$$d\ell^2 = \frac{1}{k}\left[d\eta^2 + \sinh^2\eta\, d\theta^2 + \sinh^2\eta\sin^2\theta\, d\phi^2\right]. \tag{6.94}$$

If we take the "Cartesian" coordinates related to the above via

$$x = a\sinh\eta\sin\theta\cos\phi, \qquad y = a\sinh\eta\sin\theta\sin\phi,$$
$$z = a\sinh\eta\cos\theta, \qquad u = a\cosh\eta, \tag{6.95}$$

then the three-dimensional surface with metric from (6.94) is defined by the constraint

$$x^2 + y^2 + z^2 - u^2 = -a^2, \tag{6.96}$$

so that we have a three-dimensional "hyperboloid." This time, the space has infinite volume, as you can show in Problem 6.12.

6.5.1 Stationary Geodesics and Redshift

How do particles move in the Robertson–Walker spacetime? We'll write out the equations of motion for particles and light and establish that in the former case, there exist stationary geodesic solutions. Then we can use lightlike geodesics connecting static observers to introduce the notion of redshift.

First, consider a particle with fixed $\theta = \pi/2$, $\phi = 0$ that starts from rest at $r = r_0$. The geodesic Lagrangian, in proper time parameterization, is

$$L = \frac{1}{2}m\dot{x}^\mu g_{\mu\nu}\dot{x}^\nu = \frac{1}{2}m\left[-c^2\dot{t}^2 + \frac{R(t)^2}{1-kr^2}\dot{r}^2\right] = -\frac{1}{2}mc^2. \tag{6.97}$$

Forming the equations of motion, we have

$$\ddot{t} = \frac{\dot{r}^2 R\frac{dR}{dt}}{c^2(-1+kr)},$$
$$\ddot{r} = \frac{\dot{r}}{R(-1+kr^2)}\left(kr\dot{r}R + 2\dot{t}\frac{dR}{dt}(1-kr^2)\right). \tag{6.98}$$

There is a "static" solution $r(\tau) = r_0$ with $t(\tau) = \tau$. These static geodesics leave a particle's location unchanged, and the particle's proper time coincides with the coordinate time.

As in Section 4.2, we can compute the "proper distance" (or "absolute distance") between two stationary geodesics. If we have a particle at r_0 and another at r_1, then the proper distance is the instantaneous distance between the particles, measured radially here:

$$d(t) = R(t)\int_{r_0}^{r_1}\frac{dr}{1-kr^2} = R(t)\frac{\tanh^{-1}(\sqrt{k}r)}{\sqrt{k}}\bigg|_{r=r_0}^{r_1}. \tag{6.99}$$

The important observation is that $d(t) \sim R(t)$, so that the distance between two points evolves in time according to $R(t)$.

Moving on to lightlike geodesics, we'll again consider radial separations. Start with the Lagrangian from (6.97), but now $L = 0$ since the geodesic is null for light. Let $f(r) \equiv (1 - kr^2)^{-1}$ (the details of the spatial geometry are unimportant), then dividing by $\dot{t}(\tau)$, we can write r as a function of t:

$$\left(\frac{dr}{dt}\right)^2 = \frac{c^2}{R(t)^2 f(r)}. \tag{6.100}$$

From this equation we can relate the emitted and observed frequencies of the light for "labs" at different radial locations, as we have done previously in, for example, Section 4.6 for the Schwarzschild solution. The argument (which can be found in [10]) is basically the same for any value of k in $f(r)$, and it's easiest to take $k = 0$ (you can run the argument again for $k > 0$ or $k < 0$ in Problem 6.13), so we'll set $f(r) = 1$ in what follows.

For our boundary values, suppose that light is emitted from $r = 0$ at t_e and arrives at r_o at time t_o. Taking the positive root (since $r_o > 0$) and separating variables, we have $dr = \frac{cdt}{R(t)}$, which can be integrated:

$$\int_0^{r_o} dr = \int_{t_e}^{t_o} \frac{cdt}{R(t)}. \tag{6.101}$$

The integral on the left is just r_0. Without knowing $R(t)$, we can't say much about the geodesic itself, but we can produce a general relation between the frequency of light measured at the emission and reception points, and the function $R(t)$.

Suppose that at $r = 0$ the period of the light wave is Δt_e and at r_o the period is Δt_o. Successive peaks of the wave occur at t_e and $t_e + \Delta t_e$ at $r = 0$, and at t_o and $t_o + \Delta t_o$ at r_o. Integrating $dr = \frac{cdt}{R(t)}$ between these new intervals,

$$r_o = \int_{t_e+\Delta t_e}^{t_o+\Delta t_o} \frac{cdt}{R(t)}, \tag{6.102}$$

from which by equating (6.102) with (6.101) we learn that

$$\int_{t_e}^{t_o} \frac{cdt}{R(t)} = \int_{t_e+\Delta t_e}^{t_o+\Delta t_o} \frac{cdt}{R(t)}. \tag{6.103}$$

If we assume that the periods Δt_e and Δt_o are small compared to the emission and reception times, we can Taylor expand the right-hand side of (6.103) (see Problem 6.1) to get

$$\int_{t_e}^{t_o} \frac{cdt}{R(t)} \approx \int_{t_e}^{t_o} \frac{cdt}{R(t)} + \frac{c\Delta t_o}{R(t_o)} - \frac{c\Delta t_e}{R(t_e)} \tag{6.104}$$

or

$$\frac{c\Delta t_o}{R(t_o)} = \frac{c\Delta t_e}{R(t_e)}. \tag{6.105}$$

From the period of the light, we can get the frequency as measured at each location, $f_e = 1/\Delta t_e$ and $f_o = 1/\Delta t_o$; then the ratio of the emitted frequency to the observed is

$$\frac{f_e}{f_o} = \frac{R(t_o)}{R(t_e)}. \tag{6.106}$$

This ratio tells us that the observed frequency will be less than the emitted frequency if $R(t_e) < R(t_o)$ – the light will be "redshifted" if $R(t)$ is increasing with time.

The observation that motivates this result is that most galaxies appear to be moving away from us and the light from them is redshifted. We do not exist in a privileged location in the universe (or so the cosmological principle demands), so we need a model that allows *all* points to observe redshift. The Robertson–Walker metric accommodates this assumption – every point sees every other point receding. It is traditional to quote the redshift by comparing f_e/f_o to one, so that "redshift" refers to z in $1 + z = f_e/f_o$, and then

$$z = \frac{R(t_o)}{R(t_e)} - 1. \tag{6.107}$$

Take $t_o = t_e + \Delta t$, and we'll assume that Δt is small compared to t_e; then expanding the right-hand side of (6.107) gives

$$z \approx \frac{\dot{R}(t_e)}{R(t_e)}\Delta t. \tag{6.108}$$

We can relate z to the scale factor and its derivative by noting that

$$r_0 = \int_{t_e}^{t_e+\Delta t} \frac{c\,dt}{R(t)} \approx \frac{c\Delta t}{R(t_e)} \tag{6.109}$$

for small Δt. Using this approximation in (6.108) to eliminate Δt,

$$z \approx \dot{R}(t_e)r_o/c, \tag{6.110}$$

which says that the further away a galaxy is from us (r_o), the larger its redshift, and hence its recessional speed. The observation that the speed of recession is proportional to distance is known as "Hubble's law," and we see that it emerges theoretically here (the situation is similar to the theoretical framework that Newton provided for Kepler's observations).

Each of our three options for $f(r)$ have

$$\int_0^{r_o} \sqrt{f(r)}dr \approx r_o \tag{6.111}$$

to leading order (as you will establish in Problem 6.14), so that $r_o \approx c\Delta t/R(t_e)$ no matter what form of the solution (flat, open, or closed) you choose.

6.5.2 The Einstein Tensor and Cosmological Constant

We will put the Robertson–Walker starting point into Einstein's equation, and evaluate the geometric left-hand side. That calculation will help us understand the sources available for the Robertson–Walker ansatz. Computing the Einstein tensor, its nonzero elements are

$$G_{00} = \frac{3(\dot{R}(t)^2 + c^2 k)}{c^2 R(t)^2}, \qquad \hat{G} \equiv -\frac{c^2 k + \dot{R}(t)^2 + 2R(t)\ddot{R}(t)}{c^2},$$

$$G_{rr} = g_{rr}\hat{G}, \qquad G_{\theta\theta} = g_{\theta\theta}\hat{G}, \qquad G_{\phi\phi} = g_{\phi\phi}\hat{G}. \tag{6.112}$$

The first thing to notice about the structure of these equations is that the spatial components are all of the form $\hat{G}g_{ij}$, which suggests a stress tensor that has the perfect fluid structure, with its pressure terms pg_{ij}. So in addition to its utility in describing Hubble's law, and enforcing homogeneity and isotropy, the Robertson–Walker metric has precisely the target source, a "fluid" with rest energy density ρ and pressure p.

It is here, finally, that we will introduce an additional term in Einstein's equation, the "cosmological constant." We will motivate the presence of a term proportional to $g_{\mu\nu}$ in Einstein's equation later on, in Chapter 7. For now, we state the form of the modified field equation, for constant Λ,

$$R_{\mu\nu} - \frac{1}{2}g_{\mu\nu}R + \Lambda g_{\mu\nu} = \frac{8\pi G}{c^4}T_{\mu\nu}. \tag{6.113}$$

From the point of view of "model building," what we have is a modification of general relativity, one that has a new constant attached. But we can also move the new term over to the right-hand side, writing

$$R_{\mu\nu} - \frac{1}{2}g_{\mu\nu}R = \frac{8\pi G}{c^4}\left[T_{\mu\nu} - \frac{\Lambda c^4}{8\pi G}g_{\mu\nu}\right] \tag{6.114}$$

and viewing the Λ term as a new, ubiquitous "source." Comparing with the stress tensor form in (6.59), what we have is an effective "pressure" with uniform magnitude $-\Lambda c^4/(8\pi G)$ and energy density (Λ can be positive or negative). Go back to Problem 4.14 and Problem 4.15 and see what the spherically symmetric vacuum solutions to this equation, in multiple dimensions, are.

6.6 Cosmological Field Equations

Using our modified Einstein equation from (6.113), together with the fluid source stress tensor from (6.59) (ρ and p are functions of time only, a manifestation of the assumed uniformity)

$$T^{\mu\nu} = \left(\rho(t) + \frac{p(t)}{c^2}\right)\eta^\mu\eta^\nu + p(t)g^{\mu\nu} \tag{6.115}$$

and the form of the Robertson–Walker metric

$$ds^2 = -c^2 dt^2 + R(t)^2\left[\frac{1}{1 - kr^2}dr^2 + r^2 d\theta^2 + r^2 \sin^2\theta d\phi^2\right], \tag{6.116}$$

we end up with two independent field equations. The first is just the 00 component of (6.113),

$$\frac{3(c^2 k + \dot{R}(t)^2)}{c^2 R(t)^2} - \Lambda = \frac{8\pi G \rho(t)}{c^2}, \tag{6.117}$$

one of "Friedmann's equations." For the second field equation, we'll compute the covariant divergence of the stress tensor, as we did in Section 6.4; only the zero component of the divergence is nonzero:

$$T^{0v}_{\;;v} = 0 = 3\frac{\dot{R}(t)}{R(t)}\left(\rho(t) + \frac{p(t)}{c^2}\right) + \dot{\rho}(t), \tag{6.118}$$

and you should show in Problem 6.16 that satisfying these two imply that the rest of the nonzero equations from (6.113) are also satisfied.

The first-order (6.118) can be written as

$$\frac{d}{dt}\left(\rho(t)R(t)^3\right) = -\frac{p(t)}{c^2}\frac{d}{dt}\left(R(t)^3\right), \tag{6.119}$$

where the left-hand side $\rho(t)R(t)^3$ looks like the total mass (energy), and the right-hand side is reminiscent of the work $p\,dV$. From (6.117) one can write a one-dimensional "Hamiltonian"

$$-k = \frac{\dot{R}(t)^2}{c^2} - \frac{1}{3}\left(\Lambda + \frac{8\pi G \rho(t)}{c^2}\right)R(t)^2, \tag{6.120}$$

where $-k$ is the constant "energy," $\dot{R}(t)^2$ plays the role of the kinetic energy, and the rest of the terms on the right represent an effective potential. Before we can use (6.120), we need to know $\rho(t)$ as a function of $R(t)$, and we get this relation through the equation of state. We will study two relevant equations of state: (1) $p(t) = 0$, a noninteracting dust, and (2) $p(t) = \rho(t)c^2/3$, which comes from the requirement that the stress tensor be trace-less, $T \equiv g_{\mu\nu}T^{\mu\nu} = 0$. This second choice occurs if we assume that the energy distribution in the universe is predominantly electromagnetic radiation, since the stress tensor associated with electromagnetic fields is traceless (see Section 7.4 and Problem 7.16).

Noninteracting Dust

For $p = 0$, which defines the dust, equation (6.119) tells us that $\rho(t)R(t)^3 = \alpha$, a constant. Then the effective potential in (6.120) becomes

$$V_{\text{eff}} = -\frac{1}{3}\left(\Lambda R^2 + \frac{8\pi G\alpha}{c^2 R}\right). \tag{6.121}$$

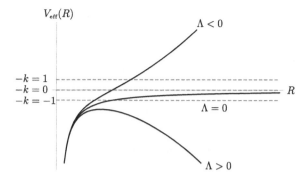

Fig. 6.5 An "energy diagram" that shows the effective potential for pressure-less dust with different signs for the cosmological constant Λ. In this diagram, motion is allowed for $-k > V_{\text{eff}}$. The three different types of spatial geometry are displayed by their constant: $-k = 1$ (open), $-k = 0$ (flat), and $-k = -1$ (closed).

Keep in mind that Λ can be either positive or negative and that without loss of generality, we can take k to be 1 (closed), 0 (flat), or -1 (open).[7] In Figure 6.5, we see the "energy diagram" that allows us to determine the dynamics of the universe as told through $R(t)$ for different choices of k and Λ. We must have $-k \geq V_{\text{eff}}$ in order for $\dot{R}^2 \geq 0$, so looking first at the $\Lambda = 0$ curve, we see that in the case of $k = 1$ (closed), there is a point at which $\dot{R} = 0$. The universe starts at $R = 0$ (to the left of the crossing), R increases in size until it reaches

$$R_c = \frac{8\pi G\alpha}{c^2}, \tag{6.122}$$

at which point $\dot{R} = 0$, and the process reverses with R decreasing back down to zero. In this case, the universe starts with a "big bang" from $R = 0$ and will eventually end with a "big crunch." Contrast that behavior with the $k = 0$ (flat) space, again at $\Lambda = 0$. In this case, the universe begins with $R = 0$, and the expansion continues forever, the curve associated with $\Lambda = 0$ never crosses the $-k = 0$ line. The same is true for the $k = -1$ (open) spatial geometry. You should try telling the story for the $\Lambda \neq 0$ cases in Problem 6.17.

Radiation

Radiation, which is thought to be the dominant form of energy in the early universe, is defined by its traceless stress tensor, $T = 0$. Using the usual perfect fluid stress tensor and the Robertson–Walker metric to compute the trace, we

---

[7] We'll use these purely numerical values for k, but from its introduction in (6.84), k must have dimension of inverse-length-squared.

find

$$T = T^{\mu\nu} g_{\mu\nu} = 3p(t) - \rho(t)c^2 = 0 \longrightarrow p(t) = \frac{1}{3}\rho(t)c^2. \qquad (6.123)$$

Inputting this expression for $p(t)$ in (6.119), we find that

$$\frac{d\rho}{dt}R^3 + 4\rho R^2 \frac{dR}{dt} = 0 \longrightarrow p(t) = \frac{\beta}{R(t)^4} \qquad (6.124)$$

with constant of integration β, from separation of variables. In this setting, the effective potential is

$$V_{\text{eff}} = -\frac{1}{3}\left(\Lambda R^2 + \frac{8\pi G\beta}{c^2 R^2}\right). \qquad (6.125)$$

Although the shift from $1/R$ to $1/R^2$ (going from pressure-less dust to radiation) is significant, it doesn't make much qualitative difference in the plot, which looks similar to the one in (6.5) (try it in Problem 6.18).

6.7 Wormholes

The Kruskal–Szekeres coordinates we encountered in Section 4.4 cover a half-plane, as shown in Figure 4.7. If we go back to (4.53), it is clear that we can cover the other half of the plane using alternate sign choices. The motivation for this full covering comes from the observation that the singularity at $r = 0$ is associated with the curve $-v^2 + u^2 = -1$ from (4.52). There are two valid curves here, for $v = \pm\sqrt{1+u^2}$, and in Figure 4.7, we have only plotted the one with $v = \sqrt{1+u^2}$. Similarly, the asymptotically flat spacetime that is found for $r \gg R_M$ has, again using (4.52), which is valid everywhere,

$$-v^2 + u^2 \approx \frac{r}{R_M}e^{r/R_M} \longrightarrow v^2 + \frac{r}{R_M}e^{r/R_M} \approx u^2, \qquad (6.126)$$

from which we conclude that $u^2 \gg v^2$ for large r/R_M. The flat spacetimes are then at $u \gg v$ and $u \ll -|v|$, so that, again, there are two choices that lead to the same geometry. We can extend the values of u and v to include the two missing quadrants from Figure 4.7 to get all four quadrants; we have already covered the first two, which we'll call I and II, in our previous discussion of Kruskal–Szekeres coordinates. Define I to be the set of points with $u \geq 0$, $v \leq u$, and $v \geq -u$. Here we take

$$v = \sqrt{-1 + \frac{r}{R_M}}e^{r/(2R_M)}\sinh\left(\frac{ct}{2R_M}\right),$$

$$\qquad\qquad\qquad\qquad\qquad\qquad\qquad\qquad\qquad\qquad (6.127)$$

$$u = \sqrt{-1 + \frac{r}{R_M}}e^{r/(2R_M)}\cosh\left(\frac{ct}{2R_M}\right),$$

which is just the pair in (4.53) for $r > R_M$ (true for region I). For region II, with $v > 0$, $u > -v$, and $u < v$, we have

$$v = \sqrt{1 - \frac{r}{R_M}} e^{r/(2R_M)} \cosh\left(\frac{ct}{2R_M}\right),$$

$$u = \sqrt{1 - \frac{r}{R_M}} e^{r/(2R_M)} \sinh\left(\frac{ct}{2R_M}\right),$$

(6.128)

the pair with $r < R_M$ from (4.53).

The two new regions, which we'll label III and IV, have similar descriptions but taking the other sign for the square roots. Region III is defined by the set of points with $u < 0$, $v > u$, and $v < -u$, where we take

$$v = -\sqrt{-1 + \frac{r}{R_M}} e^{r/(2R_M)} \sinh\left(\frac{ct}{2R_M}\right),$$

$$u = -\sqrt{-1 + \frac{r}{R_M}} e^{r/(2R_M)} \cosh\left(\frac{ct}{2R_M}\right),$$

(6.129)

the sign-flipped version of region I. Finally, region IV has $v < 0$, $u > v$, and $u < -v$, with sign-flipped versions of the transformation from region II:

$$v = -\sqrt{1 - \frac{r}{R_M}} e^{r/(2R_M)} \cosh\left(\frac{ct}{2R_M}\right),$$

$$u = -\sqrt{1 - \frac{r}{R_M}} e^{r/(2R_M)} \sinh\left(\frac{ct}{2R_M}\right).$$

(6.130)

The presence of regions III and IV, acting as separate versions of I and II, suggests the existence of two "distinct" asymptotically flat universes (they could be different asymptotically flat regions of the same universe), which we could imagine to be connected through the $r = 0$ singularity. That connection is known as a "Schwarzschild wormhole" or an "Einstein–Rosen bridge." To understand this possibility better, we will make a short digression.

6.7.1 Embedding Diagrams

The four-dimensional spacetimes we have studied often have familiar lower-dimensional elements. As an example, the Schwarzschild spacetime has a two-sphere geometry in its $\theta\phi$ subspace (fixing r and ct, you are left on a "normal" two-dimensional surface). We already know about two-spheres – we'd like to probe the novel elements of the metric, like the function $(1 - R_M/r)^{-1}$ sitting in front of the dr^2 component of the Schwarzschild line element. So we'll fix t and θ, and think about the $r\phi$ subspace. One way to visualize two-dimensional subspaces is as surfaces in a three-dimensional flat space.

Example 6.4 (One Dimension to Two). As an example to set up the idea, let's imagine a one-dimensional "space" with line element $d\ell^2 = f(x)dx^2$. We can understand the role of $f(x)$ by thinking about a curve in two dimensions. Take

$d\bar{\ell}^2 = dx^2 + dy^2$, and we'll find a functional relationship between x and y that turns $d\bar{\ell}^2$ into $d\ell^2$. For now, we'll let $y = y(x)$ be an unknown function of x; then

$$d\bar{\ell}^2 = dx^2 + y'(x)^2 dx^2 = \left(1 + y'(x)^2\right)dx^2, \tag{6.131}$$

and if we demand that this match $d\ell^2 = f(x)dx^2$, then we get

$$f(x) = \left(1 + y'(x)^2\right) \longrightarrow y(x) = \pm \int \sqrt{f(x) - 1}\, dx. \tag{6.132}$$

Take $f(x) = 1 + 4x^2$, then $y(x) = x^2$. We can draw the curve in two dimensions, $\mathbf{w}(x) = x\hat{\mathbf{x}} + x^2\hat{\mathbf{y}}$, that has (flat) length along the curve that matches the length in the modified one-dimensional space.

Following the one-dimensional example, our goal is to take a two-dimensional $r\phi$ subspace from, say, a spherically symmetric metric, and represent its deviation from a flat three-dimensional space. Because the coordinate ϕ is the usual polar angle in these spaces, it is natural to try to embed a metric with line element

$$d\ell^2 = a(r)dr^2 + r^2 d\phi^2 \tag{6.133}$$

in a three-dimensional flat space with cylindrical coordinates r, ϕ, and z, so that $d\bar{\ell}^2 = dr^2 + r^2 d\phi^2 + dz^2$. The result will be a surface that has an induced metric (see Section C.3) with line element matching the one in (6.133). Letting $z = z(r)$, the cylindrical line element is

$$d\bar{\ell}^2 = \left(1 + z'(r)^2\right)dr^2 + r^2 d\phi^2, \tag{6.134}$$

and if we demand that this match $d\ell^2$ from (6.133), then we must have

$$z(r) = \pm \int \sqrt{a(r) - 1}\, dr. \tag{6.135}$$

Taking the Schwarzschild value $a(r) = (1 - R_M/r)^{-1}$, we find

$$z(r) = 2\sqrt{R_M(r - R_M)} \tag{6.136}$$

for the positive root with $r > R_M$. You should try finding the height $z(r)$ for the interior of a uniform mass density central body in Problem 6.19.

We can take $z(r)$ and make a surface out of it ("Flamm's paraboloid" [18]) by plotting $z(r)$ versus r and rotating about the z axis as shown in Figure 6.6. The radius of the circle in the $z = 0$ plane is R_M, and we don't have a way of expressing the surface below this value. As z increases, the spacetime has increasing value of r, and as $r \to \infty$, we recover the flat spacetime. We only used positive values of z to make Figure 6.6, but the presence of the $z(r) = -2\sqrt{R_M(r - R_M)}$ branch invites us to plot a "full" version of the Schwarzschild spacetime, with two separate asymptotically flat spacetimes connected by the Einstein–Rosen bridge at $z = 0$, as shown in Figure 6.7.

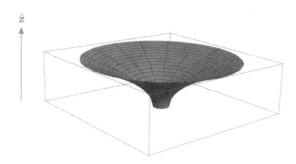

Fig. 6.6 An embedding diagram for the Schwarzschild geometry. We have taken only the positive root with $z(r)$ given by (6.136).

6.7.2 Another Model Metric

The invitation to consider the "Schwarzschild wormhole" (for a thorough description of the physics of such wormholes, see [36]) above comes from the observation about full uv plane covering in Kruskal–Szekeres coordinates. But there are other models that come from the sort of metric engineering that we are considering in this chapter. The "Ellis wormhole" [14] has the line element

$$ds^2 = -c^2dt^2 + dx^2 + \left(x^2 + \alpha^2\right)\left(d\theta^2 + \sin^2\theta d\phi^2\right) \tag{6.137}$$

for constant α. If we fix $\theta = \pi/2$ and consider the two-dimensional line element with $dt = 0$ and $d\theta = 0$,

$$d\ell^2 = dx^2 + \left(x^2 + \alpha^2\right)d\phi^2, \tag{6.138}$$

we can create an embedding diagram as above. This time, we'll let $r^2 = x^2 + \alpha^2$ to make the starting line element look more like our model in (6.133). With

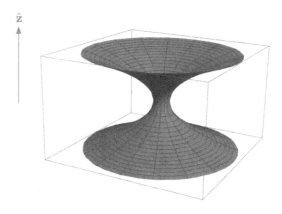

Fig. 6.7 An embedding diagram for the Schwarzschild geometry with two different asymptotically flat regions connected by an "Einstein–Rosen bridge." This time, we have used both signs in $z(r) = \pm 2\sqrt{R_M(r - R_M)}$.

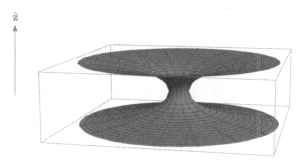

$\hat{z}$

Fig. 6.8 An embedding diagram of the Ellis wormhole with line element (6.137).

this change, the reduced line element becomes

$$d\ell^2 = \frac{r^2}{r^2 - \alpha^2} dr^2 + r^2 d\phi^2, \tag{6.139}$$

and our $a(r) = (1 - \alpha^2/r^2)^{-1}$. Putting this expression for $a(r)$ into (6.135) to find $z(r)$ for the flat cylindrical embedding, we get

$$z(r) = \alpha \log\left(\frac{r + \sqrt{r^2 - \alpha^2}}{\alpha}\right) \tag{6.140}$$

for $r > \alpha$ and assuming $z(\alpha) = 0$. The resulting embedding looks very similar to (6.7) and is shown in Figure 6.8.

6.8 Alcubierre Warp Drive

We'll close with a final piece of clever metric engineering. The Alcubierre warp drive [3] provides a mechanism for superluminal travel and is a wonderful example of creative spacetime construction even if it does appear to need negative energy to construct. We'll approach the construction in stages. In Minkowski spacetime, working in one spatial dimension with line element $ds^2 = -c^2 dt^2 + dx^2$, we know that the lightlike geodesics have $ds^2 = 0$, and so $\frac{cdt}{dx} = \pm 1$. These appear as forty five degree lines as discussed in Section A.3. At any point P in the ctx plane, we can identify the points in the future that can be influenced by P and all the points in the past that are in causal contact with P by looking at the light cone, the forward and backward boundary of the null geodesics originating at P.

But there are ways to get those lines to occur with slopes other than one on a Minkowski diagram. Consider the line element

$$ds^2 = -c^2 dt^2 + (dx - vdt)^2 \tag{6.141}$$

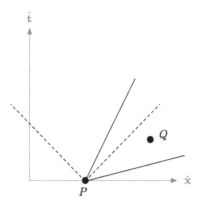

A Minkowski diagram showing the regions of influence of the point P in Minkowski spacetime (dashed line) and in the spacetime defined by the line element in (6.141) with $v = 3c$. The point Q cannot depend on events at P in the Minkowski case, but can be influenced by events at P in the spacetime defined by (6.141).

for constant v. The null geodesics of this spacetime satisfy the differential equation

$$\frac{cdt}{dx} = \frac{1}{\pm 1 + v/c}.$$ (6.142)

To see the effect of this change, take the point P at $ct = 0$ shown in Figure 6.9. Set $v = 3c$; then the null geodesics are shown as solid lines defining the light cone in this geometry. For reference, the dashed forty five degree lines show the light cones of the Minkowski spacetime. The point Q in the figure is within the light cone of P in the modified spacetime, and so events at P could influence events at Q. Not so for the Minkowski spacetime, where Q is outside of the light cone of P.

A spaceship that traveled with speed v has timelike motion in this new geometry even if $v > c$, since $ds^2 = -c^2 dt^2 < 0$. As shown in Figure 6.10, if the ship left from P, it remains within the light cone of P for all time, and we wouldn't notice anything strange happening in its local vicinity.

Suppose now that we want the new spacetime to exist only within a small region surrounding the spaceship, which has position $x_s(t) = x_0 + vt$. We can localize the effects using any smooth function $f(u)$ with $f(0) = 1$ and $f(|\epsilon|) \to 0$ for small ϵ. Then we can carve out a patch around the ship in which spacetime has line element given by (6.141) while maintaining Minkowski spacetime elsewhere:

$$ds^2 = -c^2 dt^2 + \left(dx - f\left(|x - x_s(t)|\right) v dt\right)^2.$$ (6.143)

Some representative light cones for this hybrid spacetime are shown in Figure 6.11. Away from the ship's trajectory, the light cones are the normal forty five degree lines. Within a tube of radius ϵ around the ship, the light cones are tipped so that the ship moves at speed $v > c$, but without disturbing the causal structure.

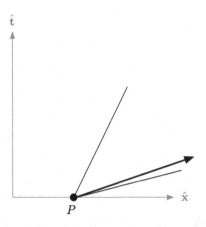

A spaceship travels at $3c$ in the geometry defined by (6.141) (with $v = 3c$). Its worldline is shown as the thick line, and the light cone at P is shown with a thin line. Anything that happens on the ship is within the light cone of both the origin P of the ship at $t = 0$ and all other points along the ship's worldline.

Finally, we can put back in the other two spatial dimensions,

$$ds^2 = -c^2dt^2 + \left(dx - vf(r_s)dt\right)^2 + dy^2 + dz^2,$$
$$r_s^2 \equiv \left(x_s(t) - x\right)^2 + y^2 + z^2. \tag{6.144}$$

This is an example of the Alcubierre warp drive metric. In the original paper, the function $f(u)$ is taken to be

$$f(u) = \frac{\tanh(\sigma(u + \epsilon)) - \tanh(\sigma(u - \epsilon))}{2\tanh(\sigma\epsilon)}, \tag{6.145}$$

where σ controls the steepness of the hyperbolic tangent functions. As $\sigma \to \infty$, the function $f(u)$ becomes 1 for $-\epsilon < u < \epsilon$ and zero for $u < -\epsilon$ and

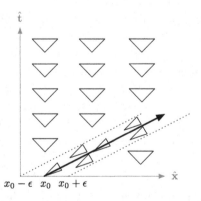

The new geometry from (6.141) is localized around a spaceship that has position $x_s(t) = x_0 + vt$ using a smooth fall-off $f(u)$ (that goes to zero for $|u| > \epsilon$) as in (6.143). The light cones for different positions, both inside and outside the ship, are shown. In this figure, the ship's speed is $v = 2c$.

$u > \epsilon$. The constant v can be made into a function of time to accommodate the acceleration and deceleration of the spaceship.

We are well beyond special relativity here; the existence of the warp drive metric requires general relativity and a very specific type of source $T_{\mu\nu}$ to generate it. You should think about the source viability by working out Problem 6.22 and Problem 6.23.

Exercises

6.1 Fill in the details (using the fundamental theorem of calculus and Taylor expansion) that give the approximation in (6.14).

6.2 Solve the set of equations in (6.45) for density $\rho(r) = \alpha r$, a linearly increasing function of radius with constant α. For a spherical mass M that has radius R with this linearly increasing density, write α in terms of M and R. What is the pressure at the center of the mass?

6.3 Redo the numerical calculation from Example 6.3 using $\gamma = 4/3$. What are the "radius" and "mass" of the Sun using this model?

6.4 Suppose you have a stress tensor for noninteracting dust, of the form $T^{\mu\nu} = \frac{u}{c^2} \frac{dx^\mu}{dt} \frac{dx^\nu}{dt}$, where u is the energy density as measured in a lab L, and $\frac{dx^\mu}{dt}$ is the velocity of the fluid as measured in L (where the zero component is c, and the spatial components are $\mathbf{v}$ in L). Find the stress tensor in a frame $\bar{L}$ related to L via a Lorentz boost in the direction of the fluid, that is, a boost in the $\hat{\mathbf{v}}$ direction with speed v (see (A.16) for the appropriate form).

6.5 Write out the nonzero components of the Einstein equations and conservation equation $T^{\mu\nu}{}_{;\mu} = 0$ for a static, spherically symmetric distribution of dust starting from the line element (6.62).

6.6 From your equations in Problem 6.5, solve the conservation, 00 and rr components of Einstein's equation for $p'(r)$, $m'(r)$ (for $m(r)$ and $\Psi(r)$ related by (6.65)), and $\Phi'(r)$ algebraically and show that all of the Einstein equations are subsequently satisfied (use the derivative of your expression for $\Phi'(r)$ to replace $\Phi''(r)$ in Einstein's equations).

6.7 Using (6.70) together with $\frac{dm(r)}{dr} = 4\pi \rho r^2$, the polytropic equation of state, and initial conditions from Example 6.3, numerically integrate to find the total mass and radius of the "Sun" in this relativistic model.

6.8 Show that the spherically symmetric metric with g_{00} and g_{rr} taken from (6.78) and (6.79) is not flat. What is the Ricci scalar for this metric?

6.9 We demanded that the Ricci scalar be constant in developing the homogenous, isotropic starting point for the Robertson–Walker metric. There are other scalars one can form, and we should check that they are also constant. Using the Robertson–Walker metric with $g_{rr} = f(r)$

from (6.83), evaluate the Kretschmann invariant $R^{\alpha\beta\mu\nu}R_{\alpha\beta\mu\nu}$. Under what condition is it a constant?

6.10 Suppose you want to construct a spacetime with constant Ricci scalar starting from the spherically symmetric line element

$$ds^2 = -a(r)c^2dt^2 + a(r)^{-1}dr^2 + r^2d\theta^2 + r^2\sin^2\theta d\phi^2. \quad (6.146)$$

What is $a(r)$ if the target is $R = \alpha$ for constant α?

6.11 Construct the conformally flat form of the metric (in which the metric can be written as a function times a flat metric) in (6.84) by considering a function $u(r)$, defining a new coordinate u.

6.12 Show that the volume of the spatial subspace, in the sense of (6.92), for the "open" line element (6.93) is infinite. How about the volume of a flat space with line element (6.85)?

6.13 Show that the result in (6.106), obtained for $k = 0$, holds for $k > 0$ and $k < 0$.

6.14 Show that $\int_0^{r_e} \sqrt{f(r)}dr \approx r_o$ for $f(r) = 1/(1 - kr^2)$ with $k > 0$ and $k < 0$, assuming that r_0 is small.

6.15 Compute the Einstein tensor for the Robertson–Walker metric starting point (6.80) with the spatial piece of the metric from (6.81).

6.16 Show that solving the two equations (6.117) and (6.118) implies a solution to all four nontrivial equations from (6.113).

6.17 Referring to Figure 6.5, what happens to the universe for each of the spatial geometries and values for Λ? For the cases that cross the constant $-k$ lines, what does the crossing point correspond to physically?

6.18 Make a plot similar to Figure 6.5 for the effective potential for radiation from (6.125) and identify the curves with $\Lambda < 0$, $\Lambda = 0$, and $\Lambda > 0$.

6.19 For the spherically symmetric metric inside a central body of uniform mass density ρ_0, make an embedding diagram using the value of g_{rr} from (6.79).

6.20 Find the Riemann tensor and Ricci tensor and scalar for the Ellis wormhole with line element given by (6.137). Is this a flat metric?

6.21 What is the energy density required to produce the Ellis wormhole with line element in (6.137)?

6.22 Does the line element in (6.141) define a flat spacetime? How about the nearly identical one found in (6.143)?

6.23 What is the energy density required to produce the Alcubierre warp drive spacetime given in (6.144) with generic $f(u)$ from (6.145)?

7 Field Theories and Gravity

We know how "point" sources generate vacuum solutions in general relativity and how particles move under the influence of those solutions from Chapter 4. In the last chapter, we saw how material particles generate spacetimes and how the Einstein field equations, together with the covariant conservation of the stress tensor, serve to constrain the properties of those sources. Our goal in this chapter is to see how more general extended sources of energy, like electromagnetic fields, generate and react to gravitational fields. To do that, we need a unified way to talk about field theories, so that we can combine E&M with gravity. We can do this using the stress tensor of the electromagnetic field, but how does that stress tensor come about, and why is it a natural source for gravity?

Just as there is a Lagrangian for point particles in classical mechanics, there are "field" Lagrangians that have similar benefits for thinking about field theories. The advantages of a Lagrangian approach are similar in the two cases. Lagrangians provide a compact, structure-revealing shorthand for a full set of dynamics. They are coordinate independent by design, ensuring that the resulting field equations are generally covariant. So although a field Lagrangian is really only the beginning of a study of field theory, which must ultimately culminate in a study of the field's equations and sources, it is a beginning that contains a wealth of information.

Most important for our purposes here is the way in which field Lagrangians enforce the universal coupling that is the hallmark of gravitational interaction. As we shall see, every field theory has a Lagrangian that depends in some way on the metric, the field of general relativity, and that dependence provides *both* the external field's source for gravity and the response of the external field to gravity. Those two elements are intertwined from the very start. Every field theory, no matter how simple, is bound up in a fundamentally nonlinear way with the spacetime it both creates and responds to. As an example, the electromagnetic field carries energy which determines a metric. But the electromagnetic field satisfies a wave equation in vacuum, and that wave equation's structure is determined by the metric ...

We'll start with a description of field Lagrangians and the associated Euler–Lagrange "field equations" for the simplest (scalar) field. In that context, we can develop notions of transformation that will ultimately lead us to the field's stress tensor with its conservation properties following from Noether's theorem. Along the way, we will encounter new players in our transformation zoo, adding "densities" to tensors and connections. Once the machinery is clear, we will show how the scalar field couples naturally to gravity. From there we will use the same approach to see how E&M interacts with gravity, which will lead us to the Reissner–Nordstrøm solution associated with charged spherically symmetric sources.

7.1 Field Lagrangian for Scalars

To begin, we'll work exclusively on a flat Minkowski background in Cartesian coordinates, so that any reference to the metric $g_{\mu\nu}$ in this section takes the form of (2.53). First, let's review the usual particle action and equations of motion to see how the upgrade to field action must go. Working in one spatial dimension, an action is an integral in time, which takes a trajectory $x(t)$ and returns a number:

$$S[x] = \int L(x, \dot{x}) dt, \tag{7.1}$$

where I've omitted the t-dependence in writing x and $\dot{x}$ to avoid clutter. The Euler–Lagrange equation comes from "extremizing" the action and reads

$$-\frac{d}{dt}\frac{\partial L}{\partial \dot{x}} + \frac{\partial L}{\partial x} = 0. \tag{7.2}$$

If our target quantity of interest is a scalar field ψ with dependence on coordinates and time (itself a coordinate, in the relativistic setting), then it makes sense to have an action that is an integral over all spacetime, with a Lagrangian that depends on ψ and all four of its derivatives democratically. A scalar field action, then, looks like (again removing the explicit coordinate dependence, $\psi(x) \to \psi$, as above)

$$S[\psi] = \int \mathcal{L}(\psi, \partial\psi) \, d^4x \tag{7.3}$$

for some "Lagrangian" $\mathcal{L}$. Similarly, the Euler–Lagrange equations should involve all derivatives equally. A natural update for (7.2), replacing temporal-derivatives with more general ones, is

$$-\partial_\mu \frac{\partial \mathcal{L}}{\partial \psi_{,\mu}} + \frac{\partial \mathcal{L}}{\partial \psi} = 0, \tag{7.4}$$

with summation implicit in the first term, as usual.

Okay, now that we know the end of the story, let's see how the field equations arise from action extremization. We want a field $\psi(x)$ – defined over a domain Ω with values given on the boundary $\partial\Omega$ – that extremizes the action S. For an ordinary, one-dimensional function $f(x)$, an "extremum" (maximum or minimum) location x_0 is defined by

$$f(x_0 + \delta x) \approx f(x_0) + \delta x f'(x_0) = f(x_0) \text{ for infinitesimal } \delta x, \tag{7.5}$$

that is, $f'(x_0) = 0$. The same is true for functionals, functions of fields like S in (7.3), although the "derivative" is harder to write out. Suppose you have an extremizing field configuration $\psi(x)$ (x refers to all four coordinates) that satisfies the boundary conditions on $\partial\Omega$. We make a "small" perturbation $\psi(x) \rightarrow \psi(x) + \delta\psi(x)$, where $\delta\psi(x)$ is itself a function over the domain with $\delta\psi(x) = 0$ on $\partial\Omega$ (so that both $\psi(x)$ and the perturbed $\psi(x) + \delta\psi(x)$ satisfy the same boundary conditions; those values are stipulated, so we do not get to change them), and look at the response of the action:[1]

$$
\begin{aligned}
S[\psi + \delta\psi] &= \int_{\Omega} \mathcal{L}(\psi + \delta\psi, \partial\psi + \partial\delta\psi)\, d^4x \\
&= \underbrace{\int_{\Omega} \mathcal{L}(\psi, \partial\psi)\, d^4x}_{=S[\psi]} + \underbrace{\int_{\Omega} \left(\frac{\partial\mathcal{L}}{\partial\psi}\delta\psi + \frac{\partial\mathcal{L}}{\partial\psi_{,\mu}}\delta\psi_{,\mu} \right) d^4x}_{\equiv \delta S}, \quad (7.6)
\end{aligned}
$$

where we have used Taylor expansion to relate $\mathcal{L}(\psi + \delta\psi, \partial\psi + \partial\delta\psi)$ to $\mathcal{L}(\psi, \partial\psi)$ and its derivatives. The second term in that expansion, defined to be δS, plays the role of $\delta x f'(x_0)$ in (7.5). We will require that $\delta S = 0$ for all $\delta\psi(x)$, taking that requirement to define the extremum of the action.

The perturbation $\delta\psi(x)$ is arbitrary (except on the boundary of $\partial\Omega$), but the derivative, $\delta\psi_{,\mu}$ is not – once $\delta\psi$ is specified, $\delta\psi_{,\mu}$ is fixed. Our goal is to vary $\delta\psi(x)$ and constrain the result, but we cannot change $\delta\psi(x)$ and its derivative independently. So we'd like to rewrite the response of the action, δS, entirely in terms of $\delta\psi$. To do this, we'll use the four-dimensional divergence theorem, together with the product rule (effectively deriving the relevant integration by parts here). First, note that

$$
\int_{\Omega} \frac{\partial}{\partial x^{\mu}} \left(\frac{\partial\mathcal{L}}{\partial\psi_{,\mu}}\delta\psi \right) d^4x = \oint_{\partial\Omega} \left(\frac{\partial\mathcal{L}}{\partial\psi_{,\mu}}\delta\psi \right) da_{\mu} = 0, \quad (7.7)
$$

since the value of $\delta\psi = 0$ on $\partial\Omega$. Using the product rule on the left, we get

$$
\int_{\Omega} \frac{\partial}{\partial x^{\mu}} \left(\frac{\partial\mathcal{L}}{\partial\psi_{,\mu}}\delta\psi \right) d^4x = \int_{\Omega} \left(\delta\psi \frac{\partial}{\partial x^{\mu}} \left(\frac{\partial\mathcal{L}}{\partial\psi_{,\mu}} \right) + \frac{\partial\mathcal{L}}{\partial\psi_{,\mu}}\delta\psi_{,\mu} \right) d^4x, \quad (7.8)
$$

and since the volume integral is zero from above, we learn that

$$
\int_{\Omega} \frac{\partial\mathcal{L}}{\partial\psi_{,\mu}}\delta\psi_{,\mu}\, d^4x = -\int_{\Omega} \delta\psi \frac{\partial}{\partial x^{\mu}} \left(\frac{\partial\mathcal{L}}{\partial\psi_{,\mu}} \right) d^4x. \quad (7.9)
$$

Using this result in the expression for δS from (7.6), we can get both terms to depend on the arbitrary $\delta\psi$,

$$
\delta S = \int_{\Omega} \left(\frac{\partial\mathcal{L}}{\partial\psi} - \frac{\partial}{\partial x^{\mu}} \left(\frac{\partial\mathcal{L}}{\partial\psi_{,\mu}} \right) \right) \delta\psi\, d^4x. \quad (7.10)
$$

---

[1] There is some compact notation in (7.6); keep in mind that

$$
\frac{\partial\mathcal{L}}{\partial\psi_{,\mu}} \equiv \frac{\partial\mathcal{L}}{\partial\left(\frac{\partial\psi}{\partial x^{\mu}}\right)},
$$

where the right-hand side indicates why we use the left-hand side in expressions like (7.6).

To get $\delta S = 0$ for all $\delta\psi$, we must set the term in parenthesis equal to zero,

$$\left(\frac{\partial\mathcal{L}}{\partial\psi} - \frac{\partial}{\partial x^{\mu}}\left(\frac{\partial\mathcal{L}}{\partial\psi_{,\mu}}\right)\right) = 0. \tag{7.11}$$

It is the arbitrary $\delta\psi$ that allows us to set the integrand to zero: imagine taking $\delta\psi$ to be a Dirac delta function that picks out a particular point in Ω; then at that point, (7.11) must hold. But we could pick $\delta\psi$ to be, sequentially, a delta function located at every point in Ω, so that (7.11) must hold throughout Ω.

Now that the structure of the action and its integrand, the Lagrangian, are in place, the question becomes: what is $\mathcal{L}$? Going back to Section 2.8, we know the most general (linear) field equation for a scalar already. So all we need to do is reverse-engineer a Lagrangian for a scalar ψ that returns $\alpha D_{\mu}D^{\mu}\psi + \beta\psi = j$ as the field equation. Since we are working on a Minkowski background for now, we want $\mathcal{L}$ with

$$-\partial_{\mu}\left(\frac{\partial\mathcal{L}}{\partial\psi_{,\mu}}\right) + \frac{\partial\mathcal{L}}{\partial\psi} = \alpha\partial_{\mu}\partial^{\mu}\psi + \beta\psi - j, \tag{7.12}$$

where j, the source function, is itself a scalar that has been provided. The first term on the right follows naturally, provided that we have

$$\frac{\partial\mathcal{L}}{\partial\psi_{,\mu}} = -\alpha\psi_{,\mu} \longrightarrow \mathcal{L} = -\frac{\alpha}{2}\psi_{,\mu}g^{\mu\nu}\psi_{,\nu}. \tag{7.13}$$

The second term could come from a Lagrangian of the form

$$\frac{\partial\mathcal{L}}{\partial\psi} = \beta\psi \longrightarrow \mathcal{L} = \frac{\beta}{2}\psi^2, \tag{7.14}$$

and the final term looks like it comes from $\mathcal{L} = -j\psi$. Since the three terms on the right in (7.12) add together, we can just add the three Lagrangian contributions to get

$$\mathcal{L} = -\frac{\alpha}{2}\psi_{,\mu}g^{\mu\nu}\psi_{,\nu} + \frac{\beta}{2}\psi^2 - j\psi. \tag{7.15}$$

The resulting field equation,

$$\alpha\partial_{\mu}\partial^{\mu}\psi + \beta\psi - j = 0 \tag{7.16}$$

is called the "Klein–Gordon equation," and the field ψ is known as a Klein–Gordon scalar field.

7.2 Coordinate Systems

The form of the action in (7.3) is correct for a Minkowski spacetime written in Cartesian coordinates. But for technical reasons (that we'll motivate below), there is a missing piece to the action integrand that is necessary, both in curvilinear coordinates and in spacetimes that are not flat. The purpose of this section is to generate an action that can be used in any geometry and with any coordinate choice. We'll address the issue first in flat spacetimes expressed in general coordinates.

Focus on the first term in the scalar Lagrangian from (7.15) and take the simplified $\mathcal{L} = \psi_{,\mu} g^{\mu\nu} \psi_{,\nu}/2$. The field equation is just

$$- \partial_\mu \left(g^{\mu\nu} \psi_{,\nu} \right) = 0, \tag{7.17}$$

where the derivative ∂_μ acts on both the metric and $\psi_{,\nu}$ terms. In Cartesian coordinates, we recover the wave equation

$$\frac{1}{c^2} \frac{\partial^2 \psi}{dt^2} - \frac{\partial \psi}{\partial x^2} - \frac{\partial \psi}{\partial y^2} - \frac{\partial \psi}{\partial z^2} = 0. \tag{7.18}$$

But what if we adopt spherical coordinates? Then (7.17) is

$$\frac{1}{c^2} \frac{\partial^2 \psi}{\partial t^2} - \frac{\partial^2 \psi}{\partial r^2} - \frac{1}{r^2} \frac{\partial^2 \psi}{\partial \theta^2} - \frac{1}{r^2 \sin^2 \theta} \frac{\partial^2 \psi}{\partial \phi^2} = 0, \tag{7.19}$$

which is not what we expect. We should have gotten the spatial Laplacian in spherical coordinates. Another way to see the problem is that both the scalar Lagrangian and the field equation involve ∂_μ instead of the D_μ that should appear in a general coordinate system, even for flat spacetimes.

We can trace the difficulty back to the Lagrangian itself. Although the action in (7.3) for scalar $\mathcal{L}$ looks fine in Cartesian coordinates, where we started, the integrand is not a scalar in general. If we transform the coordinates from Cartesian to some new, barred, set, then

$$\mathcal{L} d^4 x = \bar{\mathcal{L}} \det \left(\frac{\partial x}{\partial \bar{x}} \right) d^4 \bar{x} \neq \bar{\mathcal{L}} d^4 \bar{x}, \tag{7.20}$$

where $\frac{\partial x}{\partial \bar{x}}$ is shorthand for the matrix $\mathbb{J}$ with entries $J^\mu_\nu = \frac{\partial x^\mu}{\partial \bar{x}^\nu}$, our usual derivative of coordinates matrix. The equation above is not the transformation rule for a scalar since it involves the determinant of the matrix $\mathbb{J}$.[2] It is the volume element that throws in the factor of the determinant.

Example 7.1 (Cartesian to Cylindrical Coordinates). If we start in Cartesian coordinates and move to the cylindrical set $\{c\bar{t}, \bar{s}, \bar{\phi}, \bar{z}\}$ related to the original via $c\bar{t} = ct$, $x = \bar{s} \cos \bar{\phi}$, $y = \bar{s} \sin \bar{\phi}$, and $\bar{z} = z$, then the matrix $\mathbb{J}$ is

$$J^\mu_\nu \equiv \frac{\partial x^\mu}{\partial \bar{x}^\nu} \doteq \begin{pmatrix} 1 & 0 & 0 & 0 \\ 0 & \cos \bar{\phi} & -\bar{s} \sin \bar{\phi} & 0 \\ 0 & \sin \bar{\phi} & \bar{s} \cos \bar{\phi} & 0 \\ 0 & 0 & 0 & 1 \end{pmatrix}, \tag{7.21}$$

and the original $d^4 x = cdt\,dx\,dy\,dz$ with $d^4 \bar{x} = cd\bar{t}\,d\bar{s}\,d\bar{\phi}\,d\bar{z}$ and $d^4 x = \det(\mathbb{J}) d^4 \bar{x}$ reads

$$cdt\,dx\,dy\,dz = \bar{s}(cd\bar{t}\,d\bar{s}\,d\bar{\phi}\,d\bar{z}), \tag{7.22}$$

which we recognize as the volume element in cylindrical coordinates.

---

2 This is a "Jacobian" matrix. The term "Jacobian" is also used to refer to the determinant of $\mathbb{J}$ in an unfortunate example of operator overloading.

The necessity of the determinant of the Jacobian is clear even aside from its calculus roots. Suppose we wanted to compute the area of a disk of radius R in Cartesian coordinates,

$$A = \int_{-R}^{R} \int_{-\sqrt{R^2-x^2}}^{\sqrt{R^2-x^2}} dy\,dx = \pi R^2. \tag{7.23}$$

If we tried to compute the same area in cylindrical coordinates (really just polar since we are only working in two dimensions), without using the determinant, we would have the incorrect

$$\int_{0}^{2\pi} \int_{0}^{R} d\bar{s}\,d\bar{\phi} = 2\pi R, \tag{7.24}$$

which doesn't even have the right units. With the determinant of the Jacobian $\bar{s}$ in place, we get the correct area

$$\int_{0}^{2\pi} \int_{0}^{R} \bar{s}\,d\bar{s}\,d\bar{\phi} = \pi R^2. \tag{7.25}$$

7.2.1 Tensor Densities

Any object that transforms with factors of the determinant of the Jacobian is called a "density." You can have scalar densities and tensor densities of any rank; these transform in the usual tensorial way, but with factors of the determinant multiplying the overall transformation. The number of factors of the determinant of $\mathbb{J}^{-1}$ that appear in the transformation defines the "weight" of the density. As an example, the volume element d^4x is a "scalar density of weight 1" since $d^4\bar{x} = \det(\mathbb{J}^{-1})\,d^4x$.

If you had an object $\mathcal{A}^\mu$ that you determined transformed like

$$\bar{\mathcal{A}}^\mu = \det(\mathbb{J}^{-1})^p \frac{\partial \bar{x}^\mu}{\partial x^\nu} \mathcal{A}^\nu, \tag{7.26}$$

then we would say that $\mathcal{A}^\mu$ is a "first-rank contravariant tensor density of weight p."

One of the most useful densities is the determinant of the metric. It is a scalar density, but what is its weight? We have to determine the transformation of $g \equiv \det(g_{\mu\nu})$. Start with the metric transformation rule

$$\bar{g}_{\mu\nu} = \frac{\partial x^\alpha}{\partial \bar{x}^\mu} \frac{\partial x^\beta}{\partial \bar{x}^\nu} g_{\alpha\beta}. \tag{7.27}$$

Taking the determinant of both sides and noting that the right-hand side can be viewed as a matrix–matrix–matrix product, we have

$$\bar{g} = \det(\mathbb{J}^{-1})^{-2} g, \tag{7.28}$$

so that the metric determinant is a scalar density of weight -2.

When you multiply tensor densities together, the weights add, since it is the power of the determinant that tells us the weight. A density of weight zero has *no* factors of the determinant of the Jacobian out front. We can use the determinant of the metric to make tensors out of densities. Start by noting that $\sqrt{g}$ is a scalar density of weight -1 (taking the square root of both sides of (7.28)). Then if $\mathcal{A}^\mu$ is a tensor density of weight p, the product $\sqrt{g}^p \mathcal{A}^\mu$ is a density of weight 0, that is, just a first-rank contravariant tensor. Because of the choice of signs in our Minkowski metric, which has $g = -1$ in Cartesian coordinates, we typically use $\sqrt{-g}$ as our weight -1 factor.

Going back to d^4x, our scalar density of weight 1, we can make a scalar by multiplying by $\sqrt{-g}$, so that $\sqrt{-g}d^4x$ is a scalar. Let's verify this concretely by going back to Example 7.1. In Cartesian coordinates, $\sqrt{-g}d^4x = cdt\,dx\,dy\,dz$, and in cylindrical coordinates, $\sqrt{-\bar{g}}d^4\bar{x} = \bar{s}(cd\bar{t}\,d\bar{s}\,d\bar{\phi}\,d\bar{z})$, where the $\bar{s}$ is coming from the determinant of the metric expressed in cylindrical coordinates. So it is clear that $\sqrt{-\bar{g}}d^4\bar{x} = \sqrt{-g}d^4x$ as desired.

If you have a Lagrangian that is itself a scalar, like the one we have been thinking about, $\psi_{,\mu}g^{\mu\nu}\psi_{,\nu}/2$, then to make a scalar action, we can use $\sqrt{-g}d^4x$,

$$S[\psi] = \int \left(\frac{1}{2}\psi_{,\mu}g^{\mu\nu}\psi_{,\nu} \right) \sqrt{-g}\, d^4x. \tag{7.29}$$

There are three pieces to the action here that we can refer to separately. First, define the scalar $\mathcal{M} \equiv \psi_{,\mu}g^{\mu\nu}\psi_{,\nu}/2$; then the integrand $\mathcal{L} \equiv \mathcal{M}\sqrt{-g}$ is a scalar density of weight -1 to counter the final piece d^4x. The Lagrangian $\mathcal{L}$ is called a "Lagrangian density" to remind us of its transformation behavior. The previous section benefited from the simplifying Cartesian $\sqrt{-g} = 1$, so that absent an interest in coordinate transformation, we had $\mathcal{M} = \mathcal{L}$ there.

Let's see how the inclusion of $\sqrt{-g}$ helps make the resulting field equation truly tensorial by restoring the correct $\partial_\mu \to D_\mu$ structure. The main observation is that if you have a tensor A^μ, then

$$\sqrt{-g}A^\mu_{;\mu} = \left(\sqrt{-g}A^\mu \right)_{;\mu} = \left(\sqrt{-g}A^\mu \right)_{,\mu}. \tag{7.30}$$

Proof: It is not hard to imagine that $g_{;\mu} = 0$, since by the chain rule we have

$$g_{;\mu} = \frac{\partial g}{\partial g_{\alpha\beta}}g_{\alpha\beta;\mu} = 0, \tag{7.31}$$

using (2.138), so the first equality in (7.30) is clear. For the second equality, you will show in Problem 7.5 that $\frac{\partial g}{\partial g_{\alpha\beta}} = gg^{\beta\alpha}$, and then

$$g_{,\mu} = \frac{\partial g}{\partial g_{\alpha\beta}}g_{\alpha\beta,\mu} = gg^{\beta\alpha}\left(\Gamma^\rho_{\alpha\mu}g_{\rho\beta} + \Gamma^\rho_{\beta\mu}g_{\alpha\rho} \right) = 2g\Gamma^\rho_{\rho\mu}. \tag{7.32}$$

Using the product rule,

$$\begin{aligned} \left(\sqrt{-g}A^\mu \right)_{,\mu} &= \sqrt{-g}_{,\mu}A^\mu + \sqrt{-g}A^\mu_{,\mu} = \sqrt{-g}\left(\Gamma^\rho_{\rho\mu}A^\mu + A^\mu_{,\mu} \right) \\ &= \sqrt{-g}A^\mu_{;\mu}, \end{aligned} \tag{7.33}$$

as desired.

Let's go back to the action extremization expression we had in (7.6), where δS was obtained using Taylor expansion. With our updated action of the general form

$$S[\psi] = \int_\Omega \mathcal{M}(\psi, \partial\psi)\sqrt{-g}\, d^4x, \qquad (7.34)$$

taking $\psi \to \psi + \delta\psi$, we have

$$\delta S = \int_\Omega \left(\frac{\partial \mathcal{M}}{\partial \psi}\delta\psi + \frac{\partial \mathcal{M}}{\partial \psi_{,\mu}}\delta\psi_{,\mu}\right)\sqrt{-g}\, d^4x. \qquad (7.35)$$

Now for the integration by parts we did exclusively in the Cartesian coordinate system before, reproducing and expanding (7.7) and (7.8), we have

$$\int_\Omega \frac{\partial}{\partial x^\mu}\left(\frac{\partial \mathcal{M}}{\partial \psi_{,\mu}}\delta\psi\sqrt{-g}\right)d^4x = \oint_{\partial\Omega}\left(\frac{\partial \mathcal{L}}{\partial \psi_{,\mu}}\delta\psi\sqrt{-g}\right)da_\mu = 0, \quad (7.36)$$

and, using the product rule,

$$\begin{aligned}
&\int_\Omega \frac{\partial}{\partial x^\mu}\left(\frac{\partial \mathcal{M}}{\partial \psi_{,\mu}}\delta\psi\sqrt{-g}\right)d^4x \\
&= \int_\Omega \left(\delta\psi\frac{\partial}{\partial x^\mu}\left(\frac{\partial \mathcal{M}}{\partial \psi_{,\mu}}\sqrt{-g}\right) + \frac{\partial \mathcal{M}}{\partial \psi_{,\mu}}\delta\psi_{,\mu}\sqrt{-g}\right)d^4x.
\end{aligned} \qquad (7.37)$$

Combining these two, we get the update for (7.9):

$$\int_\Omega \frac{\partial \mathcal{M}}{\partial \psi_{,\mu}}\delta\psi_{,\mu}\sqrt{-g}\, d^4x = -\int_\Omega \delta\psi\frac{\partial}{\partial x^\mu}\left(\frac{\partial \mathcal{M}}{\partial \psi_{,\mu}}\sqrt{-g}\right)d^4x. \qquad (7.38)$$

Then the response of the action, written entirely in terms of $\delta\psi$, is

$$\delta S = \int_\Omega \left(\frac{\partial \mathcal{M}}{\partial \psi}\sqrt{-g} - \frac{\partial}{\partial x^\mu}\left(\frac{\partial \mathcal{M}}{\partial \psi_{,\mu}}\sqrt{-g}\right)\right)\delta\psi\, d^4x, \qquad (7.39)$$

and for this to vanish for all $\delta\psi$, we require the term in parentheses in the integrand to vanish as before. But now that term looks like

$$\frac{\partial \mathcal{M}}{\partial \psi}\sqrt{-g} - \left(\frac{\partial \mathcal{M}}{\partial \psi_{,\mu}}\sqrt{-g}\right)_{,\mu} = 0. \qquad (7.40)$$

Using relation (7.33), we can replace the comma derivative with a semicolon one, use the product rule to apply the covariant derivative to each term, noting that $\sqrt{-g}_{;\mu} = 0$, and cancel the common factor of $\sqrt{-g}$ to get the manifestly scalar field equation

$$\frac{\partial \mathcal{M}}{\partial \psi} - \left(\frac{\partial \mathcal{M}}{\partial \psi_{,\mu}}\right)_{;\mu} = 0, \qquad (7.41)$$

the "Euler–Lagrange" equation for the action in (7.34). Try Problem 7.6 to convince yourself that (7.41) gives you the correct field equation in spherical coordinates.

7.3 Field Stress Tensor

It should be clear from the structure of the action integrand, $\mathcal{M}\sqrt{-g}$ for scalar $\mathcal{M}$, that *any* field theory with an action description depends on the underlying metric (at the very least through $\sqrt{-g}$, but usually $\mathcal{M}$ involves factors of the metric and its inverse to make scalars out of the field and its derivatives). So far, we have used the $\mathcal{M}\sqrt{-g}$ structure only to correctly describe the field equation in different coordinate systems that still refer to the same, Minkowski, spacetime. But the formalism is more general: *any* metric will do. We could use the Schwarzschild metric and associated connection in (7.34), and we'd get the field equation governing a scalar field in the Schwarzschild geometry from (7.41). That *universal* coupling of field theories to the metric cuts both ways. The metric determines the details of the field equation, but the metric dependence in the action also couples the field theory to the gravitational field as a source.

Let's see how the scalar nature of the action leads to a conserved, second-rank tensor that automatically provides a source for gravity. Since the action is a scalar, it is unchanged by coordinate transformation: If we take $x^\mu \to \bar{x}^\mu$, then there is no change in S (which is just a number). Our plan is to look at the infinitesimal response of the action to a coordinate transformation. That response must be zero, but Taylor expansion will provide a concrete expression of that zero. The general framework and approach here provide an example of Noether's theorem, and we will discuss that connection at the end of the story.

Start with the equality of the action in the original coordinate system x and the new one:

$$S = \int_\Omega \mathcal{L}\big(\psi(x), \partial\psi(x), g(x)\big)\, d^4x = \int_{\bar\Omega} \bar{\mathcal{L}}\big(\bar\psi(\bar x), \bar\partial\bar\psi(\bar x), \bar g(\bar x)\big)\, d^4\bar x,$$
(7.42)

where $\partial\psi$ is shorthand for $\frac{\partial\psi(x)}{\partial x^\mu}$, $g(x)$ for the metric, and where the bars represent the various transformed expressions so that, for example, $\bar\partial\bar\psi$ refers to $\frac{\partial\bar\psi(\bar x)}{\partial\bar x^\mu}$.

On the right, we will reexpress the integral in terms of the original set of coordinates by transforming *just* the coordinates (not the field, its derivatives, or the metric, yet). There will be factors of the determinant of the coordinate transformation that come from the densities $d^4\bar x$ and $\bar{\mathcal{L}}$, but those will cancel out. So we have

$$S = \int \bar{\mathcal{L}}\big(\bar\psi(\bar x), \bar\partial\bar\psi(\bar x), \bar g(\bar x)\big)\, d^4\bar x = \int \mathcal{L}\big(\bar\psi(x), \bar\partial\bar\psi(x), \bar g(x)\big)\, d^4x, \quad (7.43)$$

where the bars over $\mathcal{L}$ and d^4x have been removed because the product $\mathcal{L}d^4x$ is a scalar. Since the barred variables were dummy integration variables from the start, the "old" set works just as well. The integrand on the right tells us to take the original $\mathcal{L}$, insert the transformed field, derivative and metric with the "old" coordinates as arguments in its function slots, and integrate.

To use the expression on the right in (7.43), we need to work out the response of the barred field variables and the metric, when written in terms of the original coordinates. We'll work with an infinitesimal form of the transformation. Let $\bar{x}^\alpha = x^\alpha + \epsilon f^\alpha(x)$, where f^α is a vector function of the "old" coordinates, and the dimensionless constant ϵ is much less than one. Then if we write $\psi(x) = \bar{\psi}(\bar{x}(x))$ to express the scalar nature of the transformation for ψ,[3]

$$\psi(x) = \bar{\psi}(\bar{x}) = \bar{\psi}(x + \epsilon f) \approx \bar{\psi}(x) + \epsilon\big(\partial_\alpha \bar{\psi}(x)\big) f^\alpha(x). \tag{7.44}$$

So we can think of the coordinate transformation as inducing a change in the field configuration for $\psi(x)$, writing the above as

$$\bar{\psi}(x) \approx \psi(x) - \epsilon f^\alpha(x)\partial_\alpha \bar{\psi}(x) \tag{7.45}$$

or defining the infinitesimal $\delta\psi(x) \equiv -\epsilon f^\alpha(x)\partial_\alpha \bar{\psi}(x)$, the coordinate transformation lets us think of $\bar{\psi}(x)$ as $\psi(x) + \delta\psi(x)$, where $\delta\psi(x)$ is just an arbitrary variation of the field $\psi(x)$ that vanishes on the boundary of the domain. In practice, we can accomplish that boundary requirement by taking $f^\alpha(x)$ to be zero on the boundary.

The derivative of $\bar{\psi}$ can also be expressed in this type of general variational framework. Let $A_\mu \equiv \psi_{,\mu}$. Then as a first-rank covariant tensor, we have

$$\bar{A}_\alpha(\bar{x}) = \frac{\partial x^\mu}{\partial \bar{x}^\alpha} A_\mu\big(x(\bar{x})\big). \tag{7.46}$$

We can multiply both sides of this equation by $\frac{\partial \bar{x}^\alpha}{\partial x^\rho}$ to invert the relation,

$$\frac{\partial \bar{x}^\alpha}{\partial x^\rho} \bar{A}_\alpha(\bar{x}) = A_\rho(x). \tag{7.47}$$

The preferential isolation of $A_\rho(x)$ comes because it puts $\frac{\partial \bar{x}^\alpha}{\partial x^\rho}$ on the left, and that quotient can be calculated without inversion, since both x^α and $f^\alpha(x)$ depend on the original coordinates. Writing out the pieces of (7.47), we have

$$A_\rho(x) = \big(\delta_\rho^\alpha + \epsilon f^\alpha{}_{,\rho}(x)\big)\bar{A}_\alpha(x + \epsilon f)$$
$$\approx \big(\delta_\rho^\alpha + \epsilon f^\alpha{}_{,\rho}(x)\big)\left(\bar{A}_\alpha(x) + \epsilon f^\mu(x)\frac{\partial \bar{A}_\alpha(x)}{\partial x^\mu}\right). \tag{7.48}$$

Focusing on $\bar{A}_\rho(x)$, we can invert (7.48) and keep terms at most linear in ϵ,

$$\bar{A}_\rho(x) = A_\rho(x) - \epsilon\big[f^\alpha{}_{,\rho}(x)\bar{A}_\alpha(x) + f^\mu(x)\partial_\mu \bar{A}_\rho(x)\big]. \tag{7.49}$$

On the left, we have $\bar{A}_\rho(x) \equiv \frac{\partial \bar{\psi}(x)}{\partial \bar{x}^\rho}$, the desired quantity. But on the right, inside the brackets, we have both $\bar{A}_\alpha(x)$ and $\partial_\mu \bar{A}_\rho(x)$, which involve barred-variable derivatives, and we'd like to convert those to unbarred derivatives (to

---

[3] A note on the notation below – when we write $\bar{\psi}(\bar{x})$, we mean: "take $\psi(x)$ and express the x as functions of the $\bar{x}$, $x(\bar{x})$, then put that into $\psi(x)$: $\bar{\psi}(\bar{x}) = \psi(x(\bar{x}))$." When we write $\bar{\psi}(x)$, we mean "take $\bar{\psi}(\bar{x})$ and simply send $\bar{x} \to x$."

isolate $\partial_\alpha \delta \psi$). Note that

$$\epsilon \bar{A}_\alpha(x) = \epsilon \frac{\partial \bar{\psi}(x)}{\partial \bar{x}^\alpha} = \epsilon \frac{\partial \bar{\psi}(x)}{\partial x^\rho} \frac{\partial x^\rho}{\partial \bar{x}^\alpha} \tag{7.50}$$

using the chain rule. Now while it is easy to evaluate derivatives of the barred variables in terms of the unbarred ones, since $\bar{x}^\rho = x^\rho + \epsilon f^\rho(x)$, the other direction is harder in general (which is why we have avoided it!). However, from $x^\rho = \bar{x}^\rho - \epsilon f^\rho(x)$ it is clear that the derivative of interest is $\frac{\partial x^\rho}{\partial \bar{x}^\alpha} = \delta^\rho_\alpha + O(\epsilon)$, so that, noting the ϵ outside the expressions in (7.49) (and explicitly included in (7.50)), to first order in ϵ,

$$\epsilon \bar{A}_\alpha(x) = \epsilon \frac{\partial \bar{\psi}(x)}{\partial x^\rho} \delta^\rho_\alpha + O(\epsilon^2) \approx \epsilon \frac{\partial \bar{\psi}(x)}{\partial x^\alpha}, \tag{7.51}$$

allowing us to write (7.49) as

$$\bar{A}_\rho(x) = \partial_\rho \psi(x) - \epsilon \left[f^\alpha_{,\rho}(x) \partial_\alpha \bar{\psi}(x) + f^\mu(x) \partial_\mu \partial_\rho \bar{\psi}(x) \right]. \tag{7.52}$$

Now we finally have

$$\frac{\partial \bar{\psi}(x)}{\partial \bar{x}^\rho} = \partial_\rho \psi(x) - \partial_\rho \epsilon \left[f^\alpha(x) \partial_\alpha \bar{\psi}(x) \right] = \partial_\rho \psi(x) + \partial_\rho \delta \psi(x), \tag{7.53}$$

using $\delta \psi(x) = -\epsilon f^\alpha(x) \partial_\alpha \bar{\psi}(x)$. Equation (7.53) tells us that $\bar{\partial}_\rho \bar{\psi}(x) = \partial_\rho \psi(x) + \partial_\rho \delta \psi(x)$. We could have guessed that the change in ψ induced by the change of variables flows directly to the change in $\psi_{,\rho}$, but it is reassuring (if tedious) to verify the result directly. The coordinate transformation induces a specific field variation $\delta \psi$ with associated variation of the derivative $\delta \psi_{,\mu}$ as in Section 7.1, which is good, since we will be able to use the fact that ψ extremizes the action when it satisfies its field equation to focus on the remaining term, the transformation associated with the metric.

Finally, the metric; how does it respond to the coordinate transformation? We start with the covariant transformation rule

$$\bar{g}_{\mu\nu}(\bar{x}) = \frac{\partial x^\rho}{\partial \bar{x}^\mu} \frac{\partial x^\sigma}{\partial \bar{x}^\nu} g_{\rho\sigma}\left(x(\bar{x}) \right) \tag{7.54}$$

and, as with the vector transformation above, try to isolate the original metric. This time, we multiply both sides of the equation by $\frac{\partial \bar{x}^\mu}{\partial x^\alpha} \frac{\partial \bar{x}^\nu}{\partial x^\beta}$ to get

$$g_{\alpha\beta}(x) = \frac{\partial \bar{x}^\mu}{\partial x^\alpha} \frac{\partial \bar{x}^\nu}{\partial x^\beta} \bar{g}_{\mu\nu}(x + \epsilon f). \tag{7.55}$$

Using the infinitesimal transformation and expanding the argument of $\bar{g}_{\mu\nu}(x + \epsilon f)$, we have

$$\bar{g}_{\alpha\beta}(x) \approx g_{\alpha\beta}(x) - \epsilon \left(f^\mu_{,\alpha}(x) \bar{g}_{\mu\beta}(x) + f^\nu_{,\beta} \bar{g}_{\alpha\nu}(x) + \frac{\partial \bar{g}_{\alpha\beta}(x)}{\partial x^\rho} f^\rho(x) \right) \tag{7.56}$$

to first order in ϵ. Since, in any coordinate system, $g_{\mu\nu;\alpha} = 0$, we can relate the metric derivative to its connection,

$$\epsilon \frac{\partial \bar{g}_{\alpha\beta}}{\partial \bar{x}^\rho} = \epsilon \left(\bar{\Gamma}^\sigma_{\alpha\rho} \bar{g}_{\sigma\beta} + \bar{\Gamma}^\sigma_{\beta\rho} \bar{g}_{\alpha\sigma} \right), \tag{7.57}$$

and to first order in ϵ, we have $\epsilon \frac{\partial \bar{g}_{\alpha\beta}}{\partial \bar{x}^\rho} = \epsilon \frac{\partial \bar{g}_{\alpha\beta}}{\partial x^\rho}$ (once again, the barred derivatives become unbarred ones to order ϵ, as in (7.51)). We can replace the $\partial_\rho \bar{g}_{\alpha\sigma}$ term in (7.56) with the right-hand side of (7.57), and the two connection terms can be combined with the partial derivatives of f^μ in (7.56) to turn them into covariant derivatives (try it out in Problem 7.7):

$$\bar{g}_{\alpha\beta}(x) \approx g_{\alpha\beta}(x) - \epsilon \big(f_{\alpha;\beta}(x) + f_{\beta;\alpha}(x) \big). \tag{7.58}$$

We now have an expression for the perturbation induced in the metric, $\bar{g}_{\alpha\beta}(x) \approx g_{\alpha\beta}(x) + \delta g_{\alpha\beta}(x)$ with

$$\delta g_{\alpha\beta} \equiv -\epsilon(f_{\alpha;\beta} + f_{\beta;\alpha}). \tag{7.59}$$

The result of the coordinate transformation for the metric, with both the original and transformed versions written in the same (original) coordinate system for comparison, is that

$$g_{\alpha\beta}(x) - \bar{g}_{\alpha\beta}(x) = \epsilon(f_{\alpha;\beta} + f_{\beta;\alpha}) \tag{7.60}$$

for infinitesimal transformations with arbitrary f_α. We have experience adding symmetrized vector derivatives to metrics, which was the gauge freedom from Section 2.8.2 or the linearized version in Section 3.6.1. Here we learn that the gauge freedom of general relativity is really a coordinate choice. The field equations are insensitive to these lower-rank contributions, which is fitting for field equations that are generally covariant, the same in any coordinate system.

Finally, it is time to return to (7.43) to see the effect of all of this. Replacing the barred terms with their expression as variations of the field, its derivative, and the metric, we have

$$\begin{aligned}
S &= \int_\Omega \mathcal{L}\big(\psi + \delta\psi, \partial\psi + \delta(\partial\psi), g + \delta g\big)\, d^4x \\
&\approx \underbrace{\int_\Omega \mathcal{L}(\psi, \partial\psi, g)\, d^4x}_{=S} + \underbrace{\int_\Omega \left(\frac{\partial\mathcal{L}}{\partial\psi}\delta\psi + \frac{\partial\mathcal{L}}{\partial\psi_{,\mu}}\delta\psi_{,\mu} \right) d^4x}_{=0} \\
&\quad + \underbrace{\int_\Omega \frac{\partial\mathcal{L}}{\partial g_{\mu\nu}}\delta g_{\mu\nu}\, d^4x}_{\equiv \delta S}.
\end{aligned} \tag{7.61}$$

The middle term is zero since its integrand is just the field equation for the scalar ψ, which extremizes the action for all $\delta\psi$, including that induced by the coordinate transformation. The first term is, of course, S itself, from (7.42). Then the "new" extra piece δS must be zero for all f_α.

Putting in the expression for $\delta g_{\mu\nu}$ from (7.59) and using the symmetry of the metric to combine the two terms,

$$\delta S = \epsilon \int_\Omega \left(-2 \frac{\partial\mathcal{L}}{\partial g_{\mu\nu}} \right) f_{\mu;\nu}\, d^4x. \tag{7.62}$$

The derivative of the Lagrangian is a second-rank contravariant tensor density of weight -1, so we can isolate the second-rank tensor piece of it by defining

(with signs and factors motivated by the term in parentheses above)

$$\sqrt{-g}\,T^{\mu\nu} \equiv -2\frac{\partial\mathcal{L}}{\partial g_{\mu\nu}}. \tag{7.63}$$

Using this definition in (7.62),

$$\delta S = \epsilon \int_{\Omega} \sqrt{-g}\,T^{\mu\nu} f_{\mu;\nu}\, d^4x. \tag{7.64}$$

We can use the divergence theorem and identity in (7.30) to generate an integration-by-parts relation that will flip the covariant derivative from f_μ onto $T^{\mu\nu}$. Start with the divergence theorem

$$\begin{aligned}
\int_{\Omega} \left(\sqrt{-g}\,T^{\mu\nu} f_\mu\right)_{;\nu} d^4x &= \int_{\Omega} \left(\sqrt{-g}\,T^{\mu\nu} f_\mu\right)_{,\nu} d^4x \\
&= \oint_{\partial\Omega} \sqrt{-g}\,T^{\mu\nu} f_\mu\, da_\nu = 0,
\end{aligned} \tag{7.65}$$

where we assume that f_μ vanishes on the boundary of Ω. Now using the product rule to expand the integrand on the left and noting that $\sqrt{-g}_{;\nu} = 0$ as usual, we have

$$0 = \int_{\Omega} \left(\sqrt{-g}\,T^{\mu\nu} f_\mu\right)_{;\nu} d^4x = \int_{\Omega} \sqrt{-g}\,T^{\mu\nu}_{\;\;;\nu} f_\mu\, d^4x + \int_{\Omega} \sqrt{-g}\,T^{\mu\nu} f_{\mu;\nu}\, d^4x, \tag{7.66}$$

so that, finally,

$$\delta S = -\epsilon \int_{\Omega} \sqrt{-g}\,T^{\mu\nu}_{\;\;;\nu} f_\mu\, d^4x. \tag{7.67}$$

For this to be zero for all f_μ, we must have

$$T^{\mu\nu}_{\;\;;\nu} = 0. \tag{7.68}$$

Phew! That was a lot of work, but the payoff is immense. Associated with *every* field theory that depends on the metric (which is all of them), there is a covariantly conserved symmetric tensor $T^{\mu\nu}$, just right as a source for general relativity. Noether's theorem says that the symmetry of an action leads to a conservation law. Here the symmetry of the action is the coordinate invariance, we can express the action in any coordinate system we wish. The associated conservation law is the field energy and momentum conservation encapsulated in (7.68). The work we have done here establishes Noether's theorem in this gravitational context (although the idea itself is much more broadly applicable).

From its definition we can write out the elements of this so-called "energy momentum tensor" or "stress tensor" for fields. Since $\mathcal{L} = \mathcal{M}\sqrt{-g}$, with metric dependence in both $\mathcal{M}$, the scalar piece of the Lagrangian density (which is what we are typically given), and $\sqrt{-g}$, we have

$$\boxed{T^{\mu\nu} = -\frac{2}{\sqrt{-g}}\left(\sqrt{-g}\,\frac{\partial\mathcal{M}}{\partial g_{\mu\nu}} + \mathcal{M}\frac{\partial\sqrt{-g}}{\partial g_{\mu\nu}}\right) = -\left(g^{\mu\nu}\mathcal{M} + 2\frac{\partial\mathcal{M}}{\partial g_{\mu\nu}}\right).}$$

$$\tag{7.69}$$

What would happen if a Lagrangian depended on the derivative of the metric? Why would we not expect that to happen, given what we know about general relativity?

Example 7.2 (Action Scalar). The series of approximations and substitutions made above is involved, especially when we begin to work implicitly using expansion and dropping terms of order greater than ϵ. Let's do a concrete, simple example to see how the procedure goes in practice. To keep the calculation tractable for as long as possible, we'll work in $1 + 1$ dimensions with $u \equiv ct$ for the temporal dimension and x as the spatial (the metric will take its usual Minkowski form at first). Our Lagrangian will be

$$\mathcal{L}(\psi, \partial \psi, g) = \sqrt{-g}\left(\psi_{,\mu} g^{\mu\nu} \psi_{,\nu}\right) \tag{7.70}$$

with field equation $-\frac{\partial^2 \psi}{\partial u^2} + \frac{\partial^2 \psi}{\partial x^2} = 0$. We'll work on a rectangular domain Ω with $u \in [0, U]$ and $x \in [0, X]$ for constants U and X.

As our coordinate transformation, take

$$\bar{u} = u, \qquad \bar{x} = x + \epsilon u(u - U)x(x - X)/\ell^3, \tag{7.71}$$

where ℓ is a constant with the dimension of length. This coordinate relation can be inverted exactly, as you will do in Problem 7.8, where you will also find the metric and verify the approximation in (7.58). Notice that we have chosen the transformation so that it vanishes along the boundary, ensuring that the induced scalar perturbation does as well.

For the scalar ψ that satisfies the field equation, take $\psi = \alpha u x$ for constant α. Calculating the action integral, we get

$$\begin{aligned} S[\psi] &= \int \mathcal{L}\left(\psi(x), \partial \psi(x), g(x)\right) d^2 x \\ &= \int_0^X \int_0^U \alpha^2 \left(u^2 - x^2\right) du\, dx = \frac{\alpha^2}{3} U X \left(U^2 - X^2\right). \end{aligned} \tag{7.72}$$

If we instead evaluate the Lagrangian using the transformed functions $\bar{\psi}(x)$, $\bar{\partial}\bar{\psi}(x)$, and $\bar{g}_{\mu\nu}(x)$, with the original coordinates as in (7.43), where we have used the exact transformation to generate the barred functions, then we get, expanding to first order in ϵ,

$$\begin{aligned} S[\bar{\psi}] &= \int \mathcal{L}\left(\bar{\psi}(x), \bar{\partial}\bar{\psi}(x), \bar{g}(x)\right) d^2 x \\ &= \int_0^X \int_0^U \Big[\alpha^2 \left(u^2 - x^2\right) \\ &\quad - \epsilon \frac{\alpha^2}{\ell^3} u(U - u)\left(u^2(X - 2x) + x^2(-3X + 4x)\right)\Big] du\, dx. \end{aligned} \tag{7.73}$$

It is clear that the first term here recovers all of S, since it's the same integral that we did in (7.72), which means that the second term, linear in ϵ, had better be zero. That second term is the first-order term δS from (7.61), and here we can see explicitly that it vanishes by carrying out the integration.

7.3.1 Stress Tensor for Scalar Fields

Let's work out the stress tensor for a scalar field, starting from its Lagrangian. Then we can check that the resulting tensor is covariantly conserved if ψ satisfies its field equation. The scalar piece of the Lagrangian density governing the "massive Klein–Gordon" field is

$$\mathcal{M} = \frac{1}{2}\psi_{,\alpha}g^{\alpha\beta}\psi_{,\beta} + \frac{1}{2}\mu^2\psi^2. \tag{7.74}$$

Then the field equation from (7.41) is

$$\mu^2\psi - D^\alpha D_\alpha\psi = 0. \tag{7.75}$$

Using (7.69), the stress tensor associated with this $\mathcal{M}$ is

$$T^{\mu\nu} = -\frac{1}{2}g^{\mu\nu}\left(\psi_{,\alpha}g^{\alpha\beta}\psi_{,\beta} + \mu^2\psi^2\right) - \psi_{,\alpha}\frac{\partial g^{\alpha\beta}}{\partial g_{\mu\nu}}\psi_{,\beta}. \tag{7.76}$$

It's worth taking time to compute the derivative of the metric inverse components with respect to the metric, since this type of calculation comes up a lot.

Start with the defining equation for the contravariant metric,

$$g^{\alpha\rho}g_{\rho\sigma} = \delta^\alpha_\sigma. \tag{7.77}$$

The advantage is that we know the value of δ^α_σ is independent of the metric components, since the Kronecker delta is either one or zero. So taking the derivative $\frac{\partial}{\partial g_{\mu\nu}}$ of both sides, we get zero on the right and can use the product rule on the left,

$$\frac{\partial g^{\alpha\rho}}{\partial g_{\mu\nu}}g_{\rho\sigma} + g^{\alpha\rho}\frac{\partial g_{\rho\sigma}}{\partial g_{\mu\nu}} = 0. \tag{7.78}$$

The partial derivative of the metric components with respect to themselves (ignoring the symmetry, to remain general) is $\frac{\partial g_{\rho\sigma}}{\partial g_{\mu\nu}} = \delta^\mu_\rho\delta^\nu_\sigma$. Using this result and multiplying (7.78) by $g^{\sigma\beta}$, we learn that

$$\frac{\partial g^{\alpha\beta}}{\partial g_{\mu\nu}} = -g^{\alpha\mu}g^{\nu\beta}. \tag{7.79}$$

Then the stress tensor expression from (7.76) is

$$T^{\mu\nu} = \psi_{,}{}^\mu\psi_{,}{}^\nu - \frac{1}{2}g^{\mu\nu}\left(\psi_{,\alpha}g^{\alpha\beta}\psi_{,\beta} + \mu^2\psi^2\right). \tag{7.80}$$

What happens if we take the (covariant) divergence of the stress tensor? From its derivation, we must get zero; let's see how it works out:

$$
\begin{aligned}
T^{\mu\nu}{}_{;\nu} \equiv D_\nu T^{\mu\nu} &= D^\nu\psi D_\nu D^\mu\psi + D^\mu\psi D_\nu D^\nu\psi \\
&\quad - \frac{1}{2}g^{\mu\nu}\left(2D^\alpha\psi D_\alpha D_\nu\psi + 2\mu^2\psi D_\nu\psi\right),
\end{aligned}
\tag{7.81}
$$

where we have used $\psi_{,\mu} = \psi_{;\mu} = D_\mu\psi$ to write the first derivatives of ψ. In addition, while we apply the covariant derivative using the product rule, any time it hits a metric (or metric inverse), we get zero since the covariant derivative of the metric is zero: $D_\alpha g_{\mu\nu} = 0$.

In the derivation of the conservation of the stress tensor, we required that the field ψ satisfy its field equation; otherwise the $\delta\psi$ piece of (7.61) wouldn't go away. So we need to use (7.75) to make progress evaluating the divergence of $T^{\mu\nu}$. Replacing $D_\nu D^\nu \psi$ with $\mu^2 \psi$ and expanding the expressions in (7.81),

$$D_\nu T^{\mu\nu} = D^\nu \psi D_\nu D^\mu \psi + \mu^2 \psi D^\mu \psi - D^\alpha \psi D_\alpha D^\mu \psi - \mu^2 \psi D^\mu \psi = 0, \tag{7.82}$$

the divergence of the stress tensor is, indeed, zero.

We have our first, field-theoretic, stress tensor source for use in Einstein's equation:

$$R^{\mu\nu} - \frac{1}{2} g^{\mu\nu} R = \frac{8\pi G}{c^4} \left(\psi^{,\mu} \psi^{,\nu} - \frac{1}{2} g^{\mu\nu} \left(\psi_{,\alpha} g^{\alpha\beta} \psi_{,\beta} + \mu^2 \psi^2 \right) \right). \tag{7.83}$$

This equation is interesting because it involves the unknown metric $g^{\mu\nu}$ on both the "derivative," left-hand side, and the "source," right-hand side. There is an apparent chicken-and-egg problem, where we cannot figure out the source without knowing the metric, and we cannot find the metric without knowing the source. The situation gets worse when we remember that the field equation governing ψ also depends on the metric, since the differential operator $D_\mu D^\mu$ has both metric and connection dependence hidden inside of it.

What we really need to do is solve the simultaneous nonlinear partial differential equations:

$$D^\alpha D_\alpha \psi = \mu^2 \psi,$$
$$R^{\mu\nu} - \frac{1}{2} g^{\mu\nu} R = \frac{8\pi G}{c^4} \left(\psi^{,\mu} \psi^{,\nu} - \frac{1}{2} g^{\mu\nu} \left(\psi_{,\alpha} g^{\alpha\beta} \psi_{,\beta} + \mu^2 \psi^2 \right) \right). \tag{7.84}$$

Stating things this way avoids the seemingly circular nature of the problem, without providing any insight into *how* we make progress in constructing that solution. We will return to this issue later on; it is a hallmark of the theory.

7.4 Massless Vector Field Theory

The action approach will work with fields of higher rank. Here we will consider a "massless vector field," of which the four-potential of Maxwell's electricity and magnetism is an example. Because it is the primary one of interest, we will use the terms "vector fields" and "electromagnetic fields" interchangeably as a reminder that there is a concrete, familiar physical theory here, but the same ideas apply to any other vector field theory. Note that the "massless" designation means that there is no term in the field equation that is proportional to the field itself (a technical point that comes from quantum mechanics; it is not important here).

For a vector field A_α, the action is

$$S[A] = \int_\Omega \mathcal{M}(A, \partial A)\sqrt{-g}\, d^4x, \qquad (7.85)$$

similar to (7.34), but for vector argument now. As before, $\mathcal{L} = \mathcal{M}\sqrt{-g}$ is the Lagrangian density. The field equation can be developed from action minimization, as in Section 7.2.1, but it is easy to update (7.41) to account for the additional index,

$$\frac{\partial \mathcal{M}}{\partial A_\alpha} - \left(\frac{\partial \mathcal{M}}{\partial A_{\alpha,\mu}}\right)_{;\mu} = 0. \qquad (7.86)$$

Our target field equation is the one from (2.184), and we'll start by thinking about the vacuum field equation with $J_\mu = 0$. There is no A_α term appearing on the left in (2.184), so we expect the Lagrangian to depend only on the derivative of the field. As you will verify in Problem 7.11, the scalar portion of the Lagrangian that does the job is

$$\mathcal{M} = \frac{1}{4}\beta(D_\gamma A_\nu - D_\nu A_\gamma)g^{\gamma\rho}g^{\nu\sigma}(D_\rho A_\sigma - D_\sigma A_\rho). \qquad (7.87)$$

The $D_\gamma A_\nu - D_\nu A_\gamma$ construction really comes from requiring that the vector theory make no reference to vector-disguised scalars, like the gradient $\psi_{,\alpha}$. But it is interesting that this same structure eliminates the dependence on the connection, as you showed in Problem 2.34. So we can write $\mathcal{M}$ equivalently as

$$\mathcal{M} = \frac{1}{4}\beta(\partial_\gamma A_\nu - \partial_\nu A_\gamma)g^{\gamma\rho}g^{\nu\sigma}(\partial_\rho A_\sigma - \partial_\sigma A_\rho). \qquad (7.88)$$

Define the "field strength tensor"[4] to be $F_{\gamma\nu} \equiv \partial_\gamma A_\nu - \partial_\nu A_\gamma$; then using the value of β from Problem 7.11, the scalar portion of the Lagrangian density is

$$\mathcal{M} = \frac{1}{4\mu_0}F_{\gamma\tau}F_{\rho\sigma}g^{\gamma\rho}g^{\tau\sigma}. \qquad (7.89)$$

The vector field (electromagnetic) stress tensor is obtained from (7.69) by noting the identity in (7.79),

$$T^{\mu\nu} = \frac{1}{\mu_0}\left[F^{\mu\sigma}F^\nu_{\ \sigma} - \frac{1}{4}g^{\mu\nu}F^{\rho\sigma}F_{\rho\sigma}\right]. \qquad (7.90)$$

One special property of this stress tensor is its trace in D dimensions,

$$T \equiv T^{\mu\nu}g_{\mu\nu} = \frac{1}{\mu_0}F^{\mu\sigma}F_{\mu\sigma}\left(1 - \frac{D}{4}\right). \qquad (7.91)$$

In four-dimensional spacetime, then, the trace is zero.[5]

---

[4] The vector field here is A_μ; the field strength tensor combination of derivatives of A_μ is just a convenient shorthand.

[5] This is one reason why you cannot formulate a scalar theory of gravity that is sourced by the trace of a stress tensor, since that theory cannot couple to E&M.

Let's work through the components of the stress tensor in a flat Minkowski background with Cartesian coordinates. First, we'll find the contravariant components of the field strength tensor. For the field A_μ, we interpret its contravariant components in the usual way,

$$A^\mu \doteq \begin{pmatrix} \frac{V}{c} \\ \mathbf{A} \end{pmatrix}. \tag{7.92}$$

Then the field strength tensor can be written in terms of $\mathbf{E}$ and $\mathbf{B}$ from $\mathbf{E} = -\nabla V - \frac{\partial \mathbf{A}}{\partial t}$ and $\mathbf{B} = \nabla \times \mathbf{A}$:

$$F_{\mu\nu} \doteq \begin{pmatrix} 0 & -\frac{E^1}{c} & -\frac{E^2}{c} & -\frac{E^3}{c} \\ \frac{E^1}{c} & 0 & B^3 & -B^2 \\ \frac{E^2}{c} & -B^3 & 0 & B^1 \\ \frac{E^3}{c} & B^2 & -B^1 & 0 \end{pmatrix},$$

$$F^{\mu\nu} \doteq \begin{pmatrix} 0 & \frac{E^1}{c} & \frac{E^2}{c} & \frac{E^3}{c} \\ -\frac{E^1}{c} & 0 & B^3 & -B^2 \\ -\frac{E^2}{c} & -B^3 & 0 & B^1 \\ -\frac{E^3}{c} & B^2 & -B^1 & 0 \end{pmatrix}. \tag{7.93}$$

From these and $F^\mu{}_\nu$, which you will compute in Problem 7.12, the stress tensor has the components

$$T^{00} = \frac{1}{2}\left(\frac{1}{\mu_0}B^2 + \epsilon_0 E^2\right),$$

$$T^{0i} = \frac{1}{\mu_0 c}(\mathbf{E} \times \mathbf{B})^i, \tag{7.94}$$

$$T^{ij} = -\frac{1}{\mu_0}B^i B^j - \epsilon_0 E^i E^j + \eta^{ij} T^{00},$$

with η^{ij} the spatial piece of the (inverse) metric.

These components should be familiar to you from your work in E&M. The element T^{00} is the energy density associated with the electromagnetic field, and the components T^{0i} are related to the Poynting vector $\mathbf{S} = \mathbf{E} \times \mathbf{B}/\mu_0$. From the conservation of the stress tensor, here just $T^{\mu\nu}{}_{,\nu} = 0$ (try it out in Problem 7.15), we have the set of four equations

$$\frac{\partial T^{00}}{\partial t} = -c\frac{\partial T^{0j}}{\partial x^j}, \qquad \frac{\partial}{\partial t}\left(\frac{T^{0j}}{c}\right) = -\frac{\partial T^{ij}}{\partial x^j}, \tag{7.95}$$

where the Roman indices run from $1 \rightarrow 3$. In these equations, the dual role of the components T^{0i} is clear: $cT^{0i} = (\mathbf{E} \times \mathbf{B})^i/\mu_0$ is the "current" density for the conserved "charge" T^{00}. The combination cT^{0i} is precisely the spatial "Poynting vector" $\mathbf{S} \equiv \mathbf{E} \times \mathbf{B}/\mu_0$ from E&M. In addition, T^{0i}/c (or the spatial vector $\epsilon_0 \mathbf{E} \times \mathbf{B}$) is a set of three conserved "charges" with associated current density T^{ij}. The conserved quantity in the first equation in (7.95) is energy, since it is the energy density appearing on the left. The dimension of T^{0i}/c

is that of a momentum density, so the second equation in (7.95) represents momentum conservation (upon integration over a spatial volume, of course). The current density for the momentum conservation T^{ij} has the dimension of pressure, force-per-unit-area.

To get the integral form of the differential conservation laws in (7.95), take a domain Ω with boundary $\partial\Omega$ and integrate both sides over the volume of Ω, using the divergence theorem to turn the volume integrals appearing on the right into surface integrals:

$$\frac{d}{dt}\int_{\Omega} T^{00}\, d\tau = -\oint_{\partial\Omega} cT^{0j}\, da_j, \qquad \frac{d}{dt}\int_{\Omega}\frac{T^{0j}}{c}\, d\tau = -\oint_{\partial\Omega} T^{ij}\, da_j.$$
$$(7.96)$$

The integral of T^{00} over Ω expresses the total electromagnetic energy contained in Ω. The first equation in (7.96) can be written as

$$\frac{dE}{dt} = -\oint_{\partial\Omega}\mathbf{S}\cdot d\mathbf{a}. \qquad (7.97)$$

For the second conservation statement in (7.96), the volume integral on the left gives the total electromagnetic field momentum contained in Ω, call it $\mathbf{P}$, and we can write

$$\frac{dP^i}{dt} = -\oint_{\partial\Omega} T^{ij}\, da_j. \qquad (7.98)$$

This looks like Newton's second law, so the expression on the right must be the force acting on the volume Ω. This supports the units of T^{ij} and gives us a physical interpretation for the components of $-T^{ij}$, which must be the ith component of the force per unit area acting on a surface with normal vector $\hat{\mathbf{n}}$ and $d\mathbf{a} = da\,\hat{\mathbf{n}}$. So, for $-T^{11}$, we get the x component of the force per unit area acting on a surface with normal in the $\hat{\mathbf{x}}$ direction. Similarly, $-T^{12}$ is the x component of the force per unit area acting on a surface with normal in the $\hat{\mathbf{y}}$ direction, etc. The tensor is symmetric, so that $-T^{21}$, which is officially the y component of force per unit area acting on a surface with normal in the $\hat{\mathbf{x}}$ direction, is numerically equal to $-T^{12}$.

Example 7.3 (Components of T^{ij} for Static Fields). To understand the structure of the spatial components of the stress tensor, it helps to take concrete electric and magnetic fields. To keep things simple, we'll set $\mathbf{B} = 0$ and make the electric field a constant, $\mathbf{E} = E_0(\cos\theta\hat{\mathbf{z}} + \sin\theta\hat{\mathbf{y}})$ for constant θ. Then the spatial components of the stress tensor are

$$T^{ij} \doteq \frac{\epsilon_0 E_0^2}{2}\begin{pmatrix} 1 & 0 & 0 \\ 0 & \cos(2\theta) & -\sin(2\theta) \\ 0 & -\sin(2\theta) & -\cos(2\theta) \end{pmatrix}. \qquad (7.99)$$

The nonzero component T^{11} is a little odd; we don't expect an x-directed force out of this field, but remember that in the context of a closed surface integral (think, most simply, of a cube), there will be an equal amount of $\hat{\mathbf{x}}$ surface normals and $-\hat{\mathbf{x}}$ surface normals, so the net x component will cancel. Indeed, integrated over any closed surface, *all* of these components must vanish, since for this field configuration, with $\mathbf{B} = 0$, we get no Poynting vector, the left-hand side of (7.98) is zero, and therefore so is the right.

The vacuum field equation for a vector Lagrangian, with an arbitrary metric and coordinate choice, from (7.86) is

$$D_\mu F^{\mu\nu} = 0. \tag{7.100}$$

As in (7.84), we have a combined set of nonlinear PDEs governing A_α and the metric itself through the stress tensor:

$$
\begin{aligned}
&D_\alpha F^{\alpha\beta} = 0, \\
&R^{\mu\nu} - \frac{1}{2} g^{\mu\nu} R = \frac{8\pi G}{\mu_0 c^4} \left[F^{\mu\sigma} F^\nu{}_\sigma - \frac{1}{4} g^{\mu\nu} F^{\rho\sigma} F_{\rho\sigma} \right].
\end{aligned}
\tag{7.101}
$$

Because this equation has a trace-free stress tensor source, you can show that the Ricci scalar vanishes (see Problem 7.18), leading to the simplified set

$$
\begin{aligned}
&D_\alpha F^{\alpha\beta} = 0, \\
&R^{\mu\nu} = \frac{8\pi G}{\mu_0 c^4} \left[F^{\mu\sigma} F^\nu{}_\sigma - \frac{1}{4} g^{\mu\nu} F^{\rho\sigma} F_{\rho\sigma} \right].
\end{aligned}
\tag{7.102}
$$

Here we have the equations governing a (massless) locally source-free vector field, A_μ, coupled to gravity. As with (7.84), the coupling goes both ways, with A_μ depending on the metric and $g_{\mu\nu}$ depending on the vector field.

7.5 Charged Spherical Central Body

We have obtained the field equations for gravity in the presence of free scalar or massless vector fields, and we now consider their solution. It turns out that the coupling to a vector field yields equations that are simpler than those for a scalar source, so we'll work out a solution for gravity coupled to electromagnetism. Adding sources to E&M doesn't change the stress tensor, and we'll work away from those sources in "electrovacuum," where there is no charge, so that the field equation governing A_μ is unchanged from its source-free form in (7.102).

As a source, take a sphere of mass M and charge Q. In flat spacetime, there would be an electromagnetic four-potential that was static and spherically symmetric (with $\mathbf{A} = 0$, a gauge choice for spherically symmetric sources), so that for A_μ, we would only have nonzero $A_0(r)$. We know that a spherically symmetric starting point for the metric is (4.6), and, motivated by the form of the Schwarzschild solution in (4.12), we'll start off in (4.6) with

$q(r) = 1/p(r)$.[6] With these symmetry assumptions in place, the field strength tensor has only one nonzero component, $F^{01} = A_0'(r) = -F^{10}$. In Problem 7.21, you will show that for a spherically symmetric electric field $\mathbf{E} = E(r)\hat{\mathbf{r}}$ in flat spacetime, the only nonzero component of the field strength tensor is $F^{01} = E(r)/c = -F^{10}$. This suggests that we name $A_0'(r) \rightarrow E(r)/c$ to make contact with the flat space electric field. Then the electromagnetic field equation from (7.102) only has one component

$$D_\nu F^{\mu\nu} = 0 \longrightarrow 2E(r) + rE'(r) = 0 \qquad (7.103)$$

with solution $E(r) = A/r^2$ for constant A. Notice that this result does not depend on the as yet unknown metric component $p(r)$, so we can immediately set $A = Q/(4\pi\epsilon_0)$ since that must be its value in a flat spacetime where $p(r) = 1$.

Moving on to Einstein's equation, with the known $E(r)$ in place, we get a set of four related equations. The simplest of these is the 22 component

$$R_{22} - \frac{1}{2}g_{22}R = \frac{8\pi G}{c^4}T_{22} \longrightarrow \frac{1}{2}\frac{d}{dr}\left(r^2\frac{dp(r)}{dr}\right) = \frac{GQ^2}{4\pi\epsilon_0 c^4 r^2}. \qquad (7.104)$$

From the units on both sides of the equation, we learn that the combination $GQ^2/(\epsilon_0 c^4)$ must be a length-squared, so we define the length scale R_Q by

$$R_Q^2 \equiv \frac{GQ^2}{4\pi\epsilon_0 c^4}. \qquad (7.105)$$

The solution to the ODE governing $p(r)$ in (7.104) is then

$$p(r) = 1 + \frac{B}{r} + \frac{R_Q^2}{r^2} \qquad (7.106)$$

for constant of integration B. If the central body is not charged, $R_Q = 0$, then we must recover the Schwarzschild solution (4.12), allowing us to set $B = -R_M$ with the usual $R_M \equiv 2GM/c^2$.

The metric with these constants in place is

$$g_{\mu\nu} \doteq \begin{pmatrix} -(1 - \frac{R_M}{r} + \frac{R_Q^2}{r^2}) & 0 & 0 & 0 \\ 0 & \frac{1}{1 - R_M/r + R_Q^2/r^2} & 0 & 0 \\ 0 & 0 & r^2 & 0 \\ 0 & 0 & 0 & r^2\sin^2\theta \end{pmatrix}. \qquad (7.107)$$

This metric defines the Reissner–Nordstrøm spacetime (see [37, 41]). It is associated with the exterior of a charged, static, spherically symmetric source. Work through the geodesic predictions for test particles in Problem 7.23 and Problem 7.25. What changes if the test particle is itself charged?

From (7.107) we can extract h_{00} from $g_{00} = -1 + h_{00}$,

$$h_{00} = \frac{R_M}{r} - \frac{R_Q^2}{r^2}. \qquad (7.108)$$

---

[6] That may not be good enough, and we'd have to go back and start over, but it turns out to be fine in this case.

Viewed as a metric perturbation, for large r, you can associate h_{00} with a Newtonian gravitational potential and compare with predictions made back in Section 1.5.1 in Problem 7.26.

7.6 The Einstein–Hilbert Action

Einstein's theory of general relativity also has an action, a scalar functional of the field of the theory, its derivatives, and this time, second derivatives. It is best to proceed by guessing the form of this action via a series of structural hints. We are looking for an action of the usual general form

$$S[g] = \int \mathcal{M}(g, \partial g, \partial^2 g) \sqrt{-g}\, d^4 x, \qquad (7.109)$$

where $\mathcal{M}$ is a scalar function of the metric (I've switched to the spacetime volume element $d^4 x$). We know that the field equations of general relativity are themselves nonlinear, so we expect the target scalar to involve terms higher than quadratic in the metric and its derivatives. There is a familiar, simple scalar that presents itself, the Ricci scalar. This function depends on the metric and its first and second derivatives, but in such a way that the field equations remain second order. The action

$$S[g] = \int \underbrace{R(g, \partial g, \partial^2 g) \sqrt{-g}}_{\equiv \mathcal{L}_{\text{EH}}}\, d^4 x \qquad (7.110)$$

is called the "Einstein–Hilbert" action with integrand $\mathcal{L}_{\text{EH}}$ the Einstein–Hilbert Lagrangian. You can check in that the Euler–Lagrange field equation

$$\frac{\partial}{\partial x^\mu} \frac{\partial}{\partial x^\nu} \left(\frac{\partial \mathcal{L}_{\text{EH}}}{\partial g_{\alpha\beta,\mu\nu}} \right) - \frac{\partial}{\partial x^\nu} \left(\frac{\partial \mathcal{L}_{\text{EH}}}{\partial g_{\alpha\beta,\nu}} \right) + \frac{\partial \mathcal{L}_{\text{EH}}}{\partial g_{\alpha\beta}} = 0 \qquad (7.111)$$

recovers Einstein's equation in vacuum, where $T^{\mu\nu} = 0$.

Now we can see how the universal coupling of gravity is enforced in the field theoretic actions we have considered. Any action, scalar, vector, or anything else depends on the metric, so that when we vary an action that includes the Einstein–Hilbert term, we will automatically pick up the correct source in Einstein's equation. Suppose we started with a combined scalar-gravity action

$$S[\psi, g] = \frac{c^4}{4\pi G} \int \mathcal{L}_{\text{EH}}(g, \partial g, \partial^2 g)\, d^4 x - \int \mathcal{L}(\psi, \partial \psi, g)\, d^4 x. \qquad (7.112)$$

When we vary with respect to the ψ field, upon which only the second term depends, we will recover the correct field equation (7.75) with an undetermined metric and connection. When we vary with respect to the metric itself, the field equation will be

$$\frac{c^4}{4\pi G} \left[\frac{\partial}{\partial x^\mu} \frac{\partial}{\partial x^\nu} \left(\frac{\partial \mathcal{L}_{\text{EH}}}{\partial g_{\alpha\beta,\mu\nu}} \right) + \frac{\partial}{\partial x^\nu} \left(\frac{\partial \mathcal{L}_{\text{EH}}}{\partial g_{\alpha\beta,\nu}} \right) + \frac{\partial \mathcal{L}_{\text{EH}}}{\partial g_{\alpha\beta}} \right] - \frac{\partial \mathcal{L}}{\partial g_{\alpha\beta}} = 0. \qquad (7.113)$$

The variation of the Einstein–Hilbert action gives us $-\sqrt{-g}(R^{\alpha\beta} - (g^{\alpha\beta}/2)R)$, and the $g_{\alpha\beta}$-derivative of the scalar Lagrangian density, $\mathcal{L}$, is related to the scalar stress tensor by (7.63). Putting the two together, we get

$$R^{\alpha\beta} - \frac{1}{2}g^{\alpha\beta}R = \frac{8\pi G}{c^4}T^{\alpha\beta}. \tag{7.114}$$

Clearly, *any* Lagrangian density $\mathcal{L}$ that depends on the metric, and not on its derivatives, will lead to the same form of Einstein's equation, neatly accomplishing the universal coupling of gravity to everything else. Written out, the scalar field equation and Einstein's equation leads to (7.84). If we did the same thing with a vector Lagrangian, we'd get (7.102).

7.6.1 Cosmological Constant

The claim above is that R is "the only" scalar available for making an action relevant to general relativity. That is certainly true if the target is Einstein's equation, but we can always posit the existence of a constant Λ and form a scalar Lagrangian from that. The Lagrangian density

$$\mathcal{L}_\Lambda \equiv \Lambda\sqrt{-g} \tag{7.115}$$

is available and returns $\sqrt{-g}\,g_{\mu\nu}\Lambda/2$ upon variation. So, to the Einstein–Hilbert action, we can add this cosmological constant term

$$\mathcal{L} = \frac{c^4}{4\pi G}\sqrt{-g}(R + 2\Lambda) \tag{7.116}$$

and recover the cosmological form of Einstein's equation from (6.113) when we vary with respect to $g_{\mu\nu}$ (with source terms arising, as above, from the addition of their Lagrangians to this one).

7.7 The Weyl Method

There is a clever way to develop highly symmetric solutions to Einstein's equation that proceeds directly from the action. The idea behind "Weyl's method" [54] (see [9] for a variety of useful applications) is to introduce symmetries directly into the field in the Einstein–Hilbert action, reducing the number of degrees of freedom in the field, the metric in this case, from ten to some smaller set. Then the action is varied with respect to that smaller set, leading to fewer field equations that are often easier to solve than the full set.

As an example of the procedure in a familiar context, suppose we wanted to find "the" spherically symmetric solutions to Maxwell's equation in vacuum.

Spherical symmetry for the potential A^μ means (ignoring time dependence) that $A^0 = V(r)/c$ and $\mathbf{A} = A(r)\mathbf{r}$, as in Section 4.1. The action, from (7.85) with $\mathcal{M}$ given by (7.89) and these assumptions in place, is[7]

$$\begin{aligned}
S &= \int_0^\infty \int_0^\pi \int_0^{2\pi} \frac{1}{2\mu_0}\left(B^2 - \frac{E^2}{c^2}\right)r^2 \sin\theta\, dr\, d\theta\, d\phi \\
&= \int_0^\infty \int_0^\pi \int_0^{2\pi} \frac{1}{2\mu_0}\left(-\frac{V'(r)}{c}\right)^2 r^2 \sin\theta\, dr\, d\theta\, d\phi.
\end{aligned} \tag{7.117}$$

In terms of the potential, there is no reference to $\mathbf{A}$ – that's the usual reminder that you cannot have a magnetic field associated with a spherically symmetric source. We can take $\mathbf{A} = 0$ as a gauge choice, one that leads to the same, zero, magnetic field as any other.

We can carry out the angular integrals, since the relevant remaining field $V(r)$ does not depend on θ or ϕ, giving

$$S = \int_0^\infty 2\pi \epsilon_0 \big(V'(r)r\big)^2 dr, \tag{7.118}$$

yielding a one-dimensional Lagrangian for the field $V(r)$:

$$\mathcal{L} = 2\pi \epsilon_0 \big(V'(r)r\big)^2. \tag{7.119}$$

The action is minimized by the Euler–Lagrange equations for $\mathcal{L}$,

$$-\frac{d}{dr}\left(\frac{\partial \mathcal{L}}{\partial V'(r)}\right) + \frac{\partial \mathcal{L}}{\partial V(r)} = 0 \rightarrow -4\pi \epsilon_0 r^2 V'(r) = \alpha \tag{7.120}$$

for constant α. Then we can solve for $V(r)$ by integrating:

$$V(r) = \frac{\alpha}{4\pi \epsilon_0 r} + \beta \tag{7.121}$$

for constant β. If we impose the boundary condition, $V(r) \to 0$ as $r \to \infty$, then we can set $\beta = 0$, and we recover the Coulomb potential with constant α identified with the charge.

The gravitational version proceeds in a similar fashion. For a spherically symmetric source, we assume that the metric exhibits that same spherical symmetry as in (4.6). Instead of running the metric through Einstein's equations, we form the integrand in (7.110) directly:

$$S = \int_0^r \int_0^\pi \int_0^{2\pi} \underbrace{\frac{1}{2(rpq)^2}[r^2 q p'^2 + 4p^2(q^2 + rq' - q) + rp(rp'q' - 2q(2p' + rp''))]}_{=R}$$

$$\times \underbrace{r^2 \sin\theta \sqrt{pq}}_{=\sqrt{-g}}\, dr\, d\theta\, d\phi. \tag{7.122}$$

<hr/>

[7] Technically, the action integral here is over a four-dimensional spacetime volume. I have omitted the temporal integration for simplicity and clarity (since it plays no role in the variation). If you like, put in an extra ct integration over a finite temporal domain; that will just throw in an overall factor of the total time, which won't change the field equations.

Integrating the angular dependence as in the E&M case, we get the one-dimensional Lagrangian in field variables $p(r)$ and $q(r)$:

$$\mathcal{L} = \frac{2\pi}{(rpq)^2}\left[r^2 q p'^2 + 4p^2\left(q^2 + rq' - q\right) + rp\left(rp'q' - 2q\left(2p' + rp''\right)\right)\right]r^2\sqrt{pq}.$$
(7.123)

There are two field equations here, one for p-variation and one for q. For q, we have the Euler–Lagrange equation

$$-\frac{d}{dr}\left(\frac{\partial\mathcal{L}}{\partial q'(r)}\right) + \frac{\partial\mathcal{L}}{\partial q} = 0 \rightarrow \frac{4\pi p}{(pq)^{3/2}}\left(pq - (rp)'\right) = 0. \qquad (7.124)$$

For the p equation, we need a version of the Euler–Lagrange equations appropriate for a Lagrangian that depends on second derivatives $p''(r)$. Try developing the equation in Problem 7.27; it ends up being

$$\frac{d^2}{dr^2}\left(\frac{\partial\mathcal{L}}{\partial p''(r)}\right) - \frac{d}{dr}\left(\frac{\partial\mathcal{L}}{\partial p'(r)}\right) + \frac{\partial\mathcal{L}}{\partial p(r)} = 0, \qquad (7.125)$$

which gives

$$\frac{4\pi p}{(pq)^{3/2}}\left(q^2 + rq' - q\right) = 0. \qquad (7.126)$$

We can solve the field equation in (7.124), which amounts to $pq - (rp)' = 0$ for q to get $q = (rp)'/p$. Using this in the field equation for p: $q^2 + rq' - q = 0$ gives a second-order ODE for $p(r)$,

$$\frac{r}{p}\left(2p' + rp''\right) = 0 \longrightarrow p = \frac{\alpha}{r} + \beta \qquad (7.127)$$

for constants of integration α and β. Then we get q for free,

$$q = \frac{\beta}{\beta + \alpha/r}, \qquad (7.128)$$

and we know, from experience with the weak field limit, how to set the constants α and β so as to recover the Schwarzschild spacetime.

The idea behind the Weyl method is that we can use certain types of symmetry assumptions in an action prior to varying without losing information. The symmetry of the source configuration gets passed along to fields in the field equations themselves (notably through the imposition of boundary conditions). It is, in general, not possible to use information from the field equations to simplify the action before varying. See Problem 7.28 for an example of the potential danger.

Exercises

7.1   Could you get a field equation that looked like $\Box\psi_{,\mu} = 0$ using a scalar field Lagrangian? If so, give the form of $\mathcal{L}$; if not, explain why not.

7.2   Find the field equation for the Lagrangian (in flat spacetime with Carte-
sian coordinates so that $\sqrt{-g} = 1$) $\mathcal{L} = \partial_\mu F^\mu(\psi)$, where $F^\mu(\psi)$ is a
vector function of ψ.

7.3   The Levi-Civita symbol is an example of a tensor density. In two spatial
dimensions, we have

$$\epsilon^{ij} \equiv \begin{cases} 1 & \text{if } i = 1, j = 2, \\ -1 & \text{if } i = 2, j = 1, \\ 0 & \text{else.} \end{cases} \tag{7.129}$$

Show that if the Levi-Civita symbol is to take these same numerical values in any coordinate system, then it must transform as a second-rank
tensor density of weight -1.

7.4   Carefully run the scalar Lagrangian from (7.15) through the field equations (7.11) to see that you recover the desired target, the right-hand side
of (7.12).

7.5   Using a simple 2×2 unstructured (meaning not symmetric or antisymmetric) matrix $\mathbb{M}$ with entries M_{ij} and determinant M, show that

$$\frac{\partial M}{\partial M_{ij}} = M^{ji} M, \tag{7.130}$$

where M^{ij} are the entries of $\mathbb{M}^{-1}$.

7.6   Show that (7.41) gives the correct field equation in flat spacetime,

$$D_\mu D^\mu \psi = 0 \tag{7.131}$$

starting from $\mathcal{M} = \frac{1}{2} \psi_{,\alpha} \psi^{,\alpha}$.

7.7   Show that replacing the metric derivative in (7.56) using (7.57) gives the
covariant derivative expressions in (7.58).

7.8   Working in two dimensions to keep the expressions simple, let $u \equiv ct$,
and we'll take the spatial variable x. For a rectangular domain in the ux
plane, with $u \in [0, U]$ and $x \in [0, X]$, the simplest coordinate perturbation that vanishes on the boundary is given in (7.71). By inverting the
relationship there, to get x and u as functions of $\bar{x}$ and $\bar{u}$, find the met-
ric $\bar{g}_{\mu\nu}(\bar{u}, \bar{x})$ – you can use the scalar nature of $ds^2 = dx^\mu g_{\mu\nu} dx^\nu = d\bar{x}^\mu \bar{g}_{\mu\nu} d\bar{x}^\nu$ to do this or work directly from the transformation rule for
a second-rank covariant tensor. Feel free to use a symbolic algebra pack-
age to make the calculation easier. Expand your $\bar{g}_{\mu\nu}$ to first order in ϵ
and compare with the approximate expression you get from (7.58).

7.9   Using the setup and variables from the previous problem, take the scalar
$\psi(x) = \alpha ux$ for constant α and verify that the relations in (7.45) hold
through first order in ϵ, again using a symbolic algebra program to carry
out the details.

7.10   Compute the trace $T \equiv g_{\mu\nu} T^{\mu\nu}$ of the Klein–Gordon stress tensor from
(7.80) (in D dimensions). Use this to write Einstein's equation using that
trace and your result from Problem 3.15.

7.11 Show that the scalar $\mathcal{M}$ from (7.87), when run through (7.86), gives back the left-hand side of (2.184). Find the term you have to subtract from $\mathcal{M}$ to get the source term J_α on the right side. Finally, pick β so that your field equation recovers the potential formulation of Maxwell's equation (work in Lorenz gauge).

7.12 Using the lower form for $F_{\mu\nu}$ from (7.93), find the components of the mixed field strength tensor $F^\mu{}_\nu$ (assume Minkowski spacetime here).

7.13 Write the electromagnetic field equation $D^\nu(D_\mu A_\nu - D_\nu A_\mu) = \mu_0 J_\mu$ in terms of the field strength tensor. Show that the current source J^α is covariantly conserved, $D_\alpha J^\alpha = 0$, from the field equation. This is just a reminder that the full set of Maxwell's equations is consistent with charge conservation.

7.14 The field strength tensor $F_{\mu\nu}$ embeds the electric and magnetic fields in a tensor structure. But it can be used to represent only two of Maxwell's four equations: $\partial_\mu F^{\mu\nu} = -\mu_0 J^\nu$ (in flat spacetime). The other two come from an alternate embedding, the "dual" field strength tensor: $G^{\mu\nu} \equiv (1/2)\epsilon^{\mu\nu\alpha\beta} F_{\alpha\beta}$ ($\epsilon^{\mu\nu\alpha\beta}$ is the four-dimensional Levi-Civita symbol). In terms of the Cartesian components of $\mathbf{E}$ and $\mathbf{B}$, write out the entries of $G^{\mu\nu}$. Show that $\partial_\mu G^{\mu\nu} = 0$ recovers the pair of Maxwell's equations that do not involve charge and current sources.

7.15 Show that for an electromagnetic potential A_α that satisfies its vacuum field equation, the stress tensor in (7.90) is covariantly conserved: $D_\mu T^{\mu\nu} = 0$ (work in flat spacetime, but do not assume Cartesian coordinates, that is, use the covariant derivative).

7.16 The electromagnetic stress tensor is, in a sense, defined by its tracelessness in four dimensions. Show that if you take $T^{00} = (1/2)\epsilon_0 E^2 + (1/2)B^2/\mu_0 \equiv u$, the energy density of the electromagnetic field, and call the diagonal elements $T^{xx} = T^{yy} = T^{zz} \equiv p$, then in a flat spacetime, the traceless requirement gives $p = u/3$, the "equation of state" for radiation.

7.17 Using a scalar solution to the field equation in (7.75), evaluate the stress tensor elements T^{00}, T^{0i}, and T^{ij} using (7.80) expressed in Cartesian coordinates for the plane wave solution $\psi = \psi_0 \cos(k_\mu x^\mu)$.

7.18 Show that the Ricci scalar vanishes: $R = 0$ (in four dimensions) for the field equations in (7.101). What is the Ricci scalar in dimension D not necessarily equal to four? Write Einstein's equation in terms of $R^{\mu\nu}$ (the analogue of (7.102)) in this general setting.

7.19 The Weyl tensor is defined from the Riemann tensor,

$$C_{\alpha\beta\mu\nu} = R_{\alpha\beta\mu\nu} + \frac{1}{D-2}[R_{\alpha\nu}g_{\beta\mu} - R_{\alpha\mu}g_{\beta\nu} + R_{\beta\mu}g_{\alpha\nu} - R_{\beta\nu}g_{\alpha\mu}]$$

$$+ \frac{R}{(D-1)(D-2)}[g_{\alpha\mu}g_{\beta\nu} - g_{\alpha\nu}g_{\beta\mu}]. \qquad (7.132)$$

What are the traces $C_{\alpha\mu} = g^{\beta\nu}C_{\alpha\beta\mu\nu}$ and $C = g^{\alpha\mu}C_{\alpha\mu}$? Just as the vanishing of the Riemann tensor is an indicator of flatness, the Weyl tensor is zero when a spacetime is "conformally flat," meaning that

$$ds^2 = f(t, x, y, z)\left[-c^2dt^2 + dx^2 + dy^2 + dz^2\right]. \qquad (7.133)$$

Verify that this is true (use a symbolic package if you'd like).

7.20 Use the Weyl tensor to show that the metric in (6.84) is conformally flat.

7.21 For a spherically symmetric electric field $\mathbf{E} = E(r)\hat{\mathbf{r}}$, write out the contravariant form of the field strength tensor in Cartesian coordinates using (7.93). Transform to spherical coordinates and write out the contravariant field strength tensor in those coordinates.

7.22 Using the same setup from the previous problem, write out the stress tensor in Cartesian components and then find the components in spherical. Make sure your stress tensor remains traceless in spherical coordinates.

7.23 Write out the geodesic Lagrangian in proper time parameterization for massive particles moving radially under the influence of (7.107) as we did in Section 4.3 and find the radial equation of motion. Express this equation of motion in the form $\ddot{r} = f(r)$, where the right-hand side's $f(r)$ is a function of r (and not t or its derivatives) only.

7.24 For light moving along radial geodesics in a spacetime with metric given by (7.107), what is $\frac{dr(s)}{ds}$ for an arbitrary parameter s? (Eliminate any dependence on t or its derivatives using conservation of "energy.") How about in temporal parameterization, what is $\frac{dct}{dr}$? (here we've flipped the derivative so that you could easily sketch the trajectory on a Minkowski diagram).

7.25 Working from the geodesic Lagrangian for (7.107), identify the orbital motion effective potentials for both massive particles and light, paralleling the work done in Section 4.5 for Schwarzschild.

7.26 Expand the Reissner–Nordstrøm metric in (7.107) for $r \gg R_M$ and $r \gg R_Q$. By looking at the g_{00} component identify the weak field contribution h_{00} and use that to find the effective Newtonian potential $\varphi = -h_{00}c^2/2$. How does this compare with the Newtonian prediction from (1.87)?

7.27 For a one-dimensional action of the form

$$S[f(r)] = \int_{r_0}^{r_f} L(f, f', f'')dr, \qquad (7.134)$$

find the Euler–Lagrange equations by adding a perturbation $\delta f(r)$ (that vanishes at r_0 and r_f, with vanishing derivative at those points, too) and demanding that $S[f(r) + \delta f(r)] = S[f(r)]$ to first order in $\delta f(r)$ for any $\delta f(r)$.

7.28 Using the free particle action in one dimension with Lagrangian $L = (1/2)mv^2$, find the Euler–Lagrange equations of motion, solve those, and then write the action in terms of that solution. What happens when you try to vary the action now? This is an extreme example of what can go wrong when you use equation of motion information too early in the variational procedure (precisely what the Weyl method invites us to do).

7.29 The parameterization of the metric can simplify the field equations you get from the Weyl method. Starting from the line element

$$ds^2 = -a(r)^2 b(r) c^2 dt^2 + 1/b(r) dr^2 + r^2 \left(d\theta^2 + \sin^2\theta d\phi^2 \right), \quad (7.135)$$

construct the Einstein–Hilbert Lagrangian $\mathcal{L} = R\sqrt{-g}$ and vary with respect to $a(r)$ and $b(r)$ to get the pair of field equations. Solve those up to "undetermined" constants.

7.30 For the Einstein–Hilbert Lagrangian with cosmological constant, $\mathcal{L} \sim \sqrt{-g}(R - 2\Lambda)$, use the Weyl method to find the spherically symmetric vacuum solutions in $D = 3 + 1$ and $D = 2 + 1$ dimensions. Compare with Problem 4.14 and Problem 4.15.

7.31 For the Lagrangian that combines gravity and electromagnetism,

$$\mathcal{L} = \sqrt{-g}\left(R + \sigma F^{\mu\nu} F_{\alpha\beta} g_{\mu\alpha} g_{\nu\beta} \right) \quad (7.136)$$

(for constant σ), using a spherically symmetric metric ansatz (like the one in (7.135) and a spherically symmetric four-potential with $A^0 = V(r)/c$ and $\mathbf{A} = 0$, appropriate for a static point charge (explain why $\mathbf{A} = 0$ represents a valid, general, spherically symmetric starting point for this source), carry out the Weyl method (i.e., compute the Euler–Lagrange equations for $V(r)$, $a(r)$, and $b(r)$ and solve them) to find the Reissner–Nordstrøm metric and Coulomb potential (up to constants).

Lorentz Transformations and Special Relativity

In this appendix, we review a limited but complete (for our purposes) set of observations from special relativity necessary to appreciate its role in the development of general relativity. The primary tools are the Lorentz boost in an arbitrary direction and the transformation of familiar quantities from E&M and Newtonian gravity. The review here is meant to provide context for Chapter 2, but ideas from the locally Minkowski spacetime that can be set up at a point (and its vicinity, in approximation) permeate the rest of the book. Here we'll just set the standard notation used throughout and remind the reader of some results. In particular, we'll review time dilation, length contraction, velocity addition, redshift, and the transformation of the E&M four-potential, source density, and electric and magnetic fields.

A.1 Lorentz Boost

Just as rotations are defined to be the set of transformations that leave the distance to a point in three dimensions unchanged, the Lorentz boost preserves the Minkowski length between two points in the expanded four-dimensional coordinates of spacetime. For a pair of "inertial" frames L and $\bar{L}$ moving with constant velocity relative to one another, orient the coordinate axes so that the relative motion occurs along a shared $\hat{\mathbf{x}}$ axis. We'll set $\hat{\bar{\mathbf{x}}} = \hat{\mathbf{x}}$, but when we draw the configuration, as in Figure A.1, we typically put in a visually clarifying vertical offset.

If we quickly turn on and off a light at $t = \bar{t} = 0$ when the two origins coincide, the horizontal location of the front of the light in L is $x = ct$, and in $\bar{L}$ the front is described by $\bar{x} = c\bar{t}$. This enforces the constancy of the speed of light, a well-established observation and fundamental assumption input into any theory consistent with special relativity. These relations tell us that

$$- (ct)^2 + x^2 = 0 = -(c\bar{t})^2 + \bar{x}^2, \tag{A.1}$$

and we can easily extend this to any interval, relative to the origin of the coordinate systems. So in general (i.e., not just for the front of a light signal), the ct and x coordinates in L and $\bar{L}$ are related by

$$- (c\bar{t})^2 + \bar{x}^2 = -(ct)^2 + x^2. \tag{A.2}$$

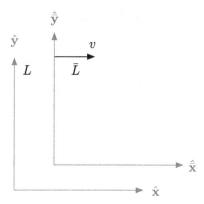

A frame (set of axes and clocks) $\bar{L}$ moves with constant speed v to the right in L. At time $t = 0$, the origins of the two coordinate systems align, and we set the clocks to $t = \bar{t} = 0$.

What transformation preserves this target invariant? Take the simplest, linear relation between the two frames,

$$c\bar{t} = Act + Bx, \qquad \bar{x} = Fct + Gx \tag{A.3}$$

for constants $\{A, B, F, G\}$. Then the requirement is

$$-(c\bar{t})^2 + \bar{x}^2 = -(c\bar{t})^2(A^2 - F^2) + (G^2 - B^2)x^2 + 2ctx(FG - AB), \tag{A.4}$$

so that we must have $FG = AB$, $A^2 - F^2 = 1$, and $G^2 - B^2 = 1$. We can satisfy these constraints by setting

$$A = G, \qquad B = F = -\sqrt{-1 + G^2}, \tag{A.5}$$

leaving G as a free parameter. To make contact with the picture in Figure A.1, pick a specific point in both coordinate systems. The origin of $\bar{L}$ is at $\bar{x} = 0$ in $\bar{L}$ and $x = vt$ in L, so that

$$0 = -\sqrt{-1 + G^2}ct + Gvt \longrightarrow G = \frac{1}{\sqrt{1 - (v/c)^2}}. \tag{A.6}$$

Define $\gamma \equiv (1 - v^2/c^2)^{-1/2}$ as usual; then the relation between coordinates in L and $\bar{L}$ that preserves (A.2) is

$$c\bar{t} = \gamma\left(ct - \frac{v}{c}x\right), \qquad \bar{x} = \gamma\left(-\frac{v}{c}ct + x\right). \tag{A.7}$$

This is the Lorentz boost relating coordinates and time in L and $\bar{L}$. The inverse transformation, relating ct and x to the coordinates in $\bar{L}$, can be obtained quickly by noting that from $\bar{L}$'s point of view, L is moving to the left with speed v,

$$ct = \gamma\left(c\bar{t} + \frac{v}{c}\bar{x}\right), \qquad x = \gamma\left(\frac{v}{c}c\bar{t} + \bar{x}\right). \tag{A.8}$$

A.2 Implications for Constant Speed

From the transformation at constant speed, we are ready to make the usual observations about time and space intervals.

A.2.1 Time Dilation and Length Contraction

First consider a clock at rest at the origin of $\bar{L}$. The clock records a time interval $\Delta \bar{t}$ while remaining at the same location relative to $\bar{L}$, $\Delta \bar{x} = 0$. In L, using the temporal transformation from (A.8),

$$c\Delta t = \gamma(c\Delta \bar{t}) \longrightarrow \Delta t = \frac{\Delta \bar{t}}{\sqrt{1 - (v/c)^2}}. \tag{A.9}$$

This equation tells us that $\Delta t > \Delta \bar{t}$. As an example, for $v = (4/5)c$, $\Delta t = (5/3)\Delta \bar{t}$. If a year passes in $\bar{L}$, a year and two thirds pass in L. This effect is known as "time dilation" and can be measured using fast moving jets. You put a clock on a jet and leave one at the airstrip. The jet flies around, and when it returns, you compare the clocks. The jet's clock has less elapsed time than the one you left on the ground.

Length measurements must be made instantaneously. If you want to measure the length of a rod moving through a lab L, you don't mark the location of the left end at some time t_1, wait until $t_2 > t_1$, mark the location of the right end and subtract to find the distance between the two ends. You measure the distance over a time interval $\Delta t = 0$. To fit the story into our L and $\bar{L}$ reference frames, if you have a rod of length $\Delta \bar{x}$ at rest in $\bar{L}$ and you want to measure its length in L, then you must do so at $\Delta t = 0$. Using these target intervals in (A.7), we learn that

$$\Delta \bar{x} = \gamma(\Delta x) \longrightarrow \Delta x = \sqrt{1 - \left(\frac{v}{c}\right)^2}\, \Delta \bar{x}, \tag{A.10}$$

so that $\Delta x < \Delta \bar{x}$. A meter stick moving at $v = (4/5)c$ has length $\Delta x = 3/5$ meter in L. This phenomenon is known as "length contraction."

A.2.2 Directions Perpendicular to the Boost

We can use length contraction to motivate the idea that the Lorentz boost doesn't change lengths that are perpendicular to the relative motion of L and $\bar{L}$. There are a variety of ways to make the case. Suppose we have a painter standing at the origin of $\bar{L}$ and moving along with his brush rolling against the wall next to him (in the xy plane). If length contraction in the y direction occurred, then the painted stripe would be one height from the painter's point of view and a different height from the point of view of a person in L. Since a painted stripe cannot be two painted stripes, we conclude that there is no length contraction in the y direction. Then the full Lorentz boost in the x direction is

$$c\bar{t} = \gamma\left(ct - \frac{v}{c}x\right), \qquad \bar{x} = \gamma\left(-\frac{v}{c}ct + x\right), \qquad \bar{y} = y, \qquad \bar{z} = z. \quad (A.11)$$

We can write the boost in matrix form by thinking of x^μ as a column vector with entries ct, x, y, and z. Then the content of (A.11) is expressed as

$$\begin{pmatrix} c\bar{t} \\ \bar{x} \\ \bar{y} \\ \bar{z} \end{pmatrix} = \begin{pmatrix} \gamma & -\frac{v}{c}\gamma & 0 & 0 \\ -\frac{v}{c}\gamma & \gamma & 0 & 0 \\ 0 & 0 & 1 & 0 \\ 0 & 0 & 0 & 1 \end{pmatrix} \begin{pmatrix} ct \\ x \\ y \\ z \end{pmatrix} \qquad (A.12)$$

with inverse

$$\begin{pmatrix} ct \\ x \\ y \\ z \end{pmatrix} = \begin{pmatrix} \gamma & \frac{v}{c}\gamma & 0 & 0 \\ \frac{v}{c}\gamma & \gamma & 0 & 0 \\ 0 & 0 & 1 & 0 \\ 0 & 0 & 0 & 1 \end{pmatrix} \begin{pmatrix} c\bar{t} \\ \bar{x} \\ \bar{y} \\ \bar{z} \end{pmatrix}. \qquad (A.13)$$

We have focused on the "standard" setup, in which the boost occurs along a shared x axis, but it is easy to see how to modify the matrix in (A.12) to accommodate a boost in the y direction,

$$\begin{pmatrix} c\bar{t} \\ \bar{x} \\ \bar{y} \\ \bar{z} \end{pmatrix} = \begin{pmatrix} \gamma & 0 & -\frac{v}{c}\gamma & 0 \\ 0 & 1 & 0 & 0 \\ -\frac{v}{c}\gamma & 0 & \gamma & 0 \\ 0 & 0 & 0 & 1 \end{pmatrix} \begin{pmatrix} ct \\ x \\ y \\ z \end{pmatrix}, \qquad (A.14)$$

and in the z direction,

$$\begin{pmatrix} c\bar{t} \\ \bar{x} \\ \bar{y} \\ \bar{z} \end{pmatrix} = \begin{pmatrix} \gamma & 0 & 0 & -\frac{v}{c}\gamma \\ 0 & 1 & 0 & 0 \\ 0 & 0 & 1 & 0 \\ -\frac{v}{c}\gamma & 0 & 0 & \gamma \end{pmatrix} \begin{pmatrix} ct \\ x \\ y \\ z \end{pmatrix}. \qquad (A.15)$$

For a boost in the arbitrary direction $\hat{\mathbf{n}}$ with speed v, define $\mathbf{v} \equiv v\hat{\mathbf{n}}$. Then the relevant transformation is

$$\begin{pmatrix} c\bar{t} \\ \bar{x} \\ \bar{y} \\ \bar{z} \end{pmatrix} = \begin{pmatrix} \gamma & -\frac{v^x}{c}\gamma & -\frac{v^y}{c}\gamma & -\frac{v^z}{c}\gamma \\ -\frac{v^x}{c}\gamma & 1 + \frac{(\gamma-1)v^x v^x}{v^2} & \frac{(\gamma-1)v^x v^y}{v^2} & \frac{(\gamma-1)v^x v^z}{v^2} \\ -\frac{v^y}{c}\gamma & \frac{(\gamma-1)v^x v^y}{v^2} & 1 + \frac{(\gamma-1)v^y v^y}{v^2} & \frac{(\gamma-1)v^y v^z}{v^2} \\ -\frac{v^z}{c}\gamma & \frac{(\gamma-1)v^x v^z}{v^2} & \frac{(\gamma-1)v^y v^z}{v^2} & 1 + \frac{(\gamma-1)v^z v^z}{v^2} \end{pmatrix} \begin{pmatrix} ct \\ x \\ y \\ z \end{pmatrix}. $$
$$(A.16)$$

A.2.3 Doppler Shift

Another important observation related to constant velocity relative motion is the Doppler shift of light. Monochromatic light is an example of a wave-like solution to Maxwell's equation and occurs with a well-defined period T related to its frequency in the usual way $f = 1/T$. Take a light source at rest in L. Light is emitted from $x = 0$ at speed c with frequency f. What is the frequency received by an observer moving away from the source, sitting at rest in $\bar{L}$ at $\bar{x} = 0$?

Suppose a maximum of the electromagnetic wave is observed by L and $\bar{L}$ at time $t = 0$ when the origins of the two frames coincide. The next time a maximum comes by the origin of L is at time $t = T$, the period of the light in L. That next peak makes it to the origin of $\bar{L}$ at time t^* in L. Since the origin of $\bar{L}$ is at $x_o = vt$, and the light travels at speed c, the origin of $\bar{L}$ and the maximum of the electric field will overlap when

$$c(t^* - T) = vt^* \longrightarrow t^* = \frac{T}{1 - v/c},\tag{A.17}$$

and the time t^* as measured in $\bar{L}$ using (A.8) (just time dilation again) gives us the period there (the first maximum occurred at 0),

$$ct^* = \gamma(c\bar{t}^*) \longrightarrow \bar{t}^* = \sqrt{1 - \left(\frac{v}{c}\right)^2}\, t^* = \sqrt{\frac{c + v}{c - v}}\, T\tag{A.18}$$

with $\bar{T} = \bar{t}^*$, so the received frequency is

$$\bar{f} = \frac{1}{\bar{T}} = \sqrt{\frac{c - v}{c + v}}\, f.\tag{A.19}$$

This is for an observer moving away form the source. For an observer moving towards the source, we just change the sign of v:

$$\bar{f} = \sqrt{\frac{c + v}{c - v}}\, f.\tag{A.20}$$

A.2.4 Velocity Addition

The fact that observers moving with constant relative velocity measure the same speed of light indicates that the usual one-dimensional velocity addition formula must be updated in special relativity. For the velocity (in one dimension, using the sign to indicate the direction) of a ball, say, relative to $\bar{L}$, $v_{B\bar{L}}$ and the velocity of $\bar{L}$ relative to L, $v_{\bar{L}L} = v$, the nonrelativistic velocity addition formula gives the ball's velocity relative to L as

$$v_{BL} = v_{B\bar{L}} + v_{\bar{L}L},\tag{A.21}$$

and if, instead of a ball, we have a light beam moving with speed c in $\bar{L}$, then that light beam has, according to this formula, speed $c + v > c$ in L.

Let's see how the Lorentz boost gives an updated expression for velocity addition. Using the same setup as above, we can write $v_{B\bar{L}} = \Delta\bar{x}/\Delta\bar{t}$ where $\Delta\bar{x}$ is the distance traveled in $\bar{L}$ over a time interval $\Delta\bar{t}$ there. We can use (A.7) to rewrite these quantities in terms of the corresponding Δt and Δx in L,

$$v_{B\bar{L}} = \frac{\Delta\bar{x}}{\Delta\bar{t}} = \frac{\gamma(-(v/c)c\Delta t + \Delta x)}{\gamma(\Delta t - (v/c^2)\Delta x)} \tag{A.22}$$

with $v_{\bar{L}L} = v$ and $v_{BL} = \Delta x/\Delta t$. We can relate the three speeds by rewriting the above in terms of these combinations,

$$v_{B\bar{L}} = \frac{v_{BL} - v_{\bar{L}L}}{1 - v_{\bar{L}L}v_{BL}/c^2}. \tag{A.23}$$

This equation can be algebraically inverted to give

$$v_{BL} = \frac{v_{B\bar{L}} + v_{\bar{L}L}}{1 + v_{B\bar{L}}v_{\bar{L}L}/c^2}. \tag{A.24}$$

The velocity addition formula here suggests that any observed speed can be at most c (put in c for all velocities on the right-hand-side). This is a more physically relevant way to see that any acceptable reference frame has speed $v \leq c$ than the warning sign present in the boost factor γ (which becomes imaginary for $v > c$).

A.3 Minkowski Diagrams

We can think of relative motion diagramatically using plots of position versus time called "Minkowski diagrams." These traditionally have time pointing vertically with position mapped in the horizontal direction, the opposite of our introductory physics experience. Because of the flipped axes, motion occurring with constant speed v is represented by a line

$$ct = \frac{c}{v}x \tag{A.25}$$

with slope c/v. So a line with a slope of 1 has $v = c$, lines with slope greater than one have $v < c$, and lines with slope less than one have $v > c$ and are not allowed to represent the motion of a particle. The three options are shown in Figure A.2. Note that I am thinking of the ct and x axes as unit (dimensionless) basis vectors, labeled $\hat{t}$ and $\hat{x}$. To plot curves on these axes, we need common units, so that the curves will be of the form $ct(x)$ versus x. When that dimensional equality is important to highlight, I'll label the axes ct and x.

An arbitrary curve can be drawn and its behavior analyzed using a Minkowski diagram. At any point along the curve, there is a line tangent to the

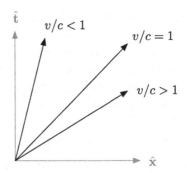

A Minkowski diagram with three types of constant speed motion, associated with lines having slope greater than one ($v < c$), equal to one ($v = c$), and less than one ($v > c$).

curve that defines an infinitesimal interval in the vicinity of a point on the curve. That interval is called "timelike" if the tangent line has a slope greater than one in absolute value, "lightlike" if the tangent has a slope magnitude of 1, and "spacelike" if the tangent line's slope is less than one in magnitude. Three example tangents, one of each type, are shown in Figure A.3.

At any point in spacetime, there is a forward "light cone" defined by rays of light emitted from the point. On a Minkowski diagram, the light cone is represented by the lines of unit slope that can be drawn in the positive and negative x directions. Some points with their light cones are shown in Figure A.4 (if you include another spatial coordinate, you get a cone by rotating the "wedges" found in Figure A.4). Any point B within the forward light cone of a point A can be causally influenced by events at A, since a signal of at most speed c can be sent from A to B. For points outside the light cone of A, like C in Figure A.4, there is no causal contact, no way for information to travel from

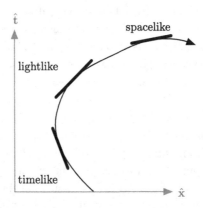

A curve on a Minkowski diagram. Lines tangent at a point define infinitesimal intervals along the curve that are characterized by the slope of the tangent line at the point: timelike intervals for slopes greater than one, lightlike for slope equal to one, and spacelike if the slope is less than one. Notice that this curve has a segment that goes backwards in time and so clearly does not represent allowable particle motion.

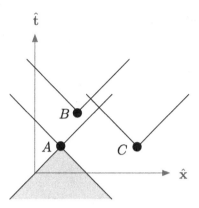

The forward light cones (in increasing t direction) for points A, B, and C. Point B is in the forward light cone of A, whereas point C is not. The shaded backwards light cone for A is shown.

A to C. We can also draw the backwards light cone; the set of all points that can influence A is just the continuation of the forward light cone backwards in time. The backwards light cone at A is shown in Figure A.4.

A.4 Proper Time

For a particle moving along a curve parameterized by s with $\mathbf{r}(s) = ct(s)\hat{\mathbf{t}} + x(s)\hat{\mathbf{x}}$ (working in one spatial, one temporal dimension), the line tangent at a point is given by $\mathbf{r}'(s)$ (primes denote the derivative with respect to a function's argument). If that line has a slope with absolute value greater than one (as it must be for particle motion), then we can make a Lorentz boost to the "rest frame" of the particle by picking our boost in the direction of $\mathbf{r}'(s)$ with speed $v = x'(s)/t'(s)$. What we have done is to generate a local $\bar{L}$ in which the particle is at rest at the origin of $\bar{L}$. Then in an infinitesimal interval, the particle moves only in the temporal direction $cd\bar{t}$. The invariance of the Minkowski length tells us that the distances in L, cdt and dx, are related to $cd\bar{t}$ by

$$ -c^2 d\bar{t}^2 = -c^2 dt^2 + dx^2, \tag{A.26} $$

and we call the time elapsed in this special rest frame the "proper time" of the particle, usually denoted τ. Rewriting the above equation in terms of the curve parameter s and calling $cd\tau$ the time interval in $\bar{L}$,

$$ -c^2 d\tau^2 = \left(-c^2 t'(s)^2 + x'(s)^2\right)ds^2. \tag{A.27} $$

Suppose you use the proper time, the time recorded in the rest frame of the particle moving along the curve, as the parameter for the curve, taking $ds = d\tau$

in (A.27). Then

$$- c^2 = -c^2 t'(\tau)^2 + x'(\tau)^2, \tag{A.28}$$

so the tangent vector has constant magnitude. Finally, note that if you use time as the parameter of the curve, then relation (A.27) provides a way to relate the proper time and coordinate time,

$$- c^2 d\tau^2 = \left(-c^2 \left(\frac{dt}{dt}\right)^2 + \left(\frac{dx}{dt}\right)^2\right) dt^2 \longrightarrow \frac{dt}{d\tau} = \frac{1}{\sqrt{1 - (\dot{x}(t)/c)^2}}, \tag{A.29}$$

or, in three dimensions,

$$\frac{dt}{d\tau} = \frac{1}{\sqrt{1 - (\mathbf{v} \cdot \mathbf{v}/c^2)}}, \qquad \mathbf{v} = \dot{x}(t)\hat{\mathbf{x}} + \dot{y}(t)\hat{\mathbf{y}} + \dot{z}(t)\hat{\mathbf{z}}, \tag{A.30}$$

with dots referring to t derivatives.

In prerelativistic mechanics, time (the coordinate time) is used almost exclusively as the parameter for describing motion curves. But in relativistic mechanics, it is sometimes useful to use proper time parameterization. As an example, take a four-dimensional curve parameterized by time,

$$\dot{x}^\mu(t) \equiv \frac{dx^\mu}{dt} = \begin{pmatrix} c \\ \dot{x}(t) \\ \dot{y}(t) \\ \dot{z}(t) \end{pmatrix} \equiv \begin{pmatrix} c \\ \mathbf{v} \equiv \frac{d\mathbf{r}}{dt} \end{pmatrix}, \tag{A.31}$$

where $d\mathbf{r}$ is the spatial infinitesimal $d\mathbf{r} \equiv dx\hat{\mathbf{x}} + dy\hat{\mathbf{y}} + dz\hat{\mathbf{z}}$. The collection $\dot{x}^\mu(t)$ does not transform nicely, since t transforms as the 0 component of a vector, and the infinitesimal dx^μ transforms as a vector (i.e., as in (A.12)) under a Lorentz boost. The proper time, however, is unchanged by a Lorentz boost,[1] so that

$$\frac{dx^\mu(\tau)}{d\tau} = \begin{pmatrix} c\frac{dt(\tau)}{d\tau} \\ \frac{dx(\tau)}{d\tau} \\ \frac{dy(\tau)}{d\tau} \\ \frac{dz(\tau)}{d\tau} \end{pmatrix} \equiv \begin{pmatrix} c\frac{dt}{d\tau} \\ \frac{d\mathbf{r}}{d\tau} \end{pmatrix}. \tag{A.32}$$

This quantity *is* a vector, meaning that its components respond to a Lorentz boost just as the components of x^μ do in (A.12) (this is the Lorentz-boost-specific definition of the more general vector definition from Section 2.1.2). Using (A.30), we can write this "four-velocity" in terms of t parameterization:

$$\frac{dx^\mu(\tau)}{d\tau} = \frac{dx^\mu}{dt}\frac{dt}{d\tau} = \frac{1}{\sqrt{1 - (\mathbf{v} \cdot \mathbf{v}/c^2)}}\begin{pmatrix} c \\ \mathbf{v} \end{pmatrix}. \tag{A.33}$$

From the four-velocity we can define the four-momentum, $p^\mu = m\frac{dx^\mu}{d\tau}$, again expressible in terms of the familiar velocity in temporal parameteriza-

---

[1] Since the boost preserves the Minkowski length, $-c^2 d\tau^2 \equiv -c^2 dt^2 + dx^2 = -c^2 d\bar{t}^2 + d\bar{x}^2$, and $d\tau$ is the same in both L and $\bar{L}$.

tion:

$$p^\mu = \frac{m}{\sqrt{1 - (\mathbf{v} \cdot \mathbf{v}/c^2)}} \begin{pmatrix} c \\ \mathbf{v} \end{pmatrix}. \tag{A.34}$$

The zero component has units of energy over c to match the momentum units of the rest of the elements. This momentum vector is sometimes called the "energy-momentum" vector, with $E = cp^0$ defining the relativistic energy of a particle, and $\mathbf{p} = m\mathbf{v}/\sqrt{1 - v^2/c^2}$ the relativistic momentum. Let's check that this vector really does transform as we expect.

The frames L and $\bar{L}$ are as shown in Figure A.1. A massive particle moves with speed $\bar{v}$ (relative to $\bar{L}$) to the right, and its energy and x-component of momentum are

$$\bar{E} = \frac{mc^2}{\sqrt{1 - \bar{v}^2/c^2}}, \qquad \bar{p} = \frac{m\bar{v}}{\sqrt{1 - \bar{v}^2/c^2}}. \tag{A.35}$$

In L the speed of the massive particle is

$$w = \frac{v + \bar{v}}{1 + v\bar{v}/c^2} \tag{A.36}$$

using the velocity addition formula from (A.24). The energy and momentum in L are

$$E = \frac{mc^2}{\sqrt{1 - w^2/c^2}}, \qquad p = \frac{mw}{\sqrt{1 - w^2/c^2}}. \tag{A.37}$$

Let

$$\gamma_u \equiv \frac{1}{\sqrt{1 - u^2/c^2}} \tag{A.38}$$

be the boost factor associated with a speed u. Then we can write the energy and momentum in $\bar{L}$ and L as

$$\bar{E} = mc^2 \gamma_{\bar{v}}, \qquad \bar{p} = m\bar{v}\gamma_{\bar{v}}, \qquad E = mc^2 \gamma_w, \qquad p = mw\gamma_w. \tag{A.39}$$

We can show algebraically that

$$\gamma_w = \gamma_v \gamma_{\bar{v}} \left(1 + \frac{v\bar{v}}{c^2} \right). \tag{A.40}$$

Using this identity in the energy and momentum relations,

$$\frac{E}{c} = mc\gamma_v \gamma_{\bar{v}} \left(1 + \frac{v\bar{v}}{c^2} \right) = \gamma_v \left(\frac{\bar{E}}{c} + \frac{v}{c}\bar{p} \right),$$

$$p = \frac{m(v + \bar{v})}{1 + v\bar{v}/c^2} \gamma_v \gamma_{\bar{v}} \left(1 + \frac{v\bar{v}}{c^2} \right) = \gamma_v \left(\bar{p} + \frac{v}{c}\frac{\bar{E}}{c} \right), \tag{A.41}$$

which looks like the inverse transformation of the coordinates (A.8) with $ct \to E/c$ and $x \to p$. Then we know that

$$\frac{\bar{E}}{c} = \gamma \left(\frac{E}{c} - \frac{v}{c}p \right) \qquad \bar{p} = \gamma \left(-\frac{v}{c}\frac{E}{c} + p \right). \tag{A.42}$$

The y and z components of the momentum are the same, $\bar{p}_y = p_y$ and $\bar{p}_z = p_z$, like $\bar{y} = y$ and $\bar{z} = z$.

Just as for the coordinates, we have the invariant

$$-(ct)^2 + x^2 + y^2 + z^2 = -(c\bar{t})^2 + \bar{x}^2 + \bar{y}^2 + \bar{z}^2, \tag{A.43}$$

we also have

$$-\left(\frac{E}{c}\right)^2 + p_x^2 + p_y^2 + p_z^2 = -\left(\frac{\bar{E}}{c}\right)^2 + \bar{p}_x^2 + \bar{p}_y^2 + \bar{p}_z^2. \tag{A.44}$$

Take $\bar{L}$ to be the instantaneous rest frame of the particle. Then $\bar{E} = mc^2$ with all the barred momenta zero. Since this is just a boost away from any other frame, the left-hand side of (A.44) has the constant value

$$-\left(\frac{E}{c}\right)^2 + \mathbf{p} \cdot \mathbf{p} = -m^2 c^2. \tag{A.45}$$

A.5 Dynamics

Relativistic dynamics proceeds from Newton's second law $\frac{d\mathbf{p}}{dt} = \mathbf{F}$ but using the relativistic momentum instead of $\mathbf{p} = m\mathbf{v}$. Then

$$\frac{d}{dt}\left(\frac{m\mathbf{v}}{\sqrt{1 - v^2/c^2}}\right) = \mathbf{F}. \tag{A.46}$$

It is in some ways an odd equation, since (1) we are only looking at the spatial portion of some larger structure, and (2) time is the parameter. If we think of the expanded four-vector setting

$$\frac{dp^\mu}{dt} = \begin{pmatrix} ? \\ \mathbf{F} \end{pmatrix}, \tag{A.47}$$

the zero component in the above is, from the four-momentum definition, $\frac{1}{c}\frac{dE}{dt}$, and we'd like to relate this to the applied force, just as we did for the spatial components using Newton's second law. Taking the temporal derivative of (A.45),

$$\frac{dE}{dt} = \frac{c^2}{E}\mathbf{p} \cdot \frac{d\mathbf{p}}{dt}. \tag{A.48}$$

Using $\frac{d\mathbf{p}}{dt} = \mathbf{F}$ and $\mathbf{p} = \mathbf{v}E/c^2$, the change in the particle's energy as a function of time is

$$\frac{dE}{dt} = \mathbf{F} \cdot \mathbf{v}, \tag{A.49}$$

the total work done by the force acting on the particle, just as in nonrelativistic mechanics. The full set of four equations is now

$$\frac{dp^\mu}{dt} \doteq \begin{pmatrix} \mathbf{F} \cdot \frac{\mathbf{v}}{c} \\ \mathbf{F} \end{pmatrix}. \tag{A.50}$$

The left-hand side of Newton's second law (A.50) does not transform like a vector. The form that we would like to have is the momentum derivative $\frac{dp^\mu}{d\tau}$, which does transform like a vector, and we can relate it to $\frac{dp^\mu}{dt}$ using (A.30) again,

$$\frac{dp^\mu}{d\tau} = \frac{dp^\mu}{dt}\frac{dt}{d\tau} = \frac{dp^\mu}{dt}\frac{1}{\sqrt{1-v^2/c^2}} \doteq \frac{1}{\sqrt{1-v^2/c^2}}\begin{pmatrix}\mathbf{F}\cdot\frac{\mathbf{v}}{c} \\ \mathbf{F}\end{pmatrix}. \tag{A.51}$$

The far right-hand side is a strange combination of force components and speed, but at least we know the whole thing transforms nicely.

Let's work out the force transformation using the vector form in (A.51) taking $\bar{L}$ to be the particle's instantaneous rest frame, so that there

$$\frac{d\bar{p}^\mu}{d\tau} \doteq \begin{pmatrix}0 \\ \bar{\mathbf{F}}\end{pmatrix}. \tag{A.52}$$

The particle is moving through the frame L. If we align the x axis with the particle's instantaneous velocity vector ($\mathbf{v} = v\hat{\mathbf{x}}$), then in L we have

$$\frac{dp^\mu}{d\tau} = \frac{1}{\sqrt{1-v^2/c^2}}\begin{pmatrix}\frac{F_x v}{c} \\ \mathbf{F}\end{pmatrix} = \gamma\begin{pmatrix}\frac{F_x v}{c} \\ \mathbf{F}\end{pmatrix}. \tag{A.53}$$

The usual Lorentz transformation for a vector then gives, for $\bar{F}_x$ and $\bar{F}_y$,

$$\bar{F}_x = \gamma\left(-\frac{v}{c}\gamma\frac{F_x v}{c} + \gamma F_x\right) = F_x, \qquad \bar{F}_y = \gamma F_y. \tag{A.54}$$

In general, the component of force that is parallel to the velocity vector of the particle at some time t has $F_\| = \bar{F}_\|$, whereas the components that are perpendicular to the instantaneous velocity vector are related to the rest-frame forces by $\mathbf{F}_\perp = \bar{\mathbf{F}}_\perp/\gamma$.

A.6 Electricity and Magnetism

E&M is a manifestly relativistic theory, and its familiar elements can be cast into simple relativistic building blocks. Start with the sources – the charge density ρ and $\mathbf{J}$ form a vector under Lorentz transformation. We'll show that this is true explicitly using our frames L and $\bar{L}$ moving with constant relative speed v along a shared x axis. Suppose in $\bar{L}$ you observe a point charge q moving with speed $\bar{v}$ to the right. Its density is $\bar{\rho} = q\delta(\bar{x} - \bar{v}\bar{t})$, and its current density is $\bar{J} = \bar{\rho}\bar{v}$. We need to be careful with the transformation of the Dirac delta function. Remember that $\delta(kx) = \delta(x)/|k|$ for constant k. So when we evaluate the delta for $\bar{x} = \gamma x$, say, we have $\delta(\bar{x}) = \delta(x)/\gamma$; it transforms as a density, not a scalar.

Moving to L, we have $\rho = q\delta(x - wt)$ and $J = \rho w$, where w is the particle's velocity relative to L from (A.36). Using the relation between coordinates provided by the Lorentz transformation, we have

$$\bar{\rho} = q\delta\big(\gamma(x - vt) - \bar{v}\gamma\big(t - v/c^2 x\big)\big)$$
$$= q\delta\left(\gamma\left(1 + \frac{v\bar{v}}{c^2}\right)(x - wt)\right) = \frac{\rho}{\gamma(1 + v\bar{v}/c^2)} \tag{A.55}$$

or

$$\rho c = \gamma\left(\bar{\rho}c + \frac{v}{c}\bar{J}\right). \tag{A.56}$$

For J, we have

$$\bar{J} = \bar{\rho}\bar{v} = \frac{\rho\bar{v}}{\gamma(1 + v\bar{v}/c^2)} = \frac{\rho}{\gamma}\frac{(w - v)}{1 - v^2/c^2} = \gamma\left(J - \frac{v}{c}\rho c\right). \tag{A.57}$$

Taken together, we recover the transformation for the density and current:

$$\bar{\rho}c = \gamma\left(\rho c - \frac{v}{c}J\right) \qquad \bar{J} = \gamma\left(J - \frac{v}{c}\rho c\right). \tag{A.58}$$

The combined vector current density J^μ has $J^0 = \rho c$ with spatial components $\mathbf{J} = \rho\mathbf{v}$. Here we have established the transformation of charge and current density for a point source, but we can add point sources together to build up any source we like. The continuity equation $\partial_\mu J^\mu = 0$ is a scalar and so holds in any coordinate system related by a boost, with the gradient operator ∂_μ undoing the transformation of J^μ.

The electric potential V and magnetic vector potential $\mathbf{A}$ also form a four-vector called A^μ (the "four-potential") with $A^0 = V/c$ and spatial components $\mathbf{A}$. Perhaps the easiest way to see this is from Maxwell's equations, these relate A^μ to its sources J^μ, which, as we have already established, is a vector,

$$\Box A^\mu = -\mu_0 J^\mu \tag{A.59}$$

for scalar $\Box \equiv \partial_\mu \partial^\mu$. Since the right-hand side is a vector, so is the left-hand side. Of course, this equation only holds in Lorenz gauge, where $\partial_\mu A^\mu = 0$, but as with the conservation of charge, once the Lorenz gauge is enforced in a frame L, it holds in any other frame. It is amusing to think of the implications for "A^μ conservation" in this gauge, where V/c plays the role of ρc^2 and $\mathbf{A}$ of $\mathbf{J}$.

Moving on to the fields, the natural derivative object associated with the potentials is the field strength tensor $F_{\mu\nu} \equiv \partial_\mu A_\nu - \partial_\nu A_\mu$; this democratically treats gradient indices and potential indices. As we saw in Section 2.8.1, it also has the advantage of eliminating dependence on Christoffel connections, which would otherwise come up in different coordinate systems. It is the upper form of component tensors that we associate with physical quantities like $\mathbf{E}$ and $\mathbf{B}$, so using two factors of the Minkowski metric, we define the contravariant $F^{\alpha\beta} = g^{\alpha\mu}g^{\beta\nu}F_{\mu\nu}$, and the entries here can be related to $\mathbf{E} = -\nabla V - \frac{\partial \mathbf{A}}{\partial t}$ and

$\mathbf{B} = \nabla \times \mathbf{A}$:

$$F^{\mu\nu} \doteq \begin{pmatrix} 0 & \frac{E^x}{c} & \frac{E^y}{c} & \frac{E^z}{c} \\ -\frac{E^x}{c} & 0 & B^z & -B^y \\ -\frac{E^y}{c} & -B^z & 0 & B^x \\ -\frac{E^z}{c} & B^y & -B^x & 0 \end{pmatrix}. \tag{A.60}$$

Since $F^{\mu\nu}$ is a second-rank contravariant tensor, it transforms with two factors of the boost matrix, itself just a way to keep track of $\frac{\partial \bar{x}^\mu}{\partial x^\alpha}$. If we denote the entries of the matrix in (A.12) (for example) by L^μ_α, then the response of $F^{\mu\nu}$ to a Lorentz transformation is

$$\bar{F}^{\mu\nu} = L^\mu_\alpha L^\nu_\beta F^{\alpha\beta}, \tag{A.61}$$

and from this rule, together with the embedding shown in (A.60), we can extract the transformation of the electric and magnetic field components.

B   Runge–Kutta Methods

In this appendix, we'll review the Runge–Kutta class of methods for solving ODEs. These are widely used and almost universally applicable, although there are better (and far worse) methods that can be used for specific problems. The method we end with here, a fourth-order Runge–Kutta (see [39] for a good discussion of the derivation of the method) is not a bad one to use when computing the geodesics, both timelike and lightlike, for the Schwarzschild and Kerr geometries. Because they are not specific to gravity, I'll present the method(s) in fairly general form. We'll continue to write expressions using Einstein summation notation, but all of the "vectors" in this appendix are simply labeled collections of functions, with no transformation property or even consistent relation between components. Roman indices will be used, again to highlight the difference with respect to other sections of the book, and will run from $1 \rightarrow n$, where n is the number of items in our collection. The index placement indicates nothing and will almost always appear down.

B.1   Problem Setup

The Runge–Kutta procedure assumes a set of first-order ODEs. Taking s to be the parameter (like time, proper time, or the parameter used for lightlike geodesic motion), we have a set of unknown target functions $\{f_i(s)\}_{i=1}^n$, their derivatives, a provided set of functions $\{G_j(s, \mathbf{f}(s))\}_{j=1}^n$, the initial values $\{h_i(0)\}_{i=1}^n$, also given (these could be given for any value of s, and we'll agree to "set the clock" at $s = 0$ for simplicity – you can move the known values to any relevant s location specific to your problem). Then the set of problems whose solution we will approximate numerically is

$$\frac{df_i(s)}{ds} = G_i\big(s, \mathbf{f}(s)\big) \text{ for } i = 1 \rightarrow n \text{ with } f_i(0) = h_i(0). \qquad \text{(B.1)}$$

The functional dependence on $\mathbf{f}$ in $G_i(s, \mathbf{f}(s))$ is meant to indicate a dependence on any/all elements of the set $\{f_i(s)\}_{i=1}^n$. It may seem like a pretty specific setup, but in fact any set of ODEs of any order can be written as a set of first-order ODEs. That idea is familiar from classical mechanics, where we move from Newton's second law, three second-order ODEs for a single particle, to a Hamiltonian description with its six first-order ODEs.

Example B.1 (Simple Harmonic Oscillator). Let's review and practice the procedure or turning multidimensional multiorder ODEs into a set of first-order ODEs with the example

$$m\ddot{x}(s) = -m\omega^2 x(s),$$

again using s as the parameter to be consistent with the rest of this section. This ODE second order in the one-dimensional position coordinate $x(s)$. We're given the initial position $x(0) = x_0$ and velocity $\dot{x}(0) = v_0$. How should we define the set of functions $\{f_i(s)\}_{i=1}^n$, and how many do we expect? The initial values provides a guide: Take $f_1(s) = x(s)$ and $f_2(s) = \dot{x}(s)$; then the ODEs are

$$\frac{d}{ds}\begin{pmatrix} f_1(s) \\ f_2(s) \end{pmatrix} = \begin{pmatrix} f_2(s) \\ -\omega^2 f_1(s) \end{pmatrix}, \qquad \begin{pmatrix} f_1(0) \\ f_2(0) \end{pmatrix} = \begin{pmatrix} x_0 \\ v_0 \end{pmatrix}. \tag{B.2}$$

In this problem, we have $G_1(s, \mathbf{f}) = f_2$ and $G_2(s, \mathbf{f}) = -\omega^2 f_1$.

Example B.2 (A Set of Higher-Order ODEs). Suppose we are given the pair of equations

$$\frac{d^3 u(s)}{ds} = 2u(s)\frac{du(s)}{ds} + v(s), \qquad \frac{d^2 v(s)}{ds^2} = v(s)\frac{d^2 u(s)}{ds^2} + u(s)\frac{dv(s)}{ds} \tag{B.3}$$

together with initial conditions

$$u(0) = a, \qquad \frac{du(s)}{ds}\bigg|_{s=0} = b, \qquad \frac{d^2 u(s)}{ds^2}\bigg|_{s=0} = c, \tag{B.4}$$

$$v(0) = d, \qquad \frac{dv(s)}{ds}\bigg|_{s=0} = e.$$

We can put these together in vector form by setting $f_1(s) = u(s)$, $f_2(s) = \frac{du(s)}{ds}$, $f_3(s) = \frac{d^2 u(s)}{ds^2}$, $f_4(s) = v(s)$, and $f_5(s) = \frac{dv(s)}{ds}$. Then the ODEs read

$$\frac{d}{ds}\begin{pmatrix} f_1(s) \\ f_2(s) \\ f_3(s) \\ f_4(s) \\ f_5(s) \end{pmatrix} = \begin{pmatrix} f_2(s) \\ f_3(s) \\ 2f_1(s)f_2(s) + f_4(s) \\ f_5(s) \\ f_4(s)f_3(s) + f_1(s)f_5(s) \end{pmatrix} \tag{B.5}$$

with $f_1(0) = a$, $f_2(0) = b$, $f_3(0) = c$, $f_4(0) = d$, and $f_5(0) = e$. The ordering of $u(s)$ and $v(s)$ and their derivatives in $\{f_i(s)\}_{i=1}^5$ is arbitrary, but once you've chosen an order, you have to stick to it.

Whatever set of ODEs you are given, start by generating the first-order set and the functions appearing on the right, $\{G_i(s, \mathbf{f})\}_{i=1}^n$.

B.2 Euler's Method

Suppose we had a solution to (B.1), a set of functions $\{f_i(s)\}_{i=1}^n$ that satisfied the ODE and initial values. At any parameter value s, we could estimate the value of the functions at a nearby $s + \Delta s$ using Taylor expansion:

$$f_i(s + \Delta s) \approx f_i(s) + \Delta s \frac{df_i(s)}{ds} = f_i(s) + \Delta s G_i\big(s, \mathbf{f}(s)\big), \qquad (B.6)$$

where we used the fact that f_i is a solution to (B.1) to replace the derivative with the function G_i evaluated at s (with its $\mathbf{f}(s)$ dependence indicating that all evaluations of the functions $\{f_i(s)\}_{i=1}^n$ appearing in $G_i(s, \mathbf{f}(s))$ are also evaluated at parameter value s).

The approximation in (B.6) is a statement about the solution of (B.1), but we can turn it around and use the observation to define a method that approximates the solution. Make a grid in s by defining a fixed Δs and then forming the set of parameter values $s_j \equiv j\Delta s$ for $j = 0, 1, 2, \ldots$. We can "project" the solution functions onto the grid, which just means evaluating them at the grid points. For the jth grid point, s_j, the approximation (B.6) is

$$f_i(s_{j+1}) \approx f_i(s_j) + \Delta s G_i\big(s_j, \mathbf{f}(s_j)\big). \qquad (B.7)$$

This equation relates the values of $\mathbf{f}$ at $s_{j+1} = s_j + \Delta s$ to the values at s_j and so represents an update of sorts.

Let's take it seriously and define the set of discrete functions $\{\bar{f}_i^j\}_{i=1}^n$ for $j = 0, 1, 2, \ldots$, where the goal is to develop a recursion relation that lets $\bar{f}_i^j$ approximate the ith solution function f_i evaluated at s_j: We want $\bar{f}_i^j \approx f_i(s_j)$. Using (B.7) as our guide, set $\bar{f}_i^0 = h_i$, so that we start off at the exactly correct initial value and then take

$$\bar{f}_i^{j+1} = \bar{f}_i^j + \Delta s G_i\big(s_j, \bar{\mathbf{f}}^j\big), \qquad (B.8)$$

where $\bar{\mathbf{f}}^j$ indicates the dependence of the function G_i on a combination of the approximations $\bar{\mathbf{f}}$ evaluated at the current s_j. We have access to those values, and so can use (B.8) as a well-defined recursion taking us from $j = 0$ to any finite integer value. The update scheme in (B.8) is known as the "explicit Euler's method."

How well does it work – does $\bar{f}_i^j$ really approximate $f_i(s_j)$? We know that at each step, an error of size Δs^2 is made, which would be the next term in the Taylor expansion appearing in (B.6), and we have dropped it in our approximation. As Δs is shrunk down, the error per step gets smaller, and the method becomes more accurate, meaning that if you started from the exact value of $\bar{f}_i^j = f_i(s_j)$ and took a step, then you'd end up with perfect agreement as $\Delta s \to 0$, $\bar{f}_i^{j+1} = f_i(s_{j+1})$. A method that has this property is called "accurate" or "consistent." But that is not enough to ensure that starting from $j = 0$, we have $\bar{f}_i^j \approx f_i(s_j)$, since we take multiple steps, and only start out with the absolutely correct values at $j = 0$. It could be that the method makes error at each step that is in the same direction. For example, it might overshoot the true value by an amount Δs^2 at each step. Then after $\sim 1/\Delta s^2$ steps, the error would be of order one, and all overestimates. We'll quantify this in a moment, but in general, a method that accumulates error over multiple steps is an "unstable" method. To get a good approximation, $\bar{f}_i^j \to f_i(s_j)$ as $\Delta s \to 0$, a method (update scheme) must be both accurate and stable. Each step must do a good job approximating the solution $f_i(s_{j+1})$, provided that it starts at $\bar{f}_i^j = f_i(s_j)$ (accuracy), and the errors cannot grow over multiple applications of the method, multiple steps (stability). We won't prove it, but this discussion sketches the motivation behind the numerical analyst's mantra: "accuracy plus stability gives convergence."[1]

Example B.3 (The Explicit Euler Method Is Unstable). The stability of a method for solving ODEs depends on the form of the ODE in addition to the details of the method. In particular, it is not always easy to determine if a numerical method is stable for a nonlinear set of ODEs. One way to think about nonlinear stability is to imagine linearizing the ODEs and applying the method to a series of linear(ized) problems.

Let's take the linear model problem, $\ddot{x}(s) = -\omega^2 x(s)$ with $x(0) = x_0$ and $\dot{x}(0) = v_0$. Written in the form of (B.1), we have $f_1(s) = x(s)$, $f_2(s) = \dot{x}(s)$, $G_1 = f_2$, $G_2 = -\omega^2 f_1$ from Example B.1. We can write the vector $\mathbf{G}$ as a matrix times the vector $\mathbf{f}(s)$:

$$\mathbf{G} \doteq \underbrace{\begin{pmatrix} 0 & 1 \\ -\omega^2 & 0 \end{pmatrix}}_{\equiv \mathbb{A}} \mathbf{f}(s). \tag{B.9}$$

Euler's method from (B.8), applied to this problem, can be similarly written in matrix form. Let $\bar{\mathbf{f}}^j$ be the vector of approximations at time level j. Then (B.8) is

$$\bar{\mathbf{f}}^{j+1} = (\mathbb{I} + \Delta s \mathbb{A})\bar{\mathbf{f}}^j, \qquad \bar{\mathbf{f}}^0 \doteq \begin{pmatrix} x_0 \\ v_0 \end{pmatrix}. \tag{B.10}$$

We can solve the recursion exactly in this example,

$$\bar{\mathbf{f}}^j = (\mathbb{I} + \Delta s \mathbb{A})^j \bar{\mathbf{f}}^0, \tag{B.11}$$

and the question of whether or not the solution has growing norm (indicating that the solution is unstable) is really a question about the eigenvalues of the matrix $\mathbb{M} \equiv \mathbb{I} + \Delta s \mathbb{A}$. If you take a random initial vector that can be decomposed into the eigenvectors of $\mathbb{M}$ and hit it with $\mathbb{M}$ over and over, as the explicit Euler method instructs us to do, then the eigenvector with the largest (absolute)

---

[1] This definition of stability is fairly restrictive and doesn't necessarily capture the methods with the best stability properties. As with most definitions of stability, it also depends on the specific set of ODEs to which it is applied. Nevertheless, it is *a* definition that is easy to understand and demonstrate with simple examples, for which there is no better motivation.

eigenvalue λ_{max} contributes $|\lambda_{max}|^j$, and each successive multiplication by $\mathbb{M}$ rotates the initial vector more and more in the direction of the eigenvector associated with λ_{max}. So, for large j, $|\bar{\mathbf{f}}^j| \sim |\lambda_{max}|^j$. If the magnitude of that largest eigenvalue is greater than one, then the solution vector $\bar{\mathbf{f}}^j$ grows in norm, and the method is unstable when applied to the linear problem.

In the present case, we can easily calculate the eigenvalues of the matrix $\mathbb{M}$; they are $\lambda_{\pm} = 1 \pm i\omega\Delta s$. The maximum absolute value of both of these is $|\lambda_{max}| = \sqrt{1 + \omega^2\Delta s^2}$, which is strictly greater than one for all values of $\Delta s > 0$. Any set of initial values will lead to a growing solution for any choice of Δs. The method is unstable when applied to this problem, according to the definition provided above.

There is a version of Euler's method that has the same level of accuracy but is also stable. The "implicit Euler's method" just replaces the evaluation $G_i(s_j, \bar{\mathbf{f}}^j)$ on the right of (B.8) with $G_i(s_{j+1}, \bar{\mathbf{f}}^{j+1})$. In the context of Example B.3, the update becomes

$$\bar{\mathbf{f}}^{j+1} = \bar{\mathbf{f}}^j + \Delta s \mathbb{A}\bar{\mathbf{f}}^{j+1} \longrightarrow (\mathbb{I} - \Delta s\mathbb{A})\bar{\mathbf{f}}^{j+1} = \bar{\mathbf{f}}^j, \quad (\text{B.12})$$

and again the recursion can be solved,

$$\bar{\mathbf{f}}^j = \left[(\mathbb{I} - \Delta s\mathbb{A})^{-1}\right]^j \bar{\mathbf{f}}^0. \quad (\text{B.13})$$

This time, it is the maximum eigenvalue of $(\mathbb{I} - \Delta s\mathbb{A})^{-1}$ that matters. The norm of the maximum eigenvalue is

$$|\lambda_{max}| = \left(1 + \omega^2\Delta s^2\right)^{-1/2}, \quad (\text{B.14})$$

and this means that the solution will *decay* for all choices of Δs. That's not obviously better than growth, but at least a decay mimics the energy loss associated with, for example, friction, and so makes physical sense.[2]

B.3 Higher-Order Runge–Kutta

We can improve the accuracy and, as it turns out, the stability of the explicit Euler method. The logic behind the improvement serves to define the Runge–Kutta class of methods, which is, roughly speaking, to achieve higher accuracy by keeping more terms in the Taylor expansion from (B.6) but organized cleverly into shifted evaluations of the functions $\{G_i(s, \mathbf{f})\}_{i=1}^n$. We'll work through the procedure to extend explicit Euler and then quote the higher-order method that is used to compute the geodesics numerically in Section 4.5.3.

---

[2] Indeed, there are damped oscillatory situations that use this numerical decay to replace some of the physical damping present in the system.

Going back to (B.7), our goal is to expand $f_i(s + \Delta s)$ beyond first order:

$$\begin{aligned}
f_i(s + \Delta s) &\approx f_i(s) + \Delta s \frac{df_i(s)}{ds} + \frac{1}{2}\Delta s^2 \frac{d^2 f_i(s)}{ds^2} \\
&= f_i(s) + \Delta s\, G_i(s, \mathbf{f}(s)) + \frac{1}{2}\Delta s^2 \frac{dG_i(s, \mathbf{f}(s))}{ds}.
\end{aligned} \tag{B.15}$$

The s-derivative of G_i depends on both the explicit reference to s found in G_i (as, for example, with time-dependent forcing in mechanics) and the implicit s lurking inside the functions $\mathbf{f}(s)$. All together, then, and using summation notation,

$$\frac{dG_i(s, \mathbf{f}(s))}{ds} = \frac{\partial G_i}{\partial s} + \frac{\partial G_i}{\partial f_j}\frac{df_j}{ds} = \frac{\partial G_i}{\partial s} + \frac{\partial G_i}{\partial f_j} G_j, \tag{B.16}$$

where we have again used the assumption that $f_i(s)$ satisfies the original ODE to rewrite its derivative in terms of G_i. The expansion in (B.15) can be written as

$$f_i(s + \Delta s) \approx f_i(s) + \Delta s \left[G_i(s, \mathbf{f}(s)) + \frac{1}{2}\Delta s \left(\frac{\partial G_i}{\partial s} + \frac{\partial G_i}{\partial f_j} G_j \right) \right], \tag{B.17}$$

where the error per step is now of order Δs^3.

To make progress, think of the Taylor expansion of the function G_i with arguments shifted off the natural evaluation point at s:

$$G_i\left(s + \frac{1}{2}\Delta s, \mathbf{f}(s) + \frac{1}{2}\Delta s\, \mathbf{G}(s, \mathbf{f}(s))\right) \approx G_i + \frac{1}{2}\Delta s \frac{\partial G_i}{\partial s} + \frac{1}{2}\Delta s \frac{\partial G_i}{\partial f_j}\frac{\partial f_j}{\partial s} \tag{B.18}$$

with all functions evaluated at s (and $\mathbf{f}(s)$ for $\mathbf{G}$). Using $\frac{df_j}{ds} = G_j(s, \mathbf{f}(s))$, the right-hand side looks just like the term in brackets in (B.17). So we could write the approximation in that equation as

$$f_i(s + \Delta s) \approx f_i(s) + \Delta s\, G_i\left(s + \frac{1}{2}\Delta s, \mathbf{f}(s) + \frac{1}{2}\Delta s\, \mathbf{G}(s, \mathbf{f}(s))\right). \tag{B.19}$$

We still have error order $O(\Delta s^3)$ in this approximation (albeit a different error from the one in (B.17)), but the form in (B.19) is more compact – it does not require us to evaluate the derivatives of $\mathbf{G}$.

As before, we use (B.17) to define a recursion that generates numerical approximations to $\mathbf{f}(s)$ on a grid with fixed spacing. The "midpoint Runge–Kutta" method is defined by

$$\begin{aligned}
\bar{f}_i^0 &= h_i, \\
\bar{f}_i^{j+1} &= \bar{f}_i^j + \Delta s\, G_i\left(s_j + \frac{1}{2}\Delta s, \bar{\mathbf{f}}^j + \frac{1}{2}\Delta s\, \mathbf{G}(s_j, \bar{\mathbf{f}}^j)\right).
\end{aligned} \tag{B.20}$$

It is difficult, with all the parentheses, to see how the update proceeds, so the update portion of the method is often defined in stages:

$$\begin{aligned}
\bar{k}_i &\equiv \Delta s\, G_i(s_j, \bar{\mathbf{f}}^j), \\
\bar{\ell}_i &\equiv \Delta s\, G_i(s_j + \Delta s/2, \bar{\mathbf{f}}^j + \bar{\mathbf{k}}/2), \\
\bar{f}_i^{j+1} &= \bar{f}_i^j + \bar{\ell}_i.
\end{aligned} \tag{B.21}$$

The intermediary $\{\bar{k}_i\}_{i=1}^n$ (abbreviated by $\bar{\mathbf{k}}$ in the evaluation of G_i) makes it clear that we are evaluating the function G_i twice, at shifted locations, to gain accuracy.

Continuing on, the most common variant is "fourth-order Runge–Kutta," which has accuracy per step of Δs^5. As always, the recursion starts from the exact initial conditions $\bar{f}_i^0 = h_i$. Written in the segmented form of (B.21) with two additional evaluations to achieve the higher accuracy, the method is defined by the update:

$$
\begin{aligned}
\bar{k}_i &\equiv \Delta s\, G_i\big(s_j, \bar{\mathbf{f}}^j\big), \\
\bar{\ell}_i &\equiv \Delta s\, G_i\big(s_j + \Delta s/2, \bar{\mathbf{f}}^j + \bar{\mathbf{k}}/2\big), \\
\bar{m}_i &\equiv \Delta s\, G_i\big(s_j + \Delta s/2, \bar{\mathbf{f}}^j + \bar{\boldsymbol{\ell}}/2\big), \\
\bar{n}_i &\equiv \Delta s\, G_i\big(s_j + \Delta s, \bar{\mathbf{f}}^j + \bar{\mathbf{m}}\big), \\
\bar{f}_i^{j+1} &= \bar{f}_i^j + \frac{1}{3}\left(\frac{\bar{k}_i}{2} + \bar{\ell}_i + \bar{m}_i + \frac{\bar{n}_i}{2}\right).
\end{aligned}
\tag{B.22}
$$

Whereas the midpoint method is unstable for the linear problem in Example B.3, the fourth-order method can be made stable by appropriate choice of Δs: $0 < \Delta s \le \sqrt{6}/\omega$. For more complicated problems, like the geodesics from Section 4.5.3, it is harder to prove that the fourth-order Runge–Kutta method is stable (or even define stability in that general setting). One quick check that you should always perform is to run the numerical solution out to some final $k\Delta s$ for integer k with step Δs and compare that with what you get by running out twice as far with step $\Delta s/2$ – if the method is stable, then you should get back similar quantities (and you can keep going by running out ten times as far with step $\Delta s/10$). This is just a rough check (it cannot establish stability), but it is a quick and harmless one.

Curvature in $D = 1, 2$

In this appendix, we'll define and discuss the notion of curvature in one and two dimensions. These definitions and the associated algorithms for computing curvature serve to build some intuition about what curvature is in the generally familiar case of spatial curves and surfaces – good discussions that formed the basis of this appendix can be found in [28, 47]. In general relativity, there are three main differences from the examples found here: (1) The dimension is higher, and curvature occurs in space-*time*; (2) The curved spacetime is not generally expressed as an embedded surface in a higher-dimensional flat space; and (3) The natural analogue of the curvatures in this appendix, the Ricci scalar, vanishes in vacuum and so does not provide a reliable description of the curvature that we are interested in (that description requires the full Riemann tensor, which is much simpler in two dimensions).

C.1 Curvature for Curves

It is easy to visually inspect a curve and determine if it is a straight line or not. But what if such visual inspection is unavailable? And how should we measure the deviation of a curve from a straight line? Motivated by the geodesics of Section 2.4.1, we'll consider parameterized curves for which $\mathbf{r}(s)$ is the vector pointing from the origin to the location of the curve at parametric value s as shown in Figure C.1. That figure displays a typical "plane curve" in which all the points lie in a plane.

At any point $\mathbf{r}(s)$ along the curve, the tangent vector is given by $\mathbf{t}(s) \equiv \mathbf{r}'(s) \equiv \frac{d\mathbf{r}(s)}{ds}$. We can choose the parameter s to be "arc length," for which $\mathbf{r}'(s) \cdot \mathbf{r}'(s) = 1$, so that $\mathbf{r}'(s)$ is a *unit* tangent vector. Taking the s derivative, we have

$$2\mathbf{r}''(s) \cdot \mathbf{r}'(s) = 0, \tag{C.1}$$

which shows that $\mathbf{r}''(s) \perp \mathbf{r}'(s)$ (as drawn in Figure C.1) for all values of s.

In general, a line in arc length parameterization has the form $\mathbf{r}(s) = \mathbf{A} + \hat{\mathbf{B}}s$, where $\mathbf{A}$ is some fixed vector, and $\hat{\mathbf{B}}$ is a unit vector (so that $\mathbf{r}'(s) \cdot \mathbf{r}'(s) = 1$). Then $\mathbf{r}'(s) = \hat{\mathbf{B}}$ and $\mathbf{r}''(s) = 0$. The line has vanishing second derivative, suggesting that $\mathbf{r}''(s)$ is a good indicator of "curvature." For our next case, take a circle of radius R with $\mathbf{r}(s) = R \cos(s/R)\hat{\mathbf{x}} + R \sin(s/R)\hat{\mathbf{y}}$. Then $\mathbf{r}'(s) =$

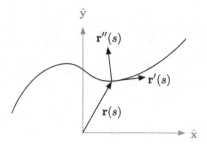

A plane curve parameterized by s.

$-\sin(s/R)\hat{\mathbf{x}} + \cos(s/R)\hat{\mathbf{y}}$ is the unit tangent vector, and

$$\mathbf{r}''(s) = -\frac{1}{R}\left[\cos(s/R)\hat{\mathbf{x}} - \sin(s/R)\hat{\mathbf{y}}\right] = -\frac{1}{R^2}\mathbf{r}(s). \qquad (C.2)$$

This time, the second derivative of $\mathbf{r}(s)$ is nonzero. As expected from (C.1), it is orthogonal to the curve's tangent vector. The "curvature vector" is defined as the second derivative in arc length parameterization, $\boldsymbol{\kappa}(s) \equiv \mathbf{r}''(s)$. The magnitude of $\mathbf{r}''(s)$ in this case is $1/R$ and is defined to be the (scalar) "curvature" of the curve. The inverse of the curvature $\rho \equiv 1/\kappa$ is called the "radius of curvature." The name makes sense for our example, with $\kappa = 1/R$ and $\rho \equiv 1/\kappa = R$ giving the radius of the circle that forms the curve. For curves that are not circles, the radius of curvature tells you the radius of the "osculating" circle that is tangent to any point along the curve and also overlaps at nearby points:[1] If you place an osculating circle of radius ρ centered at $\mathbf{r}(s) + \rho\hat{\boldsymbol{\kappa}}(s)$, then that circle will touch the curve at $\mathbf{r}(s)$, with matching tangent and curvature vectors (see examples below).

If the curve is not parameterized using arc length, then we can still develop the unit tangent vector and curvature vector $\boldsymbol{\kappa}$. Suppose the curve is parameterized by u with $\mathbf{r}(u)$ pointing from the origin to a point along the curve. Thinking of u as a function of the arc length parameter s, we can take the s derivative of $\mathbf{r}(u(s))$ using the chain rule:

$$\frac{d\mathbf{r}(u(s))}{ds} = \frac{d\mathbf{r}(u)}{du}\frac{du}{ds}, \qquad (C.3)$$

and since $\frac{d\mathbf{r}}{ds} \cdot \frac{d\mathbf{r}}{ds} = 1$ by definition, we have

$$\frac{d\mathbf{r}(u)}{du} \cdot \frac{d\mathbf{r}(s)}{du}\left(\frac{du}{ds}\right)^2 = 1 \longrightarrow \frac{du}{ds} = \frac{1}{\sqrt{\frac{d\mathbf{r}(u)}{du} \cdot \frac{d\mathbf{r}(u)}{du}}}, \qquad (C.4)$$

giving us the mapping

$$\frac{d}{ds} \rightarrow \frac{1}{\sqrt{\frac{d\mathbf{r}(u)}{du} \cdot \frac{d\mathbf{r}(u)}{du}}}\frac{d}{du}. \qquad (C.5)$$

[1]   You can make a circle of any radius "touch" a point along a curve; the osculating circle has "second-order touching," in which an infinitesimal arc overlaps the curve in the vicinity of a point.

The vector $\frac{d\mathbf{r}(u)}{du}$ is still tangent to the curve, and we can form the unit tangent vector in u parameterization:

$$\hat{\mathbf{t}}(u) \equiv \frac{\frac{d\mathbf{r}(u)}{du}}{\sqrt{\frac{d\mathbf{r}(u)}{du} \cdot \frac{d\mathbf{r}(u)}{du}}}. \tag{C.6}$$

For the curvature vector $\boldsymbol{\kappa} = \frac{d\hat{\mathbf{t}}}{ds}$, the replacement in (C.5) gives, in arbitrary u parameterization,

$$\boldsymbol{\kappa}(u) = \frac{1}{\sqrt{\frac{d\mathbf{r}(u)}{du} \cdot \frac{d\mathbf{r}(u)}{du}}} \frac{d}{du} \left(\frac{\frac{d\mathbf{r}(u)}{du}}{\sqrt{\frac{d\mathbf{r}(u)}{du} \cdot \frac{d\mathbf{r}(u)}{du}}} \right). \tag{C.7}$$

This expression reduces correctly in the $u = s$ case, for which $\mathbf{r}' \cdot \mathbf{r}' = 1$.

Example C.1 (Curvature of a Parabola). Take the parabolic curve $\mathbf{r}(u) = u\hat{\mathbf{x}} + ku^2\hat{\mathbf{y}}$ for constant $k > 0$. Running this position vector through (C.7) gives the curvature vector

$$\boldsymbol{\kappa}(u) = \frac{1}{(1 + 4k^2u^2)^2} \left(-4k^2u\hat{\mathbf{x}} + 2k\hat{\mathbf{y}} \right) \tag{C.8}$$

with radius of curvature

$$\rho(u) \equiv \frac{1}{\kappa(u)} = \frac{1}{2k}\left(1 + 4k^2u^2\right)^{3/2}. \tag{C.9}$$

A plot of the parabola and the osculating circles centered at $\mathbf{r}(u) + \rho(u)\hat{\boldsymbol{\kappa}}(u)$ with radius $\rho(u)$ for two different values of u are shown in Figure C.2.

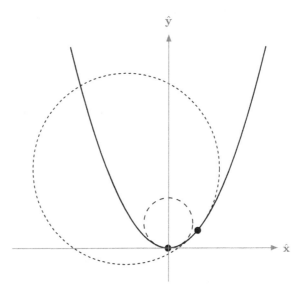

Fig. C.2 The parabolic curve together with the tangent circles at two points marked with dots.

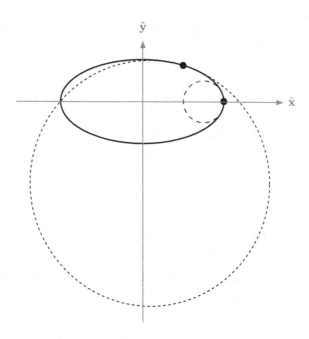

The ellipse with parametric representation $\mathbf{r}(u) = a\cos(u)\hat{\mathbf{x}} + b\sin(u)\hat{\mathbf{y}}$. At the two dots, the circles with the same curvature are shown, tangent to the curve.

Example C.2 (Curvature of an Ellipse). For the elliptical curve with $\mathbf{r}(u) = a\cos(u)\hat{\mathbf{x}} + b\sin(u)\hat{\mathbf{y}}$ (for constants a and b), the curvature vector from (C.7) is

$$\boldsymbol{\kappa}(u) = -\frac{ab}{(b^2\cos^2(u) + a^2\sin^2(u))^2}\left(b\cos(u)\hat{\mathbf{x}} + a\sin(u)\hat{\mathbf{y}}\right). \qquad (C.10)$$

From the magnitude of this vector, the radius of curvature is

$$\rho(u) = \frac{1}{\kappa(u)} = \frac{1}{ab}\left(b^2\cos^2(u) + a^2\sin^2(u)\right)^{3/2}. \qquad (C.11)$$

The curve, together with two of its osculating circles, centered at $\mathbf{r}(u) + \rho(u)\hat{\boldsymbol{\kappa}}(u)$ with radius $\rho(u)$, is shown in Figure C.3. We can check the radius of curvature in the case $a = b = R$, a circular reduction. In that case, we correctly recover $\rho(u) = R$. The other degenerate case that we can evaluate easily from (C.11) is the $b = 0$ line segment. A line has zero curvature and so an infinite radius of curvature, precisely what we get from (C.11).

C.2 Frenet–Serret Equations

In arc length parameterization, the tangent and its derivative form an orthogonal pair. There is a third vector that we can define to complete an orthonormal basis (in three dimensions) defined by the curve $\mathbf{r}(s)$:

$$\hat{\mathbf{b}}(s) \equiv \hat{\mathbf{t}}(s) \times \hat{\boldsymbol{\kappa}}(s) \tag{C.12}$$

(where $\hat{\boldsymbol{\kappa}}$ is the unit curvature vector). The s derivatives of these vectors are related to each other and form a nicely closed set of equations. We'll take derivatives of the definitions and project onto the basis $\{\hat{\mathbf{t}}, \hat{\boldsymbol{\kappa}}, \hat{\mathbf{b}}\}$ to develop the set.

We get the derivative of the tangent vector for free from the definition of curvature,

$$\hat{\mathbf{t}}' = \kappa \hat{\boldsymbol{\kappa}} \tag{C.13}$$

(using primes to denote the derivative – all vectors are in arc length, s-parameterization here).

Taking the s derivative of both sides of the definition of $\hat{\mathbf{b}}$ in (C.12) gives

$$\hat{\mathbf{b}}' = \hat{\mathbf{t}}' \times \hat{\boldsymbol{\kappa}} + \hat{\mathbf{t}} \times \hat{\boldsymbol{\kappa}}' = \hat{\mathbf{t}} \times \hat{\boldsymbol{\kappa}}', \tag{C.14}$$

where the first term in the middle expression vanishes ($\hat{\mathbf{t}}' \parallel \hat{\boldsymbol{\kappa}}$). For the unit vector $\hat{\mathbf{b}} \cdot \hat{\mathbf{b}} = 1$, we have $\hat{\mathbf{b}}' \cdot \hat{\mathbf{b}} = 0$, so that no component of $\hat{\mathbf{b}}'$ points in the $\hat{\mathbf{b}}$ direction. The component in the $\hat{\mathbf{t}}$ direction is

$$\hat{\mathbf{b}}' \cdot \hat{\mathbf{t}} = \left(\hat{\mathbf{t}} \times \hat{\boldsymbol{\kappa}}'\right) \cdot \hat{\mathbf{t}} = 0, \tag{C.15}$$

and there is no component of $\hat{\mathbf{b}}'$ in the $\hat{\mathbf{t}}$ direction, so we conclude that

$$\hat{\mathbf{b}}' = -\sigma \hat{\boldsymbol{\kappa}}; \tag{C.16}$$

this vector points in the $\hat{\boldsymbol{\kappa}}$ direction with s-parameterized magnitude determined by σ (the minus sign above is convention).

Finally, the derivative of $\hat{\boldsymbol{\kappa}}(s)$ can be obtained from its projections onto $\hat{\mathbf{t}}$ and $\hat{\mathbf{b}}$. The derivative of $\hat{\boldsymbol{\kappa}} \cdot \hat{\mathbf{t}} = 0$ gives

$$\hat{\boldsymbol{\kappa}}' \cdot \hat{\mathbf{t}} = -\hat{\boldsymbol{\kappa}} \cdot \hat{\mathbf{t}}' = -\kappa. \tag{C.17}$$

The derivative of $\hat{\boldsymbol{\kappa}} \cdot \hat{\mathbf{b}} = 0$ tells us

$$\hat{\boldsymbol{\kappa}}' \cdot \hat{\mathbf{b}} = -\hat{\boldsymbol{\kappa}} \cdot \hat{\mathbf{b}}' = \sigma, \tag{C.18}$$

and we know there is no $\hat{\boldsymbol{\kappa}}$ in $\hat{\boldsymbol{\kappa}}'$ since $\hat{\boldsymbol{\kappa}} \cdot \hat{\boldsymbol{\kappa}} = 1$ (take the s derivative to get $\hat{\boldsymbol{\kappa}}' \cdot \hat{\boldsymbol{\kappa}} = 0$). Putting these pieces together, we learn that

$$\hat{\boldsymbol{\kappa}}' = -\kappa \hat{\mathbf{t}} + \sigma \hat{\mathbf{b}}. \tag{C.19}$$

The set of three equations is known as the Frenet–Serret equations

$$\hat{\mathbf{t}}' = \kappa \hat{\boldsymbol{\kappa}}, \qquad \hat{\mathbf{b}}' = -\sigma \hat{\boldsymbol{\kappa}}, \qquad \hat{\boldsymbol{\kappa}}' = -\kappa \hat{\mathbf{t}} + \sigma \hat{\mathbf{b}}. \tag{C.20}$$

Given a curve in arc length parameterization $\mathbf{r}(s)$, we can get the curvature $\kappa(s)$. We can also find the value of $\sigma(s)$, called the "torsion," from the equation for $\hat{\mathbf{b}}'$ – what is our interpretation of this function? If $\sigma = 0$, then $\hat{\mathbf{b}}$ is a constant vector, so $\hat{\mathbf{t}}(s) \times \hat{\boldsymbol{\kappa}}(s)$ is a constant. That constant cross product means that the curve is planar (in the plane orthogonal to $\hat{\mathbf{b}}$), as you can see in the example shown in Figure C.1, where $\hat{\mathbf{t}}(s)$ and $\hat{\boldsymbol{\kappa}}(s)$ have a constant cross product, pointing out of the page: $\hat{\mathbf{b}} = \hat{\mathbf{z}}$. The other direction holds as well: If the curve is planar, then the torsion vanishes.

Example C.3 (Torsion for a Helix). We'll compute the torsion for a helical curve defined by

$$\mathbf{r}(s) = R\cos\left(\frac{s}{\sqrt{R^2+v^2}}\right)\hat{\mathbf{x}} + R\sin\left(\frac{s}{\sqrt{R^2+v^2}}\right)\hat{\mathbf{y}} + \frac{vs}{\sqrt{R^2+v^2}}\hat{\mathbf{z}} \quad (C.21)$$

for constants R and v. The curve is in arc length parameterization, and its tangent is

$$\hat{\mathbf{t}}(s) = \mathbf{r}'(s)$$
$$= \frac{1}{\sqrt{R^2+v^2}}\left(-R\sin\left(\frac{s}{\sqrt{R^2+v^2}}\right)\hat{\mathbf{x}} + R\cos\left(\frac{s}{\sqrt{R^2+v^2}}\right)\hat{\mathbf{y}} + v\hat{\mathbf{z}}\right).$$
$$(C.22)$$

The curvature vector is

$$\boldsymbol{\kappa}(s) = \hat{\mathbf{t}}'(s) = -\frac{R}{R^2+v^2}\left(\cos\left(\frac{s}{\sqrt{R^2+v^2}}\right)\hat{\mathbf{x}} + \sin\left(\frac{s}{\sqrt{R^2+v^2}}\right)\hat{\mathbf{y}}\right)$$
$$(C.23)$$

with magnitude

$$\kappa(s) = \frac{R}{R^2+v^2}, \quad (C.24)$$

a constant. The vector $\hat{\mathbf{b}}(s)$ for this curve is

$$\hat{\mathbf{b}}(s) = \hat{\mathbf{t}}(s) \times \hat{\boldsymbol{\kappa}}(s)$$
$$= \frac{1}{\sqrt{R^2+v^2}}\left(v\sin\left(\frac{s}{\sqrt{R^2+v^2}}\right)\hat{\mathbf{x}} - v\cos\left(\frac{s}{\sqrt{R^2+v^2}}\right)\hat{\mathbf{y}} + R\hat{\mathbf{z}}\right).$$
$$(C.25)$$

To find the torsion, we need the derivative of $\hat{\mathbf{b}}(s)$:

$$\mathbf{b}'(s) = \frac{v}{R^2+v^2}\underbrace{\left(\cos\left(\frac{s}{\sqrt{R^2+v^2}}\right)\hat{\mathbf{x}} + \sin\left(\frac{s}{\sqrt{R^2+v^2}}\right)\hat{\mathbf{y}}\right)}_{=-\hat{\boldsymbol{\kappa}}(s)}, \quad (C.26)$$

and then from $\hat{\mathbf{b}}'(s) = -\sigma\hat{\boldsymbol{\kappa}}(s)$ (note that the minus sign in the definition cancels the minus sign appearing in (C.26)) we can pick out the torsion

$$\sigma = \frac{v}{R^2+v^2}. \quad (C.27)$$

In the case of a circle in the xy plane, where $v = 0$, we recover $\sigma = 0$, that is, the torsion vanishes, as expected.

C.3 Curvature for Surfaces

A two-dimensional surface can be parameterized using two surface coordinate u^1 and u^2. In a generalization of the curve description depicted in Figure C.1, the vector $\mathbf{X}(u^1, u^2)$ points from the origin to the points on the surface as shown in Figure C.4.

At any point, we can compute the vectors pointing in the direction of increasing u^1 and u^2 coordinate values. As usual, these are given by the derivatives of $\mathbf{X}$ with respect to the surface coordinates,

$$\mathbf{X}_1 \equiv \frac{\partial \mathbf{X}}{\partial u^1}, \qquad \mathbf{X}_2 \equiv \frac{\partial \mathbf{X}}{\partial u^2}. \tag{C.28}$$

For a two dimensional surface, these two directions must be linearly independent (else we have a curve). We can use these two vectors, both tangent to the surface, to define a vector that is normal to the surface, $\mathbf{n}(u^1, u^2) = \mathbf{X}_1(u^1, u^2) \times \mathbf{X}_2(u^1, u^2)$, and the unit normal is $\hat{\mathbf{n}}(u^1, u^2) = \mathbf{n}(u^1, u^2)/n(u^1, u^2)$.

How should we go about thinking of the curvature of this *surface*? We'd like a description that gives zero curvature for a plane, just as the curvature κ is zero for a line. For nonzero surface curvature, we'd further like to recover the interpretive elements of κ and ρ from curves. These targets suggest that we define surface curvature in terms of curvature for curves. Which family of curves should we consider? One way to proceed is to take the curves defined by the "normal sections" of the surface at a point. To get the normal section at a point P on the surface, you intersect a plane with the surface at point P such that the plane has unit normal perpendicular to the surface normal at P. The intersection of this plane with the surface defines a curve, call it C. The "normal curvature" at P is the curvature of this curve, evaluated at P. Referring to Figure C.5, we can see the closed surface with a plane intersecting it. The

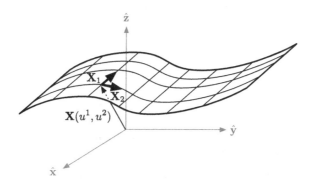

Fig. C.4 A generic two-dimensional surface. The vector $\mathbf{X}(u^1, u^2)$ points from the origin to points on the surface. In this figure, the lines on the surface represent constant coordinate values u^1 and u^2. At any point, the vectors $\mathbf{X}_1 \equiv \frac{\partial \mathbf{X}}{\partial u^1}$ and $\mathbf{X}_2 \equiv \frac{\partial \mathbf{X}}{\partial u^2}$ form a linearly independent set.

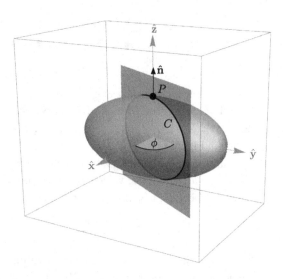

Fig. C.5 A closed two-dimensional surface with an intersecting plane at azimuthal angle ϕ. The intersection defines a one-dimensional curve C, and we can compute the curvature of that curve at the point P.

curve C that defines the intersection is outlined. In the figure, the point P that we have in mind lies on the positive $\hat{\mathbf{z}}$ axis, with $\hat{\mathbf{n}} = \hat{\mathbf{z}}$. Then the set of normal section planes is parameterized by the azimuthal angle ϕ, giving the related set of intersection curves indexed by $\phi = 0 \to \pi$.

In particular cases, we can compute the curvature of the family of normal section curves directly from the specification of the surface. Then all we have to do is identify the vector $\mathbf{r}(u)$ that defines the intersection curve and compute $\kappa(u)$ from (C.7). In the next example, we'll carry out that calculation for a particular surface.

Example C.4 (Curvature for a Specific Surface). Suppose we take the blimp-shaped surface shown in Figure C.5. The surface can be parameterized by y and an angle θ defined in the xz plane with $\theta = 0$ giving points with $x = 0$:

$$\mathbf{X}(y, \theta) = \frac{b}{a}\sqrt{a^2 - y^2}(\sin\theta\hat{\mathbf{x}} + \cos\theta\hat{\mathbf{z}}) + y\hat{\mathbf{y}}, \qquad (C.29)$$

where a and b are parameters, $y \in [-a, a]$, and $\theta \in [0, 2\pi]$. To describe the surface beyond its pictorial form, consider the curve made up of points on the surface with $\theta = 0$. These points lie in the yz plane and form an ellipse with semimajor and semiminor axes a and b (we don't know which are major and minor until the values have been given). Another curve of interest is formed by the set of points on the surface with $y = 0$. These lie in the xz plane, and the curve is a circle of radius b.

Using the azimuthal angle ϕ (depicted in Figure C.5), the elliptical curve associated with $\theta = 0$ above has $\phi = \pi/2$, whereas the curve with $y = 0$, the circle of radius b in the xz plane, has $\phi = 0$. We can compute the normal sections for these two values of ϕ. For $\phi = \pi/2$, the curve is parameterized by

y with $\theta = 0$ in (C.29):

$$\mathbf{r}(y) = y\hat{\mathbf{y}} + \sqrt{b^2 - y^2 b^2/a^2}\,\hat{\mathbf{z}}. \tag{C.30}$$

Computing the curvature vector using (C.7) gives the curvature

$$\kappa(y) = \frac{a^4 b}{(a^4 - a^2 y^2 + b^2 y^2)^{3/2}}. \tag{C.31}$$

For $\phi = 0$, taking $y = 0$ in (C.29), the curve is

$$\mathbf{r}(\theta) = b(\sin\theta\,\hat{\mathbf{x}} + \cos\theta\,\hat{\mathbf{z}}), \tag{C.32}$$

which we already know has curvature $1/b$.

To find the curvature of the normal section curves at point P for an arbitrary value of the angle ϕ, it is easiest to switch from the y and θ parameterization of (C.29) to one in terms of y and z. In the new surface coordinates, we have

$$\mathbf{X}(y, z) = \frac{b}{a}\left(\pm\sqrt{a^2\left(1 - \frac{z^2}{b^2}\right) - y^2}\,\hat{\mathbf{x}} + y\hat{\mathbf{y}} + z\hat{\mathbf{z}}\right) \tag{C.33}$$

for $y \in [-a, a]$ and $z \in [-b, b]$. The advantage here is that the plane at angle ϕ can also be parameterized by y and z via $\mathbf{X}_\phi(y, z) = y\cot\phi\,\hat{\mathbf{x}} + y\hat{\mathbf{y}} + z\hat{\mathbf{z}}$, so the intersection of the plane $\mathbf{X}_\phi$ with the surface $\mathbf{X}$ is easy to compute. The curvature vector of the intersecting curve, parameterized by z and evaluated at the point P where $z = b$, is

$$\kappa(z = b) = -\left(\frac{1}{b}\cos^2\phi + \frac{b}{a^2}\sin^2\phi\right)\hat{\mathbf{z}}, \tag{C.34}$$

and the curvature is the magnitude of this vector as always. Each value of ϕ gives a new value for $\kappa(z)$, as expected. Taking the ϕ derivative of $\kappa(b)$ and setting it equal to zero gives the maximum and minimum values of the curvature for all normal sections – these values are called the "principal normal curvatures." In this case, the maximum and minimum occur at $\phi = 0$ and $\phi = \pi/2$, the values that we have already computed separately above. For this pair of values, the radii of curvature for the circles that are tangent to the surface at P are b and a^2/b, respectively.

The procedure for finding the curvature values at a point on the surface demonstrated in the previous example can be difficult to carry out. One of the main obstacles is the parameterization of the curves themselves, which can make computing the curvature tedious at best. We can streamline the calculation by noting that we don't need the full intersection curve C to find its curvature at a point. A curve that lies on the surface can be expressed using the surface coordinates u^1 and u^2. If we use a parameter p to identify the points on the curve in the surface, $u^1(p)$ and $u^2(p)$, then the curve takes a parameterized form $\mathbf{r}(p) = \mathbf{X}(u^1(p), u^2(p))$. The expression for the tangent vector $\mathbf{r}'(p)$ in terms of $\mathbf{X}(u^1(p), u^2(p))$ is

$$
\begin{aligned}
\mathbf{r}'(p) &= \frac{d\mathbf{X}(u^1(p), u^2(p))}{dp} = \frac{\partial \mathbf{X}}{\partial u^1}\frac{du^1(p)}{dp} + \frac{\partial \mathbf{X}}{\partial u^2}\frac{du^2(p)}{dp} \\
&= \mathbf{X}_1\frac{du^1(p)}{dp} + \mathbf{X}_2\frac{du^2(p)}{dp}.
\end{aligned}
\tag{C.35}
$$

To find the curvature, we need the unit tangent vector. The squared length of the tangent vector is

$$
\begin{aligned}
\mathbf{r}'(p) \cdot \mathbf{r}'(p) &= \mathbf{X}_1 \cdot \mathbf{X}_1\left(\frac{du^1(p)}{dp}\right)^2 + 2\mathbf{X}_1 \cdot \mathbf{X}_2\frac{du^1(p)}{dp}\frac{du^2(p)}{dp} \\
&\quad + \mathbf{X}_2 \cdot \mathbf{X}_2\left(\frac{du^2(p)}{dp}\right)^2,
\end{aligned}
\tag{C.36}
$$

and this expression can be written in a familiar quadratic form. Define the "first fundamental form" by its elements, $g_{ab} \equiv \mathbf{X}_a \cdot \mathbf{X}_b$ for $a, b = 1, 2$. We can interpret g_{ab} as a metric on the surface, since it gives the lengths of vectors found there. The squared length by (C.36) becomes

$$
\mathbf{r}'(p) \cdot \mathbf{r}'(p) = g_{ab}\frac{du^a(p)}{dp}\frac{du^b(p)}{dp}.
\tag{C.37}
$$

In arc length parameterization, which we'll continue to highlight with parameter s, the defining feature is $\mathbf{r}'(s) \cdot \mathbf{r}'(s) = 1$, which tells us that

$$
g_{ab}\frac{du^a(s)}{ds}\frac{du^b(s)}{ds} = 1,
\tag{C.38}
$$

and the curvature is given by $\kappa(s) = \mathbf{r}''(s)$. Again moving to the surface description of the curve, we can take the derivative of $\mathbf{r}'(s) = \mathbf{X}_a\frac{du^a(s)}{ds}$ from (C.35) (employing summation notation) to get

$$
\mathbf{r}''(s) = \frac{d\mathbf{X}_a}{ds}\frac{du^a(s)}{ds} + \mathbf{X}_a\frac{d^2u^a(s)}{ds^2} = \frac{d^2\mathbf{X}_a}{du^b}\frac{du^a(s)}{ds}\frac{du^b(s)}{ds} + \mathbf{X}_a\frac{d^2u^a(s)}{ds^2}.
\tag{C.39}
$$

For the normal curvature we computed in Example C.4, the curvature vector points in the direction normal to the surface at the point P of interest, $\mathbf{r}''(s) \sim \hat{\mathbf{n}}$. Picking out that direction from the curvature, the normal curvature is given by $\kappa_n \equiv \mathbf{r}''(s) \cdot \hat{\mathbf{n}}$. Since $\mathbf{X}_1$ and $\mathbf{X}_2$ are tangent to the surface at any point, the second term on the right in (C.39) goes away when we dot the curvature into the normal (i.e. $\hat{\mathbf{n}} \cdot \mathbf{X}_a = 0$):

$$
\kappa_n = \hat{\mathbf{n}} \cdot \mathbf{X}_{ab}\frac{du^a(s)}{ds}\frac{du^b(s)}{ds} \qquad \text{with } \mathbf{X}_{ab} \equiv \frac{\partial^2\mathbf{X}}{\partial u^a \partial u^b}.
\tag{C.40}
$$

In κ_n the expression involving the quadratic derivatives of u is similar to the structure on the right in (C.37). The symmetric $\hat{\mathbf{n}} \cdot \mathbf{X}_{ab} \equiv h_{ab}$ is called the "second fundamental form" and goes along with g_{ab}.

To compute κ_n using (C.40), we still must be able to specify curves in arc length parameterization. Again, our goal is a description of the normal curvature that doesn't rely on a particular curve parameterization (or even the curves themselves, beyond their structural information at a point). From the equation for normal curvature, written in terms of the second fundamental form,

$$\kappa_n = h_{ab} \frac{du^a(s)}{ds} \frac{du^b(s)}{ds}, \tag{C.41}$$

the easiest way to remove the parameter dependence is to multiply on the left by the left-hand side of (C.38) and on the right by the right-hand side of (C.38), giving

$$\kappa_n \underbrace{g_{ab} \frac{du^a(s)}{ds} \frac{du^b(s)}{ds}}_{=1} = h_{ab} \frac{du^a(s)}{ds} \frac{du^b(s)}{ds}. \tag{C.42}$$

This form is manifestly reparameterization-invariant. We don't even really need the parameter at all. At a point P, any parameterization of the curve leads to the same values of $du^1 = \frac{du^1(p)}{dp} dp$ and $du^2 = \frac{du^2(p)}{dp} dp$ for use in (C.42). Thinking of that individual point, the normal curvature expression from (C.42) gives

$$\kappa_n = \frac{h_{ab} du^a du^b}{g_{xy} du^x du^y}. \tag{C.43}$$

Now the coordinate differentials aren't attached to a particular curve at all. Any curve going through the point P has a unit tangent direction there, $\mathbf{r}'(p)/\sqrt{\mathbf{r}'(p) \cdot \mathbf{r}'(p)}$, which depends only on the ratio of du^1 to du^2. So instead of identifying the set of all curves that go through the point P with curvature direction parallel to $\hat{\mathbf{n}}$, we can get the normal curvature by considering just variations of the ratio du^1/du^2. The tangent directions, in vector form, are given by $\mathbf{t} = du^1 \mathbf{X}_1 + du^2 \mathbf{X}_2$ (for the linearly independent vectors $\mathbf{X}_1$ and $\mathbf{X}_2$ at P as in Figure C.4). We can finally write the normal curvature entirely in terms of objects that can be computed using only the surface definition and a given point,

$$\kappa_n = \frac{h_{ab} t^a t^b}{g_{xy} t^x t^y}. \tag{C.44}$$

In Example C.4, we found the maximum and minimum values of the normal curvature by taking the derivative of the curvature of the family of normal section curves with respect to that family's parameter ϕ. The expression for normal curvature in (C.44) can be similarly extremized by taking its derivative with respect to the tangent directions $\mathbf{t}$,

$$\frac{\partial \kappa_n}{\partial t^q} = \frac{2 h_{aq} t^a}{g_{xy} t^x t^y} - \frac{h_{ab} t^a t^b}{(g_{xy} t^x t^y)^2} 2 g_{xq} t^x, \tag{C.45}$$

and setting this equal to zero,

$$h_{aq} t^a - \kappa_n g_{aq} t^a = 0. \tag{C.46}$$

Define the matrix with entries h_{ab} to be $\mathbb{H}$ and the matrix with entries g_{ab} to be $\mathbb{G}$. Then (C.46) becomes $(\mathbb{H} - \kappa_n \mathbb{G})\mathbf{t} = 0$ or

$$\mathbb{G}(\mathbb{G}^{-1}\mathbb{H} - \kappa_n \mathbb{I})\mathbf{t} = 0 \longrightarrow (\mathbb{G}^{-1}\mathbb{H})\mathbf{t} = \kappa_n \mathbf{t}. \tag{C.47}$$

The second version of this equation tells us that the extremizing directions **t** and associated normal curvature values κ_n are solutions to the eigenvalue problem for the matrix $\mathbb{G}^{-1}\mathbb{H}$, which is determined entirely by components of the surface geometry in the vicinity of the point P of interest.

We finally have a way to compute the principal normal curvatures. It still requires some 2×2 matrix manipulation to carry out, and in part for that reason, people tend to form the easily obtainable linear combinations $\kappa_{\min} + \kappa_{\max}$, called the "mean curvature" (especially if you throw in a factor of $1/2$), and $\kappa_{\min}\kappa_{\max}$, the "Gauss curvature." These are the sums and products of the eigenvalues of the matrix $\mathbb{G}^{-1}\mathbb{H}$. We know that the trace of a matrix is the sum of its eigenvalues and that its determinant is the product of its eigenvalues, and then the relevant combinations are

$$\kappa_{\min} + \kappa_{\max} = \text{Tr}\left(\mathbb{G}^{-1}\mathbb{H}\right), \qquad \kappa_{\min}\kappa_{\max} = \text{Det}\left(\mathbb{G}^{-1}\mathbb{H}\right). \tag{C.48}$$

Example C.5 (Curvature for a Sphere of Radius r). For a sphere, using the usual polar and azimuthal angle parameterizations, the vector pointing from the origin to points on the sphere is

$$\mathbf{X}(\theta, \phi) = r(\sin\theta \cos\phi \hat{\mathbf{x}} + \sin\theta \sin\phi \hat{\mathbf{y}} + \cos\theta \hat{\mathbf{z}}) \tag{C.49}$$

with the directions of increasing θ and ϕ given by

$$\begin{aligned}
\mathbf{X}_1 &\equiv \frac{\partial \mathbf{X}}{\partial \theta} = r\cos\theta(\cos\phi\hat{\mathbf{x}} + \sin\phi\hat{\mathbf{y}}) - r\sin\theta\hat{\mathbf{z}}, \\
\mathbf{X}_2 &\equiv \frac{\partial \mathbf{X}}{\partial \phi} = r\sin\theta(-\sin\phi\hat{\mathbf{x}} + \cos\phi\hat{\mathbf{y}}),
\end{aligned} \tag{C.50}$$

leading immediately to the first fundamental form

$$g_{pq} \doteq \begin{pmatrix} r^2 & 0 \\ 0 & r^2\sin^2\theta \end{pmatrix}, \tag{C.51}$$

which you should compare with the metric for the surface of a sphere in (3.12). All points are equivalent on the sphere, so we can pick P to be anywhere we like.[2] Take P to be the point $r\hat{\mathbf{y}}$ with $\theta = \pi/2$ and $\phi = \pi/2$; then $\hat{\mathbf{n}} = \hat{\mathbf{y}}$, and the second fundamental form is

$$h_{pq} \doteq \begin{pmatrix} -r\sin\theta\sin\phi & r\cos\theta\cos\phi \\ r\cos\theta\cos\phi & -r\sin\theta\sin\phi \end{pmatrix}. \tag{C.52}$$

The eigenvalues of $\mathbb{G}^{-1}\mathbb{H}$, evaluated at P, are $\kappa_1 = -1/r$ and $\kappa_2 = -1/r$. The minus signs in this case have to do with the surface normal conventions.[3] These "principal directions of curvature" are the same, which is not surprising for the sphere. The mean curvature is $-2/r$, with Gauss curvature $1/r^2$.

---

[2] There are inconvenient choices. For example, if you take P to be at the north pole, then the entries of h_{pq} are not well-defined, since for $\theta = 0$, ϕ itself is not well-defined. Try computing h_{pq} in this case to see what happens.

[3] The relative signs of curvature components matter, and tell us about the convexity or concavity at a point.

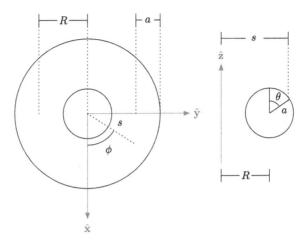

Fig. C.6 These figures define the coordinates θ and ϕ of the torus. On the left, a top down view of a torus with distance from the center of the coordinate system to the middle of the circular tube R and tube radius a. On the right, a cross-section view of the tube. The angle ϕ is measured relative to the x axis, and θ is measured relative to vertical within the torus tube. The auxiliary variable $s \equiv R + a \sin \theta$, a cylindrical distance to the z axis, is shown in both figures.

Example C.6 (Gauss and Mean Curvature for the Surface from Example C.4). We can compute the fundamental forms and curvatures directly from (C.29). The entries of the induced metric are

$$g_{pq} \doteq \begin{pmatrix} \frac{a^4 - a^2 y^2 + b^2 y^2}{a^2(a^2 - y^2)} & 0 \\ 0 & b^2(1 - \frac{y^2}{a^2}) \end{pmatrix}. \tag{C.53}$$

For the second fundamental form, taking $\hat{\mathbf{n}} = \hat{\mathbf{z}}$, appropriate for the point P in Figure C.5, we have

$$h_{pq} \doteq \begin{pmatrix} -\frac{ab \cos \theta}{(a^2 - y^2)^{3/2}} & \frac{by \sin \theta}{a\sqrt{a^2 - y^2}} \\ \frac{by \sin \theta}{a\sqrt{a^2 - y^2}} & -\frac{b}{a}\sqrt{a^2 - y^2} \cos \theta \end{pmatrix}. \tag{C.54}$$

The eigenvalues of the matrix $\mathbb{G}^{-1}\mathbb{H}$, evaluated at $y = 0$ and $\theta = 0$, are $\kappa_1 = -1/b$ and $\kappa_2 = -b/a^2$. The mean curvature (trace of $\mathbb{G}^{-1}\mathbb{H}$) is the sum $-1/b - b/a^2$, and the Gauss curvature (determinant of $\mathbb{G}^{-1}\mathbb{H}$) is the product $1/a^2$.

Example C.7 (Curvature for a Torus). A torus can be parameterized by an azimuthal angle ϕ measured with respect to the x axis and a local polar angle measured from the north pole of the "tube" circle. For a torus with large radius R and tube radius a, the coordinate definitions are shown in Figure C.6. A point on the surface of the torus has the vector location

$$\mathbf{X}(\theta, \phi) = (R + a \sin \theta)(\cos \phi \hat{\mathbf{x}} + \sin \phi \hat{\mathbf{y}}) + a \cos \theta \hat{\mathbf{z}}. \tag{C.55}$$

In this example, we'll compute the curvature everywhere using the surface normal constructed from the cross product of the local derivative vectors

$\mathbf{X}_1(\theta, \phi)$ and $\mathbf{X}_2(\theta, \phi)$. These vectors, together with the unit normal, are

$$\mathbf{X}_1 = a \cos \theta (\cos \phi \hat{\mathbf{x}} + \sin \phi \hat{\mathbf{y}}) - a \sin \theta \hat{\mathbf{z}},$$

$$\mathbf{X}_2 = (R + a \sin \theta)(- \sin \phi \hat{\mathbf{x}} + \cos \phi \hat{\mathbf{y}}),$$

$$\hat{\mathbf{n}} \equiv \frac{\mathbf{X}_1 \times \mathbf{X}_2}{\sqrt{(\mathbf{X}_1 \times \mathbf{X}_2) \cdot (\mathbf{X}_1 \times \mathbf{X}_2)}} = \sin \theta (\cos \phi \hat{\mathbf{x}} + \sin \phi \hat{\mathbf{y}}) + \cos \theta \hat{\mathbf{z}}. \tag{C.56}$$

Using these, we can construct the first and second fundamental forms at any point on the surface of the torus,

$$g_{pq} \doteq \begin{pmatrix} a^2 & 0 \\ 0 & (R + a \sin \theta)^2 \end{pmatrix}, \qquad h_{pq} \doteq \begin{pmatrix} -a & 0 \\ 0 & -(R + a \sin \theta) \sin \theta \end{pmatrix}. \tag{C.57}$$

Forming the matrix $\mathbb{G}^{-1}\mathbb{H}$ and computing its eigenvalues, we get

$$\kappa_1 = -\frac{1}{a}, \qquad \kappa_2 = -\frac{\sin \theta}{R + a \sin \theta}, \tag{C.58}$$

leading to the mean curvature

$$\kappa_1 + \kappa_2 = -\left(\frac{1}{a} + \frac{\sin \theta}{R + a \sin \theta} \right) \tag{C.59}$$

and Gauss curvature

$$\kappa_1 \kappa_2 = \frac{\sin \theta}{a(R + a \sin \theta)}. \tag{C.60}$$

Notice that the Gauss curvature does change sign from the outside of the torus, where $\sin \theta > 0$ (and hence so is the curvature), to the inside, where $\sin \theta < 0$. This indicates that the relative sign of the Gauss curvature tells us whether or not the surface is bending away from the local unit normal (positive) or towards the unit normal (negative).

References

[1] www.ligo.caltech.edu

[2] B. P. Abbott et al., "Observation of gravitational waves from a binary black hole merger," *Phys. Rev. Lett.* **116** (6) p. 061102 (2016).

[3] Miguel Alcubierre, "The warp drive: hyper-fast travel within general relativity," *Classical Quantum Gravity* **11** pp. L73–L77 (1994).

[4] H. Bondi, F. A. E. Pirani & I. Robinson, "Gravitational waves in general relativity. III. Exact plane waves," *Proc. R. Soc. Lond. A* **251** pp. 519–533 (1959).

[5] Robert H. Boyer & Richard W. Lindquist, "Maximal analytic extension of the Kerr metric," *J. Math. Phys.* **8** pp. 265–281 (1967).

[6] H. A. Buchdahl, "General relativistic fluid spheres," *Phys. Rev.* **116** (4) pp. 1027–1034 (1959).

[7] S. Chandrasekhar, *The Mathematical Theory of Black Holes*, Oxford University Press, 1992.

[8] S. Chandrasekhar, *An Introduction to the Study of Stellar Structure*, Dover Publications, 2010.

[9] S. Deser & B. Tekin, "Shortcuts to high symmetry solutions in gravitational theories," *Classical Quantum Gravity* **20** pp. 4877–4884 (2003).

[10] Ray d'Inverno & James Vickers, *Introducing Einstein's Relativity: A Deeper Understanding*, 2nd ed., Oxford University Press, 2022.

[11] F. W. Dyson, A. S. Eddington & C. Davidson, "A determination of the deflection of light by the Sun's gravitational field, from observations made at the total eclipse of May 29, 1919," *Philos. Trans. R. Soc. A* **220**, pp. 291–333 (1920).

[12] A. S. Eddington, "A comparison of Whitehead's and Einstein's formulae," *Nature* **113** p. 192 (1924).

[13] A. Einstein, *The Principle of Relativity*, Dover Publications, 1952. This is "A Collection of Original Papers on the Special and General Theory of Relativity" including papers by H. A. Lorentz, H. Weyl, and H. Minkowski, in addition to A. Einstein. Notes by A. Sommerfeld.

[14] Homer G. Ellis, "Ether flow through a drainhole: a particle model in general relativity," *J. Math. Phys.* **14** pp. 104–118 (1973).

[15] C. W. F. Everitt et al., "Gravity Probe B: final results of a space experiment to test general relativity," *Phys. Rev. Lett.* **106** p. 221101 (2011).

[16] C. W. F. Everitt, B. Muhlfelder, C. M. Will & A. S. Silbergleit (eds.), "Focus issue: Gravity Probe B," *Classical Quantum Gravity* (2015).

[17] David Finkelstein, "Past–future asymmetry of the gravitational field of a point particle," *Phys. Rev.* **119** pp. 965–967 (1958).

[18] L. Flamm, "Beiträge zur Einsteinischen Gravitationstheorie," *Phys. Z.* **17** pp. 448–454 (1916). Translation and Republication of: "Contributions to Einstein's Theory of Gravitation," *Gen. Relativ. Gravit.* **47** (6) (2015).

[19] A. P. French, *Special Relativity*, W. W. Norton & Company, 1968.

[20] Herbert Goldstein, Charles S. Poole & John Safko, *Classical Mechanics*, 3rd ed., Pearson Education, 2001.

[21] David J. Griffiths, *Introduction to Electrodynamics*, 4th ed., Cambridge University Press, 2017.

[22] David J. Griffiths, David Derbes & Richard B. Sohn (eds.), *Sidney Coleman's Lectures on Relativity*, Cambridge University Press, 2022.

[23] James B. Hartle, *Gravity: An Introduction to Einstein's General Relativity*, Cambridge University Press, 2021.

[24] Thomas J. R. Hughes & Jerrold E. Marsden, *A Short Course in Fluid Mechanics*, Mathematics Lecture Series 6, Publish or Perish, Inc., 1976.

[25] R. A. Hulse & J. H. Taylor, "Discovery of a pulsar in a binary system," *Astrophys. J.* **195**, pp. L51–L53 (1975).

[26] John David Jackson, *Classical Electrodynamics*, 3rd ed., Wiley, 1998.

[27] Roy P. Kerr, "Gravitational field of a spinning mass as an example of algebraically special metrics," *Phys. Rev. Lett.* **11** (5) pp. 237–238 (1963).

[28] Erwin Kreyszig, *Differential Geometry*, Dover Publications, 1991.

[29] M. D. Kruskal, "Maximal extension of Schwarzschild metric," *Phys. Rev.* **119** pp. 1743–1745 (1960).

[30] Cornelius Lanczos, *The Variational Principles of Mechanics*, 4th ed., Dover Publications, 1986.

[31] L. D. Landau & E. M. Lifshitz, *The Classical Theory of Fields*, 4th ed., Butterworth-Heinemann, 1980.

[32] P.-S. Laplace, *Exposition du système du monde, tome 2*, Imprimerie du Cercle-Social, Paris, 1796. English translation by H. H. Harte, *The System of the World, Volume 2*, Legare Street Press, 2022.

[33] John Michell, "On the means of discovering the distance, magnitude, &c. of the fixed stars, in consequence of the diminution of the velocity of their light, in case such a diminution should be found to take place in any of them, and such other data should be procured from observations, as would be farther necessary for that purpose," *Philos. Trans. R. Soc.* **74** pp. 35–57 (1784).

[34] L. M. Milne-Thomson, *Theoretical Hydrodynamics*, 5th ed., Dover Publications, 2013.

[35] Charles W. Misner, Kip S. Thorne & John Archibald Wheeler, *Gravitation*, Princeton University Press, 1973.

[36] Michael S. Morris & Kip S. Thorne, "Wormholes in spacetime and their use for interstellar travel: a tool for teaching general relativity," *Am. J. Phys.* **56** pp. 395–412 (1988).

[37] G. Nordstrøm, "On the energy of the gravitational field in Einstein's theory," *Proc. K. Ned. Akad. Wet.* **20** pp. 1238–1245 (1918).

[38] R. V. Pound & G. A. Rebka, Jr., "Gravitational red-shift in nuclear resonance," *Phys. Rev. Lett.* **3** (9) pp. 439–441 (1959).

[39] William H. Press, Saul A. Teukolsky, William T. Vetterling & Brian P. Flannery, *Numerical Recipes: The Art of Scientific Computing*, 3rd ed., Cambridge University Press, 2007.

[40] Edward M. Purcell & David J. Morin, *Electricity and Magnetism*, 3rd ed., Cambridge University Press, 2013.

[41] H. Reissner, "Über die Eigengravitation des elektrischen Feldes nach der Einsteinschen Theorie," *Ann. Phys.* **355** (9) pp. 106–120 (1916).

[42] H. P. Robertson, "Kinematics and world-structure," *Astrophys. J.* **82** pp. 284–301 (1935).

[43] Bernard Schutz, *A First Course in General Relativity*, 3rd ed., Cambridge University Press, 2022.

[44] K. Schwarzschild, "Über das Gravitationsfeld eines Massenpunktes nach der Einsteinschen Theorie," *Sitzungsberichte der Königlich Preussischen Akademie der Wissenschaften zu Berlin* pp. 189–196 (1916). "On the Gravitational Field of a Mass Point According to Einstein's Theory," English translation by S. Antoci & A. Loinger, arXiv:physics/9905030 (1999).

[45] K. Schwarzschild, "Über das Gravitationsfeld einer Kugel aus inkompressibler Flüssigkeit nach der Einsteinschen Theorie," *Sitzungsberichte der Königlich Preussischen Akademie der Wissenschaften zu Berlin* pp. 424–434 (1916). "On the Gravitational Field of a Sphere of Incompressible Fluid According to Einstein's Theory," English translation by S. Antoci, arXiv:physics/9912033 (1999).

[46] J. Soldner, "Über die Ablenkung eines Lichtstrahls von seiner geradlinigen Bewegung durch die Attraktion eines Weltkörpers, an welchem er nahe vorbeigeht," *Astronomisches Jahrbuch für das Jahr 1804* pp. 161–172 (1801). With commentary and English translation in Stanley L. Jaki, "Johann Georg von Soldner and the Gravitational Bending of Light, with an English Translation of His Essay on It Published in 1801," *Found. Phys.* **8** pp. 927–950 (1978).

[47] J. L. Synge & A. Schild, *Tensor Calculus*, Dover Publications, 1978.

[48] G. Szekeres, "On the singularities of a Riemannian manifold," *Publ. Math. Debrecen* **7** pp. 285–301 (1960). Reprinted in *Gen. Relativ. Gravit.* **34** pp. 2001–2016 (2002).

[49] Kip S. Thorne & Roger D. Blandford, *Modern Classical Physics: Optics, Fluids, Plasmas, Elasticity, Relativity and Statistical Physics*, Princeton University Press, 2017.

[50] U. I. Uggerhøj, R. E. Mikkelsen & J. Faye, "The young centre of the Earth," *Eur. J. Phys.* **37** p. 035602 (2016).

[51] Robert M. Wald, *General Relativity*, University of Chicago Press, 1984.

[52] J. Weber, *General Relativity and Gravitational Waves*, Interscience Publishers, Inc., New York, 1961. Dover edition: Dover Publications, 2013.

[53] Rainier Weiss, in D. J. Muehlner, Rainier Weiss, "Gravitation Research," Massachusetts Institute of Technology, Research Laboratory of Electronics, Quarterly Progress Report, April 15, 1972.

[54] H. Weyl, "Zur Gravitationstheorie," *Ann. Phys.* **54** pp. 117–145 (1917).

[55] Andrew Zangwill, *Modern Electrodynamics*, Cambridge University Press, 2012.

[56] Ya. B. Zel'dovich & I. D. Novikov, *Stars and Relativity*, Dover Publications, 2011.

Index

Printed in the United States
by Baker & Taylor Publisher Services